SYNCHRONOUS MACHINES

(Their Theory, Stability, and Excitation Systems)

SYNCHRONOUS MACHINES

(Their Theory, Stability, and Excitation Systems)

Mulukutla S. Sarma

Northeastern University

GORDON AND BREACH SCIENCE PUBLISHERS

New York London Paris

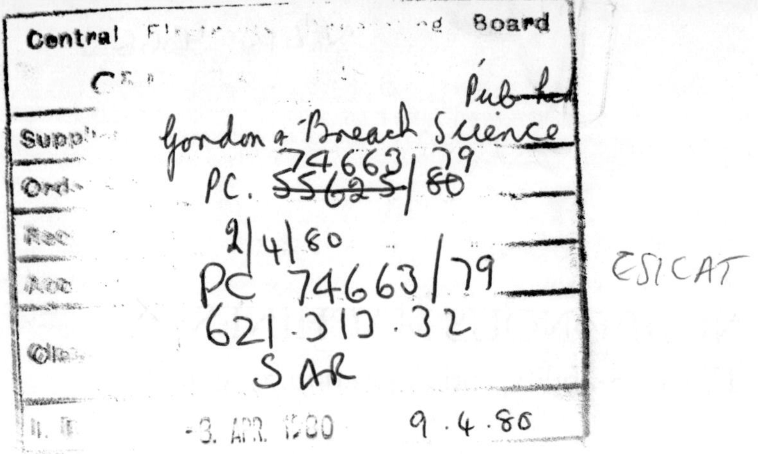

Gordon and Breach, Science Publishers, Inc.
One Park Avenue
New York, NY10016

Gordon and Breach Science Publishers Ltd.
42 William IV Street
London WC2 4DF

Gordon & Breach
7-9 rue Emile Dubois
Paris 75014

Library of Congress Cataloging in Publication Data

Sarma, Mulukutla S 1938-
Synchronous machines.

Bibliography: p.
Includes index.
1. Electric machinery, Synchronous. I. Title.
TK2731.S27 621.313'3 78-13920
ISBN 0-677-03930-1

Library of Congress catalog card number 78-13920. ISBN 0 677 03930 1.

Printed in Malta by Interprint Limited

Dedicated to:

my wife, Savitri Devi, without whose patience, understanding and encouragement this book would not have been written.

PREFACE

During the thirties and forties most universities in the United States used to offer good electric-machinery courses and power-engineering programs well-suited to the needs of those times. In fact, some of the present courses in the general field of electric power are remnants from those days. Also, a large percentage of today's electric-machine designers and power-system engineers are of pre-World War II vintage. As most of these engineers are retiring through natural attrition, the manufacturers of electric power apparatus and so also the electric utility industry have been facing somewhat of a manpower shortage. The whole field of energy is in the midst of a crisis with a new set of technological problems due to ever-increasing power density, limited fuel resources, complex and costly environmental considerations, competition from outside countries, and the controversial nuclear revolution.

In 1978 only a very few universities in the United States have well-established and modern programs in electric power apparatus and systems. Most universities offer neither well-balanced undergraduate/graduate courses nor graduate research opportunities in the field. At a time when several new textbooks are annually published in relatively new areas such as computer sciences, control theory, solid-state electronics, etc., a professor offering courses in electric power apparatus must be content with textbooks which were published ten to twenty-five years ago; this is particularly true at the graduate level.

The electric utility industry in the United States is the largest industry not only in the country but also in the world from the viewpoint of a total plant and equipment investment of about 100 billion dollars, and an annual production figure of about 2 trillion kilowatthours, accounting for more than one-third of the total production of electric energy in the world. More nuclear plant capacity than the rest of the world combined is under development and construction in the United States. With the advent of this development and the general energy crisis, there has been renewed interest on the part of the universities in developing modern energy engineering programs and establishing distinguished power professorships supported by the manufacturing and utility industries. The urgent need for updating the power programs and production of highly competent engineers capable of working on the frontiers of fast-developing, competitive technology has been recognized.

The synchronous machine may be cited as the single most important basic component of an electric energy system. For economic reasons, the size of a single generator unit is of the order of 500 to 1000 MW. Conversion of energy from mechanical to electrical form and vice-versa remains a major aspect of our current engineering practice. All the benefits of gigantic computational robots and sophisticated automatic controls would shrink to insignificance if energy-conversion equipment did not operate satisfactorily in order to assure uninterrupted power supply within set limits of voltage and frequency. It is therefore very important for all electric power engineers to have a sound knowledge of energy conversion principles.

A synchronous machine constitutes a multitude of windings characterized by time-varying self-inductances and mutual inductances. Also, saturation and magnetic nonlinearity tend to complicate matters considerably.

Physical behavior may often become obscure in the light of mathematical analysis unless one is very careful.

The main objective of this book is to present a unified development of the fundamental coupled-circuit theory of the transient performance of synchronous machines. Steady-state theory is deduced as a special case. Transient performance under balanced and unbalanced short-circuit conditions, as well as short-circuit torques have been analyzed. Various aspects of saturation and flux distribution in synchronous machines are also discussed. Formulation for computer-aided analysis from an electromagnetic field point of view is put forth. Relevant stability studies and excitation systems are also presented to make it a complete and comprehensive text on the subject. In addition, the philosophy of modeling, generator protection as well as trends in future development are briefly discussed. Emphasis is on a more or less rigorous mathematical development which is sound enough from a practical engineering point of view and on presenting a fundamental physical understanding of the machine so that the reader will be equipped with the concepts required to extend the theory as he or she needs it. It is, after all, the fundamental concepts that underlie creative engineering and become the most valuable and permanent part of a student's background.

The material that has been scattered in a few books and hundreds of technical papers has been brought together under one cover to make it available for students in a comprehensive manner. Recent topics such as flux distribution by the latest numerical methods, subsynchronous resonance and modeling techniques have been included in a textbook for the first time. The material of this text is the result of a gradual development to meet the needs of the author's classes taught at universities in the United States and India over the past fifteen years. In developing this text extensive use has been made of technical papers, most of which were published in the IEEE Transactions by R. H. Park, R. E. Doherty, C. A. Nickle, C. F. Wagner, L. A. Kilgore, S. B. Crary, C. Concordia, S. H. Wright, R. D. Evans, E. W. Kimbark, W. A. Lewis, A. W. Rankin, G. Kron, E. Clarke, E. A. Erdélyi, and many others. Without the developments made by these outstanding engineers and several IEEE committees, this textbook could not have been written. Profound influence of the earlier books written by C. Concordia, W. A. Lewis, E. W. Kimbark, A. E. Fitzgerald and some others, as well as the notes developed by H. B. Palmer, C. C. Young and N. Simons is gratefully acknowledged. The theory presented in this book is the culmination of the work of many engineers over a period of about fifty years. Acknowledgement of sources can therefore be made only through the bibliography.

This textbook may be used by machine designers and practicing power system engineers who are directly concerned with the prediction of machine performance under different conditions of operation. In the environment of a classroom in an educational institution, this text may be adopted for a two-semester sequence course at graduate level. The first half of it may also be offered as a senior undergraduate elective. The reader would be most comfortable if he or she had a usual undergraduate first course in rotating electric machinery. The reader is expected to have taken the usual undergraduate introductory circuit courses and to have been exposed to linear differential equations, Laplace transforms, and matrices.

The motivation for developing and publishing this text at this time, besides the urgent need for such a book on the subject, may best be described in terms of the following objectives:

1. A textbook that is student-oriented, comprehensive and up-to-date on the subject with consistent notation and necessary detailed explanation.
2. A text that can easily be adapted to the classroom environment, and that can also be used for independent study by a practicing engineer.
3. A text based on the material that is actually taught over the past fifteen years at different universities including Northeastern University and that incorporates student feedback.
4. A text containing the problems which have all actually been assigned to students at one time or another, and which have been solved in detail.
5. A textbook that would be within the reach of almost all the readers interested in the subject within the Unites States as well as other countries of the world, in spite of the rising cost of paper, printing, and publishing.

Assistance and encouragement were received from many sources during the preparation of the manuscript. The harmonious climate in our department provided the proper academic setting, due in great measure to the energetic, diplomatic, and intelligent leadership of our power-system engineering program director, Dr. James M. Feldman, and our department chairman, Dr. Harold R. Raemer. Finally, several typists and illustrators have worked on the manuscript in various stages and I would like to thank them as a group. I would also like to take this opportunity to thank my wife and children for their confidence and constant encouragement.

Mulukutla S. Sarma

Boston, Massachusetts

SYNCHRONOUS MACHINES

(Their Theory, Stability, and Excitation Systems)

by

Mulukutla S. Sarma

TABLE OF CONTENTS

CHAPTER I

GENERAL CONSIDERATIONS

1.1 Physical Description of a Synchronous Machine

A synchronous machine is a rotating apparatus, which is usually a part of a power system. It essentially consists of two elements: a set of armature coils and a field structure in relative motion. For constructional convenience and economic reasons, the armature winding is normally located in the stator, and the rotor contains the field winding. When the field is supplied with the direct current and is rotated by a prime mover, alternating voltages are induced in the armature coils. A three-phase machine will be considered in this text because it is most commonly used in practice and because most electric power is generated as three-phase power.

The number of poles of a synchronous machine is determined by the mechanical speed and the electric frequency at which the machine is intended to operate. The synchronous speed of a synchronous machine is that speed at which the machine normally runs under balanced, steady-state conditions, and is given by

$$n = \frac{120f}{p} \qquad (1.1.1)$$

where n is the speed in rev/min, f is the frequency in hertz, and p is the number of poles.

Hydraulic turbines operate at relatively low speeds and therefore a relatively large number of poles is required to produce the desired frequency, which is 60Hz for most power systems in the United States. A salient-pole construction is characteristic of hydroelectric-generator rotors, as it is better suited mechanically to the situation. On the other hand, steam and gas turbines operate best at relatively high speeds. These turbine-generator rotors are commonly of 2- or 4-pole cylindrical-rotor construction, and are made from a single steel forging or from several forgings. Typical salient-pole and cylindrical rotors are shown in Figures 1.1.1 and 1.1.2.

The disposition of the armature coils and the magnetic-flux paths for two- and four-pole machines are sketched in Figures 1.1.3 and 1.1.4. Since the angle included in one pole pair is 360 electrical degrees, the angle θ in electrical units is then related to the mechanical angle θ_m through the number of poles P as follows:

$$\theta = \frac{P}{2}\,\theta_m \tag{1.1.2}$$

There must be as many complete sets of armature coils as the number of pairs of poles, symmetrically distributed around the stator, and each set consists of three coils that are 120 electrical degrees apart. A typical stator of a synchronous machine with its armature conductors is shown in Figure 1.1.5. The steady-state voltages produced under balanced load conditions are always 120° apart in phase regardless of the speed of rotation of the field. The three-phase windings are either star- or delta-connected. For most of the modern high-voltage machines, they are invariably connected in star from the viewpoint of insulation requirements and availability of the neutral.

The magnetic flux produced by the field winding under

Figure 1.1.1. Salient-Pole Rotor of 2250 kva, 2400/4160 volt, 150 rpm Synchronous Generator viewed from coupling end.
(Photo courtesy of Allis-Chalmers Corporation).

Figure 1.1.2. Completely Wound Direct-Cooled Rotor of a 150 MW, 0.85 pf, 3600-rpm, Turbogenerator prior to mounting the coil-end retaining rings. (Photo courtesy of Allis-Chalmers Power Systems, Inc.).

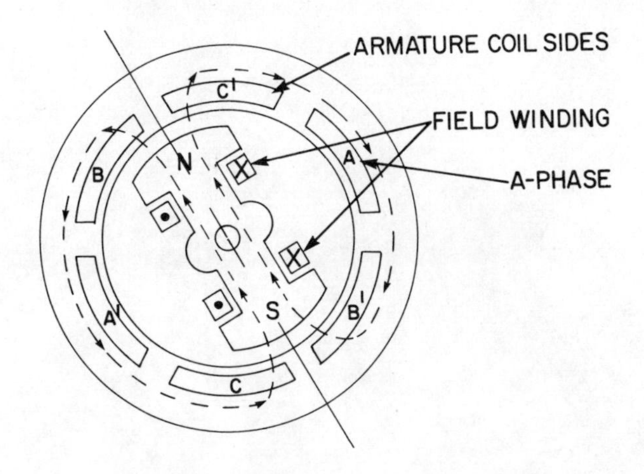

Figure 1.1.3. Disposition of Armature coils in a 2-pole Synchronous Machine.

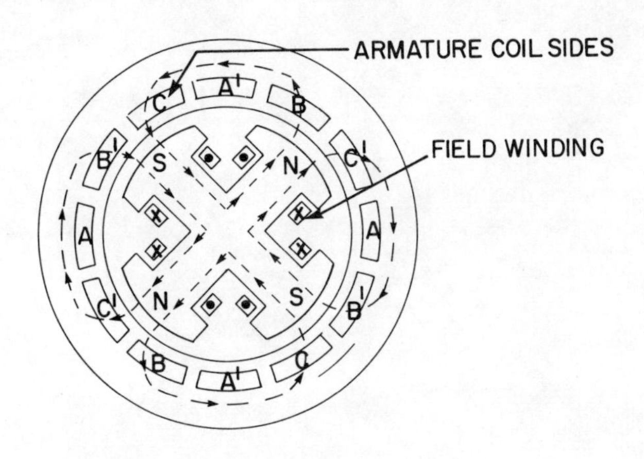

Figure 1.1.4. Disposition of Armature Coils in a 4-pole Synchronous Machine.

Figure 1.1.5. Completed Stator with Water Cooled Armature Winding. (Photo courtesy of General Electric Company).

normal operating conditions is rotating at synchronous speed with respect to the stator. The fundamental component of the magnetic field of the balanced three-phase armature currents rotates at synchronous speed and so is stationary with respect to the field. The magnetic fields of stator and rotor are constant in amplitude and stationary with respect to each other, and a steady electromagnetic torque is produced from the tendency of the two magnetic fields to align themselves. In the case of a generator this torque opposes the mechanical torque applied from the prime mover. In a motor the direction of rotation is determined by the electromagnetic torque and the speed voltages that are produced act in opposition to the applied voltages. Electromagnetic torque and rotational voltages, that are essential for electromechanical energy conversion, are produced in both generators and motors.

The magnetic flux produced by the field winding is rotating with respect to the stator windings and its supporting magnetic structure; so the voltages are produced in the iron as well as in the armature coils. Thus it becomes necessary to laminate the stator iron in order to break up the eddy-current paths and thereby minimize the eddy-current losses. An assembly of stator-core laminations is shown in Figure 1.1.6. The field structure on the other hand sustains principally a constant flux only and so it does not have to be laminated throughout. However, the space harmonics of the armature mmf* contribute to rotor-

---

*See for details any undergraduate text on electrical machines, as for example, Fitzgerald, A.E., et al, <u>Electric Machinery</u> (Ch. 3, and App. B), 3rd Edition, McGraw-Hill Book Company, New York, 1971, or Matsch, L.W., <u>Electromagnetic and Electromechanical Machines</u> (Ch. 5), Intext Educational Pub., New York, 1972.

Figure 1.1.6. Assembly of Generator Stator Core Laminations Awaiting Assembly of Winding (Photo courtesy of General Electric Company).

surface eddy-current losses, which make it desirable to laminate the surface of rotor iron at least wherever possible. Most of the salient-pole machines have laminated pole faces.

Many salient-pole synchronous machines are equipped with damper or amortisseur windings, which consist usually of a set of copper or brass bars set in pole-face slots and connected together at the ends of the machine. A sketch of damper bars in shown in Figure 1.1.7. Amortisseur windings have some effect on stability and serve several useful functions such as the starting of synchronous motors as induction motors, damping rotor oscillations, reducing overvoltages under certain short-circuit conditions, and to aid in synchronizing the machine. The cylindrical rotor of a turbine generator may be considered as equivalent to an amortisseur of infinitely many circuits.

The construction of synchronous machines may vary from one to another in several aspects. While the hydroelectric generators are commonly of the vertical type, most of the others including turbine-generators are of the horizontal type. The constructional details also depend greatly on the type of cooling methods employed. The cooling problem is particularly serious in large turbine generators, where economy, mechanical requirements, transportation limitations, erection and assembly problems demand compactness, especially for the rotor forging. Rather elaborate systems of cooling ducts have to be provided to ensure that the cooling medium removes most effectively the heat arising from the losses. Closed ventillating systems are commonly used for moderate sizes of machines. In the case of most modern large turbo-generators, water-cooled stators (the conductors have hollow passages through which cooling water

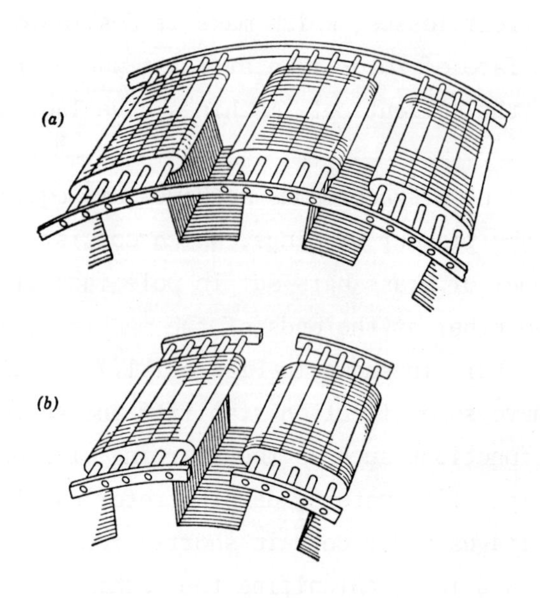

Figure 1.1.7. Sketch of Damper Bars
(a) Connected
(b) Non-Connected.

(*Reproduced by permission of the author:* E. W. Kimbark, Power System Stability: Synchronous Machines, p. 215, Dover Publications, Inc., New York, 1956/1968).

in direct contact with the copper conductors is circulated) and hydrogen-cooled rotors are most common, while water-cooling is employed in both the stator and rotor in a few cases. Sectional views of two synchronous machines with different cooling arrangements are shown in Figure 1.1.8.

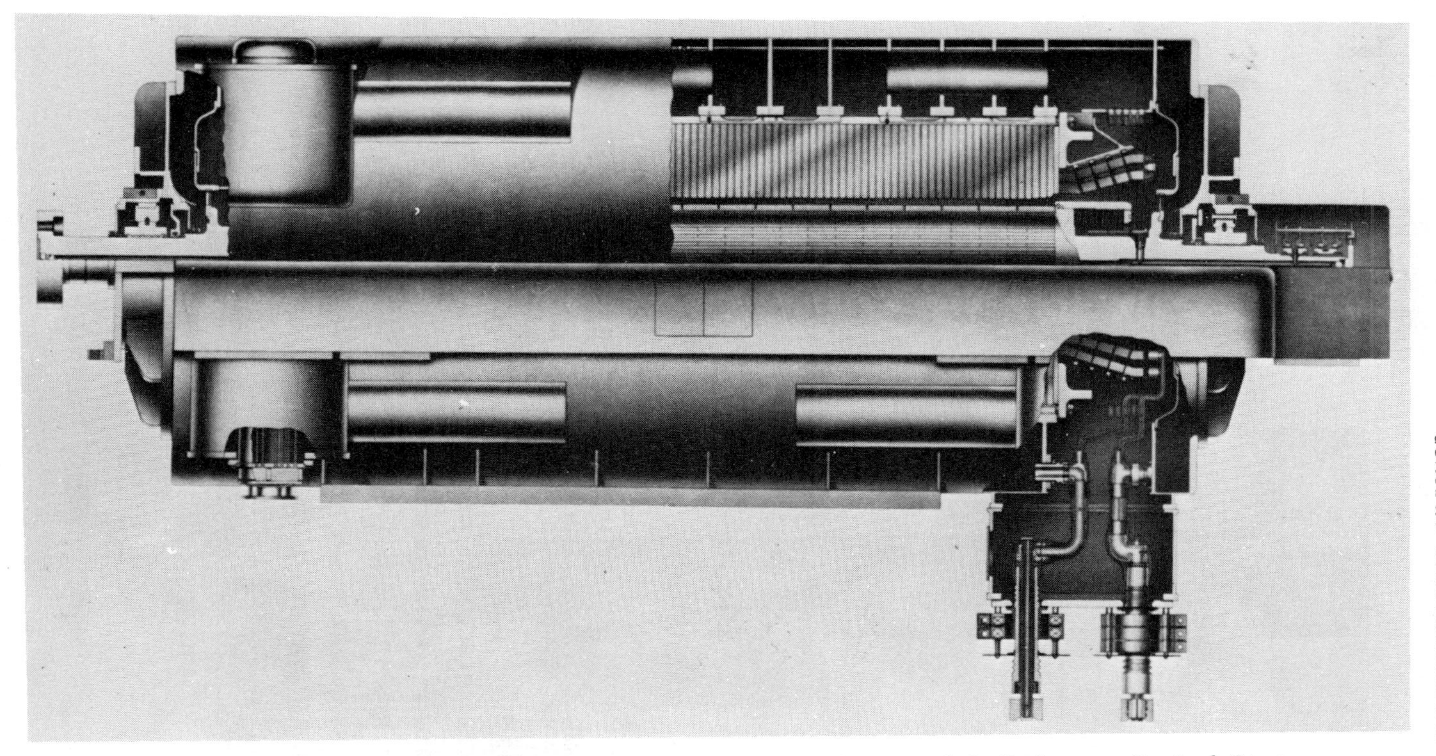

Figure 1.1.8(a). Partial cross-section of 3600-rpm Generator with Hydrogen-Cooled Stator and Rotor.
(Photo courtesy of General Electric Company).

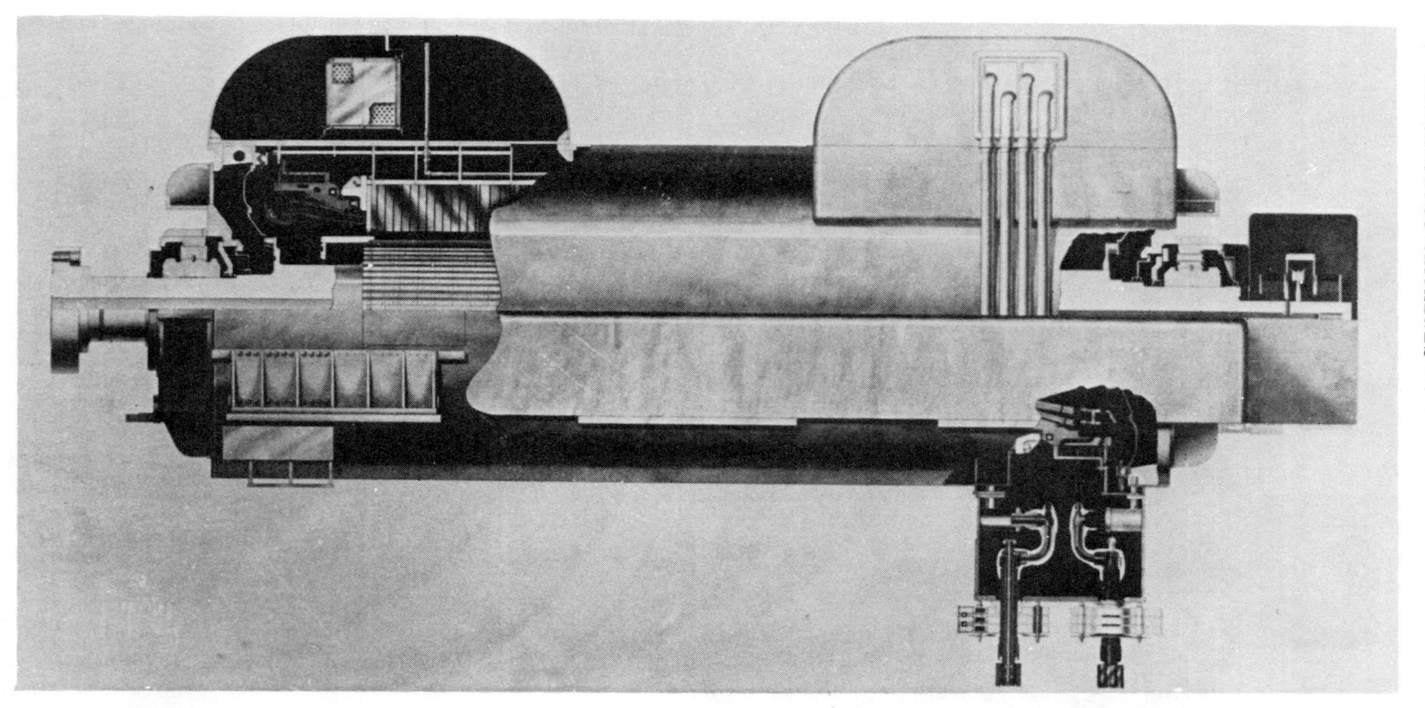

Figure 1.1.8(b). Partial Cross-section of an 1800-rpm Liquid-Cooled Generator (Photo courtesy of General Electric Company).

1.2 Field and Armature Windings

The windings of both field and armature are distributed around the periphery of the machine. The field winding of a salient-pole synchronous machine is located on the salient poles, as in the case of slow-speed engine-driven or water-wheel-driven generators. It is distributed around the poles, successive poles being wound in opposite directions so that the poles alternate magnetically north and south. On the other hand, for high-speed steam- or gas-turbine-driven cylindrical-rotor generators, the field winding is arranged in the slots of the rotor. The terminals of the field winding are brought to two collector or slip rings attached to, but insulated from, the shaft. The field current is brought to the rotor through the brushes, usually of carbon, running on these rings.

The armature winding consisting of a number of coils is distributed in the slots of the stator punchings around the periphery of the machine to correspond to the desired number of poles and phases. A 3-turn single coil of armature winding as viewed from the shaft is shown in Figure 1.2.1. A coil comprises of two coil sides which lie in two different slots of the machine. The number of slots spanned by each coil is known as the coil pitch, which may also be expressed in terms of electrical degrees, mechanical degrees, or a fraction of a full pole pitch. Two- or double-layer windings only will be considered here, as they are commonly used for all synchronous machines except the smallest ones. In such an arrangement, one coil side of each coil lies in the bottom-half of a slot, while the other coil side occupies the top-half of a slot. Thus each slot contains two coil sides of two different coils insulated

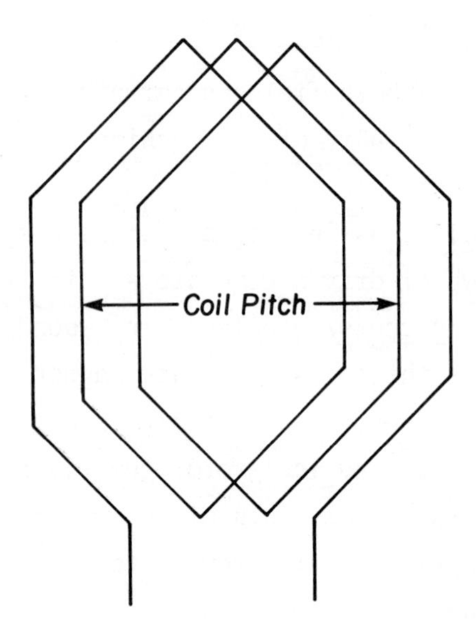

Figure 1.2.1. A 3-Turn Single Coil of Armature Winding.

from each other.

The armature windings may be classified as integral-slot windings and fractional-slot windings, depending on whether the number of slots per pole per phase (spp) is an integer or an improper fraction. The fractional-slot windings offer great flexibility of design and manufacture, and allow a wider latitude in the choice of the number of slots that can be used in the armature punchings. If all individual coils span a full pole pitch, the winding is termed as a full-pitch winding. On the other hand, a fractional-pitch, or chorded, winding is one in which any coil spans only a fraction (that is usually less than one) of a pole pitch. A few examples of windings will now be considered.

A flattened layout of a three-phase, two-pole, full-pitch winding distributed over 12 slots (2 spp) is shown in Figure 1.2.2. It may be seen that the coil sides occupying the top and bottom of each slot will always belong to the same phase in the case of full-pitch windings. All phase belts are alike when an integral number of slots per pole per phase is used. For a normal three-phase machine, the peripheral angle subtended by a phase belt is 60 electrical degrees. Positive sequence abc will be considered.

Figure 1.2.3 shows the winding layout of a 3-phase, 2-pole, fractional-pitch (5/6 of a pole pitch) winding chorded by 1 slot pitch or 30 electrical degrees, with 2 slots per pole per phase. As is typical of fractional-pitch windings, the coil sides occupying the top and bottom of some slots are of different phases. Individual phase groups are still displaced by 120 electrical degrees from the groups in other phases so that three-phase voltages are produced. Chorded windings will be found* to decrease the harmonic content of both the voltage and mmf waves significantly, lower the generated voltage a little, and somewhat shorten the end connections.

Details of the group connections connecting in series several coils of each group, inter-pole connections interconnecting several groups, and the end leads to be brought out for star or delta connection are not shown here and are left to the imagination of the reader.* Series-parallel group connections are possible and are sometimes used in a

---

*See for details any undergraduate text on electrical machines, such as Say, M.G., Alternating Current Machines, (Ch. 3), A Halsted Press Book, John Wiley & Sons, New York, 1976.

Slot Number	1	2	3	4	5	6	7	8	9	10	11	12
Bottom Layer	a	a	-c	-c	b	b	-a	-a	c	c	-b	-b
Top Layer (nearest to air gap)	a	a	-c	-c	b	b	-a	-a	c	c	-b	-b

phase belt (slots 1–2); 1 pole pitch (slot 1 to slot 7)

NOTE: The signs indicate the instantaneous directions of current flow, in and out of the paper.

Figure 1.2.2. A Flattened Layout of a 3-phase, 2-pole, Full-pitch Winding with spp = 2.

Slot Number	1	2	3	4	5	6	7	8	9	10	11	12
Bottom Layer	a	a	-c	-c	b	b	-a	-a	c	c	-b	-b
Top Layer	a	-c	-c	b	b	-a	-a	c	c	-b	-b	a

|←5/6 of a full pole pitch→|

Figure 1.2.3. A Flattened Layout of a 3-phase, 2-pole, Fractional-pitch Winding with spp = 2.

few machines, making sure that each of the parallel circuits will have the same voltage and current.

Next let us consider an example of a fractional-slot double-layer armature winding of a 3-phase, 10-pole, 42-slot synchronous machine. The corresponding layout is shown partially in Figure 1.2.4 with a coil span of 20/21 of a full pole pitch. The details given below should suffice to understand and develop the layout of any other fractional-slot winding also.

The slots per pole per phase is given by

$$\text{spp} = \frac{42}{10 \times 3} = \frac{7}{5} = \frac{m}{n} \text{ (say)} \tag{1.2.1}$$

where n is the smallest number of poles for which the coil grouping repeats, in this case 5, and m is the number of coils per phase in each repeatable group of n poles, in this case 7.

The pole span for this example measured in number of slots is given by

$$\text{pole pitch} = \frac{42}{10} = 4\ 1/5 \text{ slots} \tag{1.2.2}$$

Considering the coil pitch to be 4 slots arbitrarily, the fractional pitch works out for our example as

$$\text{fractional pitch} = \frac{4}{4\ 1/5} = \frac{4 \times 5}{21} = \frac{20}{21} \tag{1.2.3}$$

The slot pitch in electrical degrees may be calculated as

$$\text{slot pitch} = \frac{180}{4\ 1/5} = \frac{180 \times 5}{21} = \frac{300}{7} = 42\ 6/7° \tag{1.2.4}$$

Referring to Figure 1.2.4, it becomes a simple matter then

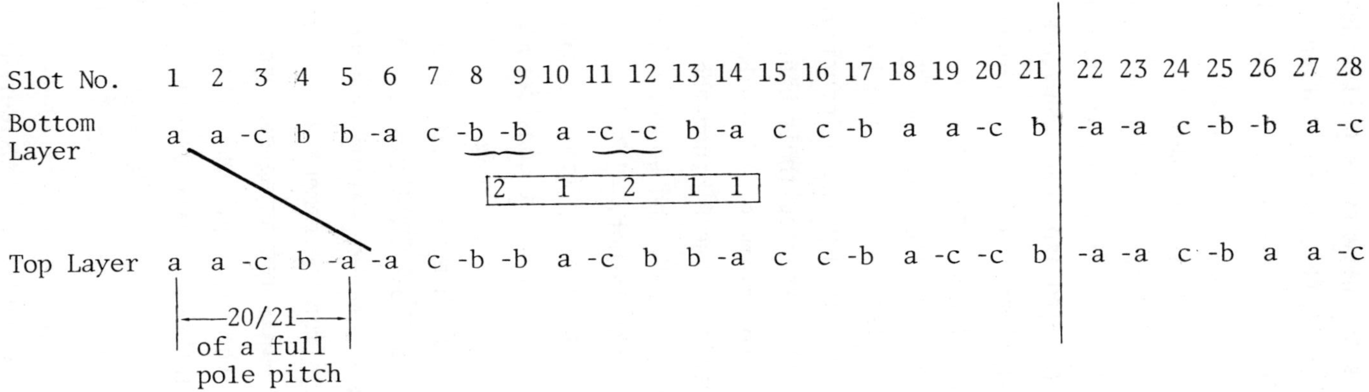

Slot No.	1	2	3	4	5	6	7	8	9	10	11	12	13	14	15	16	17	18	19	20	21	22	23	24	25	26	27	28
Bottom Layer	a	a	-c	b	b	-a	c	-b	-b	a	-c	-c	b	-a	c	c	-b	a	a	-c	b	-a	-a	c	-b	-b	a	-c
								2		1	2		1	1														
Top Layer	a	a	-c	b	-a	-a	c	-b	-b	a	-c	b	b	-a	c	c	-b	a	-c	-c	b	-a	-a	c	-b	a	a	-c

20/21 of a full pole pitch

Figure 1.2.4. Fractional-Slot, Double-Layer Armature Winding Layout of a 3-phase, 10-pole, 42-slot Synchronous Machine.

to assign the phases to the various coil sides. Starting with the bottom layer, coil side in slot number 1 is assigned phase a. Each succeeding coil side in the bottom layer will be assigned to phase a, if the angle between it and the first coil side in slot number 1 lies between 0-60, 180-240, 360-420, etc., electrical degrees. Phase c allocation is given to those coil sides, if the angle lies between 60-120, 240-300, 420-480, etc.; phase b is assigned to those coil sides, if the angle lies between 120-180, 300-360, 480-540, etc. The sign needs to be alternated whenever the phase belt changes in order to make the sequence come out as positive. Once the bottom-layer coil sides in all slots are given phase allocation, the top-layer coil sides can be assigned their phases by considering the coil pitch. For example, if the bottom layer in slot number 1 belongs to phase a, then the top layer in slot 5 will be assigned "-a", and so on. Thus, it is possible to complete the layout.

A few points are noteworthy from the resultant layout of the above example we have just considered. The winding repeats itself after 5 poles (n of Eq. 1.2.1) or 21 slots, of course with the sign changed. That is the reason why a partial layout only is shown in Figure 1.2.4. On further examination one can see the repeating sequence of 21211, disregarding the sign for the moment as it can be taken care of eventually. This repeating sequence is known as the index number of the winding. Once this is known, it is fairly easy to obtain the complete layout of the winding by assigning proper phases and relative signs. Table 1.2.1 shows the mechanics of arriving at the index number of any fractional-slot winding; the figures shown in there correspond to the example we have been considering.

Table 1.2.1

Mechanics of Arriving at the Index Numbers of a Fractional-Slot Winding

Coil No. = N	1	2	3	4	5	6	7	8	9	10	11
(N - 1) n	0	5	10	15	20	25	30	25	40	45	50
Value of k	k=0	k=0	k=0	k=1	k=2	k=2	k=3	k=4	k=5	k=5	
(N - 1) n - km	0	5	10	8	6	11	9	7	5	10	
-m			-7	-7		-7	-7	-7		-7	
Remainder (Positive)			3	1		4	2	0		3	
Assigned to phase	a		-c	b		-a	c	-b		a	
Number of Coils (Index Number)	2		1	2		1	1				

NOTE: 1. k is an integer taking values of 1, 2, 3, and so on, starting from k = 0.

2. m = 7 and n = 5 in our example.

3. Each time m can be subtracted from [(N - 1) n - km] with a positive remainder, the corresponding coil is assigned to the next phase (with proper sign), and the value of k is increased by 1 for the next coil.

(*Adapted by permission of the author:* Lewis, W. A., The Principles of Synchronous Machines, p. 6-23, Illinois Institute of Technology, Chicago, 1959].

All the coils of a repeatable group are always connected in series. The voltage generated in a fractional-slot winding (with spp = m/n) is the same as for the integral-slot winding with m coils per phase per pole and a slot pitch of (60/m) electrical degrees. It can be shown* that the distribution factors (which are introduced in the next few pages) for a fractional-slot winding are also the same as for an integral-slot winding having m coils per pole per phase.

The pitch factors** for fractional-pitch integral-slot windings of 3-phase machines are given by

$$k_{pn} = \sin\left\{\frac{n(\text{coil span in Electrical Degrees})}{2}\right\} \qquad (1.2.5)$$

where n denotes the order of the harmonic to which it applies. Pitch factors that are likely to be encountered for the usual slot combinations are given in Table 1.2.2. It may be noticed that the harmonics are substantially reduced in most cases, by comparing the pitch factors for the harmonics with the pitch factor for the fundamental. Third harmonic voltages and multiples of the third harmonic (as the ninth, fifteenth, etc.), caused by corresponding harmonic components in the flux-density wave will not appear in the line-to-line voltages of normally designed either star- or delta-connected three-phase machines. Some difficulty may occasionally be caused by such harmonics in the

---

* See Lewis, W.A., The Principles of Synchronous Machines, p. 6.19, Art. 6.8, Illinois Institute of Technology, Chicago, 1959.

** See for details of derivation any undergraduate text on electrical machines cited under Bibliography.

Table 1.2.2

Pitch Factors for Fractional-pitch Integral-slot Windings Three-phase Machines

Slots per Pole per Phase	Slots per Pole	Coil Pitch			Pitch Factor k_{pn} for n =					
		Slots	Fractional	Decimal	1 (Fund.)	3	5	7	9	11
1	3	3	1	1.0	1.0	-1.0	1.0	-1.0	1.0	-1.0
		2	2/3	0.667	0.8660	0.0000	-0.8660	0.8660	0.0000	-0.8660
2	6	5	5/6	0.833	0.9659	-0.7071	0.2588	0.2588	-0.7071	0.9659
		4	2/3	0.667	0.8660	0.0000	-0.8660	0.8660	0.0000	-0.8660
3	9	8	8/9	0.889	0.9848	-0.8660	0.6428	-0.3420	0.0000	0.3420
		7	7/9	0.778	0.9397	-0.5000	-0.1736	0.7660	-1.0000	0.7660
		6	2/3	0.667	0.8660	0.0000	-0.8660	0.8660	0.0000	-0.8660
4	12	11	11/12	0.916	0.9914	-0.9239	0.7934	-0.6088	0.3827	-0.1305
		10	5/6	0.833	0.9659	-0.7071	0.2588	0.2588	-0.7071	0.9659
		9	3/4	0.750	0.9239	-0.3827	-0.3827	0.9239	-0.9239	0.3827
		8	2/3	0.667	0.8660	0.0000	-0.8660	0.8660	0.0000	-0.8660
5	15	14	14/15	0.933	0.9945	-0.9511	0.8660	-0.7431	0.5878	-0.4067
		13	13/15	0.867	0.9782	-0.8090	0.5000	-0.1045	-0.3090	0.6691
		12	4/5	0.800	0.9511	-0.5878	0.0000	0.5878	-0.9511	0.9511
		11	11/15	0.733	0.9136	-0.3090	-0.5000	0.9782	-0.8090	0.1045
		10	2/3	0.667	0.8660	0.0000	-0.8660	0.8660	0.0000	-0.8660
6	18	17	17/18	0.945	0.9962	-0.9659	0.9063	-0.8192	0.7071	-0.5736
		16	8/9	0.889	0.9848	-0.8660	0.6428	-0.3420	0.0000	0.3420
		15	5/6	0.833	0.9659	-0.7071	0.2588	0.2588	-0.7071	0.9659
		14	7/9	0.778	0.9397	-0.5000	-0.1736	0.7660	-1.0000	0.7660
		13	13/18	0.722	0.9063	-0.2588	-0.5736	0.9962	-0.7071	-0.0872
		12	2/3	0.667	0.8660	0.0000	-0.8660	0.8660	0.0000	-0.8660
7	21	20	20/21	0.952	0.9972	-0.9749	0.9307	-0.8660	0.7817	-0.6800
		19	19/21	0.905	0.9888	-0.9010	0.7331	-0.5000	0.2226	0.0746
		18	6/7	0.857	0.9749	-0.7817	0.4338	0.0000	-0.4340	0.7820
		17	17/21	0.810	0.9556	-0.6236	0.0750	0.5000	-0.9009	0.9889
		16	16/21	0.762	0.9309	-0.4338	-0.2949	0.8660	-0.9749	0.5632
		15	5/7	0.714	0.9009	-0.2222	-0.6236	1.0000	-0.6233	-0.2229
		14	2/3	0.667	0.8660	0.0000	-0.8660	0.8660	0.0000	-0.8660

(*Reproduced by permission of the author:* Lewis, W. A., The Principles of Synchronous Machines, p. 6.7, I. I. T., Chicago, 1959.)

line-to-neutral voltages of star-connected machines; in such cases two-thirds pitch may be used. Usually the most significant harmonics to be minimized by the use of fractional-pitch windings are the fifth and seventh. Higher-frequency harmonics than the ninth are usually so small that little attention is required except in rare cases.

The distribution factors* for integral-slot windings are given by

$$k_{dn} = \frac{\sin (nm\nu/2)}{m \sin (n\nu/2)} \tag{1.2.6}$$

where n denotes the order of the harmonic to which it applies; m is the number of slots per phase belt in which the winding is distributed; and ν is the slot pitch in electrical degrees. For common three-phase windings $(m\nu/2)$ will be equal to 30 electrical degrees and so Eq. (1.2.6) may be modified accordingly. Table 1.2.3 gives the distribution factors for integral-slot windings of three-phase machines corresponding to different numbers of coils per phase belt. Distributing the winding results in a marked reduction in the harmonic content of the voltage.

The rms generated voltage per phase for a concentrated winding having N_{ph} turns in series per phase is given by a familiar expression, which is derived in any basic book on rotating machines:

$$|E_1| = \frac{2\pi \; f \; N_{ph} \; \phi_1}{\sqrt{2}} = 4.44 \; f \; N_{ph} \; \phi_1 \tag{1.2.7}$$

---

*See for details of derivation any undergraduate text on electrical machines cited under Bibliography.

Table 1.2.3

Distribution Factors for Integral-slot Windings

Three-phase Machines

Number of Coils per Phase Belt m	Distribution Factor k_{dn} for n =					
	1 (Fund.)	3	5	7	9	11
1	1.0000	1.0000	1.0000	1.0000	1.0000	1.0000
2	0.9660	0.7071	0.2588	-0.2588	-0.7071	-0.9660
3	0.9601	0.6667	0.2176	-0.1774	-0.3333	-0.1774
4	0.9579	0.6533	0.2053	-0.1575	-0.2706	-0.1261
5	0.9569	0.6472	0.2000	-0.1495	-0.2472	-0.1095
6	0.9560	0.6440	0.1972	-0.1453	-0.2357	-0.1017
7	0.9558	0.6421	0.1955	-0.1429	-0.2291	-0.09743
∞	0.9549	0.6366	0.1910	-0.1364	-0.2122	-0.08681

(*Reproduced by permission of the author:* Lewis, W. A., The Principles of Synchronous Machines, p. 6.10, I. I. T., Chicago, 1959).

where subscript 1 denotes the fundamental, f the frequency, and ϕ_1 the fundamental field flux per pole. When both the distribution factor and pitch factor for the distributed, fractional-pitch winding are applied, the rms voltage per phase is modified as

$$|E_1| = \left(\frac{2\pi}{\sqrt{2}}\right) k_{d1}\, k_{p1}\, f\, N_{ph}\, \phi_1 \tag{1.2.8}$$

for the fundamental and

$$|E_n| = \left(\frac{2\pi}{\sqrt{2}}\right) k_{dn}\, k_{pn}\, k_{fn}\, f\, N_{ph}\, \phi_1 \tag{1.2.9}$$

for the nth harmonic; k_{fn} is the coefficient depending on the nature of the rotor surface, corresponding to the nth harmonic. Typical values of k_{fn} for the odd harmonics through the eleventh range from 0.01 to 0.06. The ratio of the voltage magnitude of any harmonic to the fundamental is then given by

$$\frac{|E_n|}{|E_1|} = \frac{k_{dn}\, k_{pn}\, k_{fn}}{k_{d1}\, k_{p1}} \tag{1.2.10}$$

For normal designs and machine proportions, the ratios (k_{dn}/k_{d1}), (k_{pn}/k_{p1}), and the coefficient k_{fn} are each much less than unity, so that their product results in a very small quantity; with the result the harmonics in the terminal voltage may usually be neglected under normal conditions.

The rms value of the complete voltage wave will be given by

$$|E| = |E_1| \sqrt{1 + \frac{|E_3|^2}{|E_1|^2} + \frac{|E_5|^2}{|E_1|^2} + \ldots} \qquad (1.2.11)$$

Since the value of the radical will be very nearly unity, the rms value of the complete voltage is very nearly the same as the rms value of the fundamental alone, when the harmonics are small in comparison with the fundamental. Thus, for practical purposes, the rms value of the generated voltage per phase will be given by

$$|E_{ph}| = 4.44\ k_{d1}\ k_{p1}\ f\ N_{ph}\ \phi_1 \qquad (1.2.12)$$

where k_{d1} is the distribution factor for the fundamental, k_{p1} the pitch factor for the fundamental, f the frequency, N_{ph} is the number of turns in series per phase, and ϕ_1 the fundamental field flux per pole.

This section is intentionally kept concise to provide the necessary background. For additional details and elaborate explanation, the reader is referred to the book by Lewis quoted under Bibliography.

1.3 Mathematical Description of a Synchronous Machine

The assumptions on which the model and the present analysis are based are listed below:

(1) The stator windings are sinusoidally distributed around the periphery along the air gap as far as all mutual effects with the rotor are concerned.

(2) The effect of the stator slots on the variation of any of the rotor inductances with rotor angle

is neglected.

(3) Saturation may be neglected at least for the present.

Justification for any assumption comes from the comparison of performance calculated on that basis with actual performance obtained by test. It will be found that the first two assumptions are quite reasonable, while the third one is not, as saturation does significantly affect certain aspects of the machine performance.

The electrical performance equations of a synchronous machine will now be developed from a coupled-circuit viewpoint. The generator convention for polarities will be adopted so that positive current corresponds to generator action. By Faraday's law, the voltage induced in the stator coil is

$$e = -\frac{d\lambda}{dt} \tag{1.3.1}$$

where λ is the flux linkage with the stator coil and the minus sign associated with Faraday's law implies generator reference directions. That is, while the flux linking the coil is decreasing, an emf will be induced in it in a direction to try to produce a current which would tend to prevent the flux linking it from decreasing.

Based on the above, one can write the voltage relations for the armature or stator circuits as follows:

$$e_a = p\lambda_a - ri_a \tag{1.3.2}$$

$$e_b = p\lambda_b - ri_b \tag{1.3.3}$$

$$e_c = p\lambda_c - ri_c \tag{1.3.4}$$

where

e_a = terminal voltage of phase a

λ_a = total flux linkage of phase a

i_a = current in phase a

a,b,c are the three phases

r = the resistance of each armature winding, assumed to be the same for the three phases

p = the derivative operator d/dt; t is time.

As for the rotor field circuit, whenever the flux linkage is subject to change, a voltage drop in the direction of positive current will occur in the coil, and the resistance of the field circuit also gives rise to a voltage drop in the same direction, so that the voltage equation may be written as

$$e_f = p\lambda_f + r_f i_f \tag{1.3.5}$$

The symbols e, λ and i have the same meaning as indicated earlier, and the subscript f denotes the circuit in question.

From the physical description of the synchronous machine, the field winding has its axis in line with the pole axis, and assuming reasonably that the rotor magnetic paths and all of its electric circuits are symmetrical about both the pole and interpolar axes for a salient-pole machine, one may choose conveniently the pole axis as direct axis and the interpolar axis as quadrature axis, the angle between the two axes being 90 electrical degrees. The quadrature axis is taken as 90 electrical degrees ahead of the direct axis in the direction of normal rotor rotation. Figure 1.3.1 shows the schematic diagram of phase windings and field winding along with the notations adopted for voltages and currents, as well as the location of the direct and quadrature axes. A salient two-pole machine has been chosen for simplicity and convenience.

A diagram of damper circuits is given in Figure 1.3.2 (flattened layout is chosen for convenience), showing the direct-axis circuits numbered as 1d, 2d, etc., and quadrature-axis circuits numbered as 1q, 2q, etc. The symmetrical choice of the rotor circuits has the virtue of making all mutual inductances and resistances between direct- and quadrature-axis rotor circuits equal to zero. However, the amortisseur circuits are resistance-coupled as well as inductance coupled. The voltage equations of direct-axis and quadrature-axis amortisseur circuits will then be given by:

$$0 = p\lambda_{1d} + r_{11d}i_{1d} + r_{12d}i_{2d} + \cdots \tag{1.3.6}$$

$$0 = p\lambda_{2d} + r_{21d}i_{1d} + r_{22d}i_{2d} + \cdots \tag{1.3.7}$$

etc.

$$0 = p\lambda_{1q} + r_{11q}i_{1q} + r_{12q}i_{2q} + \cdots \tag{1.3.8}$$

$$0 = p\lambda_{2q} + r_{21q}i_{1q} + r_{22q}i_{2q} + \cdots \tag{1.3.9}$$

etc.

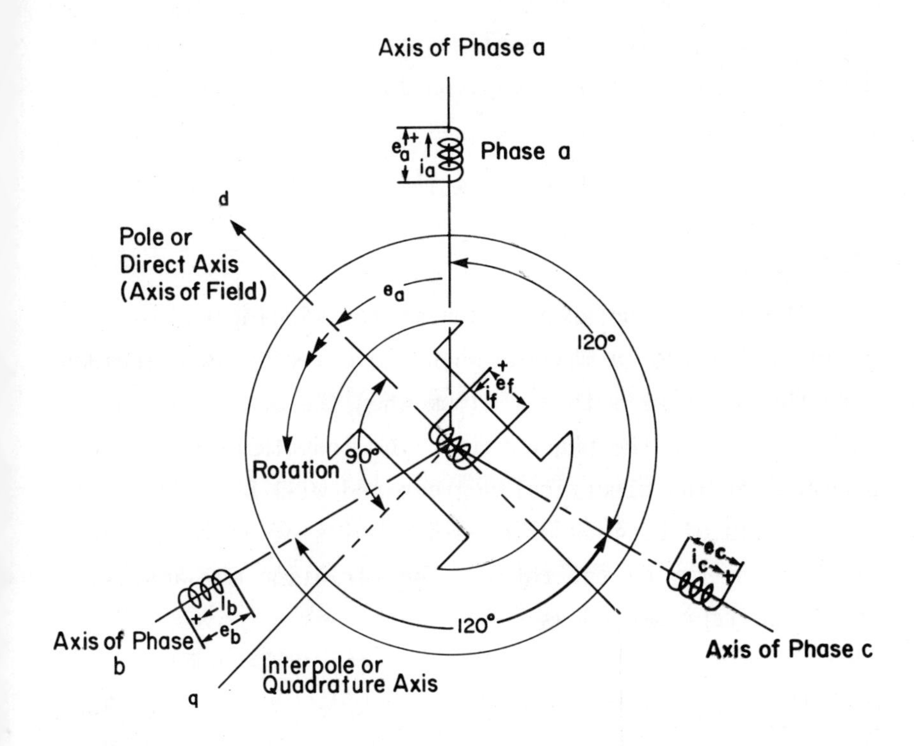

Figure 1.3.1. Diagram Showing Conventions.

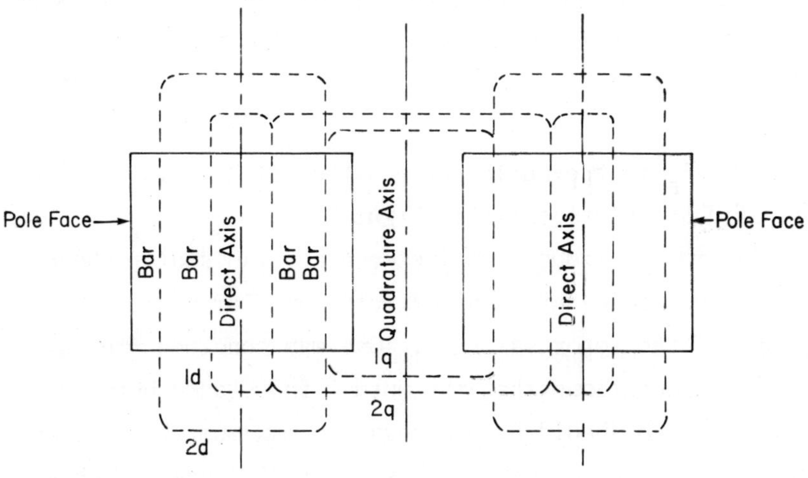

Figure 1.3.2. Diagram of damper circuits.

Since the field axis aligns itself with the pole or direct axis, Eq. (1.3.5) may consistently be written as

$$e_{fd} = p\lambda_{fd} + r_{fd}i_{fd} \tag{1.3.10}$$

Inductances:

The self-inductance of any stator winding varies periodically from a maximum when the direct axis coincides with the phase axis to a minimum when the quadrature axis is in line with the phase axis. The inductance will have a period of 180 electrical degrees and will be expressible by a series of cosines of even harmonics of angle, in view of the symmetry of the rotor. The variation may approximately be represented as

$$\ell_{aa} = L_{aa0} + L_{aa2} \cos 2\theta \tag{1.3.11}$$

$$\ell_{bb} = L_{aa0} + L_{aa2} \cos 2(\theta - 120°) \tag{1.3.12}$$

$$\ell_{cc} = L_{aa0} + L_{aa2} \cos 2(\theta + 120°) \tag{1.3.13}$$

where $\theta = \theta_a$ is the angle between the direct axis and the axis of phase a, as shown in Figure 1.3.1.

The mutual inductances between any two stator phases are also periodic functions of rotor angular position because of the rotor saliency. One may conclude from symmetry considerations that the mutual inductance between phases a and b should have a negative maximum when the pole axis is lined up 30° behind phase a or 30° ahead of phase b, and a negative minimum when it is midway between the two phases.

It can be shown* that the variable part of the mutual inductance is of exactly the same magnitude as that of the variable part of the self-inductance, and that the constant part has a magnitude of very nearly half that of the constant part of the self-inductance. Thus the variations of stator mutual inductances may be represented as

$$\ell_{ab} = \ell_{ba} = -[L_{ab0} + L_{aa2} \cos 2(\theta + 30°)] \qquad (1.3.14)$$

$$\ell_{bc} = \ell_{cb} = -[L_{ab0} + L_{aa2} \cos 2(\theta - 90°)] \qquad (1.3.15)$$

$$\ell_{ca} = \ell_{ac} = -[L_{ab0} + L_{aa2} \cos 2(\theta + 150°)] \qquad (1.3.16)$$

All the rotor self-inductances, ℓ_{ffd}, ℓ_{11d}, ℓ_{22d}, ℓ_{11q}, etc., are constants since the effects of stator slots and of saturation are neglected.

Mutual inductance between any rotor direct- and any quadrature-axis circuit vanishes. All mutual inductances between any two circuits both in the direct axis, and between any two circuits both in the quadrature axis are constant.

$$\ell_{fd1q} = \ell_{fd2q} = \ell_{1d1q} = \ell_{1d2q} = \ell_{1qfd} = \ell_{1q1d}$$
$$= \ell_{1q2d} = 0 \quad , \text{etc.} \qquad (1.3.17)$$

$$\ell_{f1d} = \ell_{1fd} = \text{constant} \quad , \text{etc.} \qquad (1.3.18)$$

---

*See Concordia, C., Synchronous Machines, (Ch. 2), John Wiley & Sons, Inc., New York, 1951.

Next let us consider the mutual inductances between stator and rotor circuits, which are periodic functions of rotor angular position. Since only the space-fundamental component of the flux produced will link the sinusoidally distributed stator, all stator-rotor mutual inductances will vary sinusoidally, reaching a maximum when the two coils in question align. Thus their variations may be written as:

$$\ell_{afd} = \ell_{fda} = L_{afd} \cos \theta \tag{1.3.19}$$

$$\ell_{bfd} = \ell_{fdb} = L_{afd} \cos (\theta - 120°) \tag{1.3.20}$$

$$\ell_{cfd} = \ell_{fdc} = L_{afd} \cos (\theta + 120°) \tag{1.3.21}$$

$$\ell_{a1d} = \ell_{1da} = L_{a1d} \cos \theta \tag{1.3.22}$$

$$\ell_{b1d} = \ell_{1db} = L_{a1d} \cos (\theta - 120°) \tag{1.3.23}$$

$$\ell_{c1d} = \ell_{1dc} = L_{a1d} \cos (\theta + 120°), \text{ etc.} \tag{1.3.24}$$

$$\ell_{a1q} = \ell_{1qa} = L_{a1q} \cos (\theta + 90°) = -L_{a1q} \sin \theta \tag{1.3.25}$$

$$\ell_{b1q} = \ell_{1qb} = -L_{a1q} \sin (\theta - 120°) \tag{1.3.26}$$

$$\ell_{c1q} = \ell_{1qc} = -L_{a1q} \sin (\theta + 120°), \text{ etc.} \tag{1.3.27}$$

Flux-Linkage Relations:

The rotor flux-linkage equations are given below:

$$\lambda_{fd} = \ell_{ffd}i_{fd} + \ell_{f1d}i_{1d} + \ell_{f2d}i_{2d} + \cdots + \ell_{f1q}i_{1q} +$$

$$+ \ell_{f2q}i_{2q} + \cdots - \ell_{fda}i_a - \ell_{fdb}i_b - \ell_{fdc}i_c \qquad (1.3.28)$$

$$\lambda_{1d} = \ell_{1fd}i_{fd} + \ell_{11d}i_{1d} + \ell_{12d}i_{2d} + \cdots + \ell_{1d1q}i_{1q} +$$

$$+ \ell_{1d2q}i_{2q} + \cdots - \ell_{1da}i_a - \ell_{1db}i_b - \ell_{1dc}i_c \qquad (1.3.29)$$

$$\lambda_{1q} = \ell_{1qfd}i_{fd} + \ell_{1q1d}i_{1d} + \ell_{1q2d}i_{2d} + \cdots + \ell_{11q}i_{1q} +$$

$$+ \ell_{12q}i_{2q} + \cdots - \ell_{1qa}i_a - \ell_{1qb}i_b - \ell_{1qc}i_c \qquad (1.3.30)$$

The above equations may be rewritten as follows after substituting the inductance variations that have been obtained earlier:

$$\lambda_{fd} = -L_{afd}[i_a \cos \theta + i_b \cos (\theta - 120°) +$$

$$+ i_c \cos (\theta + 120°)] + L_{ffd}i_{fd} + L_{f1d}i_{1d} +$$

$$+ L_{f2d}i_{2d} + \cdots \qquad (1.3.31)$$

$$\lambda_{1d} = -L_{a1d}[i_a \cos \theta + i_b \cos (\theta - 120°) +$$

$$+ i_c \cos (\theta + 120°)] + L_{1fd}i_{fd} + L_{11d}i_{1d} +$$

$$+ L_{12d}i_{2d} + \cdots \qquad (1.3.32)$$

$$\lambda_{1q} = +L_{a1q}[i_a \sin\theta + i_b \sin(\theta - 120°) +$$

$$+ i_c \sin(\theta + 120°)] + L_{11q}i_{1q} + L_{12q}i_{2q} + \cdots \tag{1.3.33}$$

The form of the above equations suggests the introduction of new variables for simplification:

$$i_d = K[i_a \cos\theta + i_b \cos(\theta - 120°) + i_c \cos(\theta + 120°)] \tag{1.3.34}$$

$$i_q = -K[i_a \sin\theta + i_b \sin(\theta - 120°) +$$

$$+ i_c \sin(\theta + 120°)] \tag{1.3.35}$$

where K could be any constant chosen for convenience. i_d and i_q may be seen to be proportional to the components of mmf in the direct and quadrature axes respectively, produced by the resultant of all three armature currents i_a, i_b and i_c. For balanced phase currents of any given maximum magnitude, the maximum values of i_d and i_q can be made of the same magnitude. Under the balanced conditions, the maximum magnitude of any one of the phase currents will then be given by $\sqrt{i_d^2 + i_q^2}$. In order to achieve this, a value of 2/3 will be assigned to the constant K in the above equations.

It would appear as the next logical step to eliminate the old variables i_a, i_b and i_c in favor of the newly introduced variables. If three currents are to be eliminated, three substitute variables will be required in general. Hence, we need to introduce another new variable i_o, which is the conventional zero-phase sequence current of the symmetrical-component theory.

$$i_o = +\frac{1}{3}(i_a + i_b + i_c) \tag{1.3.36}$$

We have now established the transformation from the "abc" phase variables to the "dqo" variables, which may be expressed as follows in a matrix notation:

$$\begin{bmatrix} i_d \\ i_q \\ i_o \end{bmatrix} = \begin{bmatrix} \frac{2}{3}\cos\theta & \frac{2}{3}\cos(\theta - 120°) & \frac{2}{3}\cos(\theta + 120°) \\ -\frac{2}{3}\sin\theta & -\frac{2}{3}\sin(\theta - 120°) & -\frac{2}{3}\sin(\theta + 120°) \\ \frac{1}{3} & \frac{1}{3} & \frac{1}{3} \end{bmatrix} \begin{bmatrix} i_a \\ i_b \\ i_c \end{bmatrix} \tag{1.3.37}$$

The above may be rewritten as given below for the sake of simplicity:

$$[i_B] = [A][i_p] \tag{1.3.38}$$

where i_B are the Blondel components, i_p the phase components, and A the Blondel transformation matrix. The same transformation matrix can be applied to the flux linkages as well as the voltages just as it has been applied to the currents.

It may be pointed out here that one can find the phase components from the Blondel components through inverse Blondel transformation.

$$[i_p] = [A]^{-1}[i_B] \tag{1.3.39}$$

where $[A]^{-1}$ can be seen to be

$$[A]^{-1} = \begin{bmatrix} \cos\theta & -\sin\theta & 1 \\ \cos(\theta - 120°) & -\sin(\theta - 120°) & 1 \\ \cos(\theta + 120°) & -\sin(\theta + 120°) & 1 \end{bmatrix} \qquad (1.3.40)$$

Going back to the rotor flux-linkage equations (1.3.31) to (1.3.33), and substituting the new variable currents, one obtains:

$$\lambda_{fd} = -\frac{3}{2} L_{afd} i_d + L_{ffd} i_{fd} + L_{f1d} i_{1d} + \cdots \qquad (1.3.41)$$

$$\lambda_{1d} = -\frac{3}{2} L_{a1d} i_d + L_{1fd} i_{fd} + L_{11d} i_{1d} + \cdots \qquad (1.3.42)$$

$$\lambda_{1q} = -\frac{3}{2} L_{a1q} i_q + L_{11q} i_{1q} + L_{12q} i_{2q} + \cdots \qquad (1.3.43)$$

The above equations contain the inductances that are all constant, as indicated by the capital L's.

Next the armature flux-linkage relations can be written as:

$$\lambda_a = -\ell_{aa} i_a - \ell_{ab} i_b - \ell_{ac} i_c + \ell_{afd} i_{fd} + \ell_{a1d} i_{1d} +$$

$$+ \ell_{a2d} i_{2d} + \cdots + \ell_{a1q} i_{1q} + \ell_{a2q} i_{2q} + \cdots \qquad (1.3.44)$$

$$\lambda_b = -\ell_{ba} i_a - \ell_{bb} i_b - \ell_{bc} i_c + \ell_{bfd} i_{fd} + \ell_{b1d} i_{1d} +$$

$$+ \ell_{b2d} i_{2d} + \cdots + \ell_{b1q} i_{1q} + \ell_{b2q} i_{2q} + \cdots \qquad (1.3.45)$$

$$\lambda_c = -\ell_{ca} i_{ca} - \ell_{cb} i_b - \ell_{cc} i_c + \ell_{cfd} i_{fd} + \ell_{c1d} i_{1d} +$$

$$+ \ell_{c2d} i_{2d} + \cdots + \ell_{c1q} i_{1q} + \ell_{c2q} i_{2q} + \cdots \qquad (1.3.46)$$

The inductance variations that have been obtained earlier may now be substituted in the above equations. The new variables of flux linkages (λ_d, λ_q and λ_o) may now be introduced in terms of the phase variables (λ_a, λ_b and λ_c), and relatively simple relations given below may be obtained after considerable simplification, the unexciting details of which are left out to the student as a desirable exercise:

$$\lambda_d = -(L_{aao} + L_{abo} + \frac{3}{2} L_{aa2})\, i_d + L_{afd} i_{fd} + L_{a1d} i_{1d} + \\ + L_{a2d} i_{2d} + \cdots \qquad (1.3.47)$$

$$\lambda_q = -(L_{aao} + L_{abo} - \frac{3}{2} L_{aa2})\, i_q + L_{a1q} i_{1q} + \\ + L_{a2q} i_{2q} + \cdots \qquad (1.3.48)$$

$$\lambda_o = -(L_{aao} - 2L_{abo})\, i_o \qquad (1.3.49)$$

λ_d and λ_q may be interpreted as the flux linkages in coils moving with the rotor and centered over the direct and quadrature axes respectively. The equivalent direct-axis moving armature circuit can be seen to have the self-inductance

$$L_d = L_{aao} + L_{abo} + \frac{3}{2} L_{aa2} \qquad (1.3.50)$$

which is known as the direct-axis synchronous inductance. The equivalent quadrature-axis moving armature circuit has the self-inductance

$$L_q = L_{aao} + L_{abo} - \frac{3}{2} L_{aa2} \qquad (1.3.51)$$

which is known as the quadrature-axis synchronous inductance. There is also an equivalent zero-sequence axis coil, which is completely separated magnetically from all the other coils and which has the self-inductance

$$L_o = L_{aao} - 2L_{abo} \tag{1.3.52}$$

which is known as the zero-sequence inductance.

Armature Voltage Equations:

The new voltages e_d, e_q and e_o may now be defined in the same manner as the currents and flux linkages through Blondel transformation. The equations (1.3.2) to (1.3.4) may then be substituted for e_a, e_b and e_c; the new currents i_d, i_q and i_o may be introduced in order to eliminate the phase quantities i_a, i_b and i_c. Then the armature voltage relations could be rewritten as

$$e_d = p\lambda_d - \lambda_q p\theta - ri_d \tag{1.3.53}$$

$$e_q = p\lambda_q + \lambda_d p\theta - ri_q \tag{1.3.54}$$

$$e_o = p\lambda_o - ri_o \tag{1.3.55}$$

From a physical standpoint the manipulations through the transformation correspond to the specification of the armature quantities along axes fixed to the rotor and rotating with speed, $p\theta$, with respect to the stator. One can therefore naturally expect to find the generated- or speed-voltages as well as the induced voltages produced by the rotating flux linkages. The representation of a synchronous machine in terms of direct-axis, quadrature-axis and zero-sequence windings is given in Figure 1.3.3.

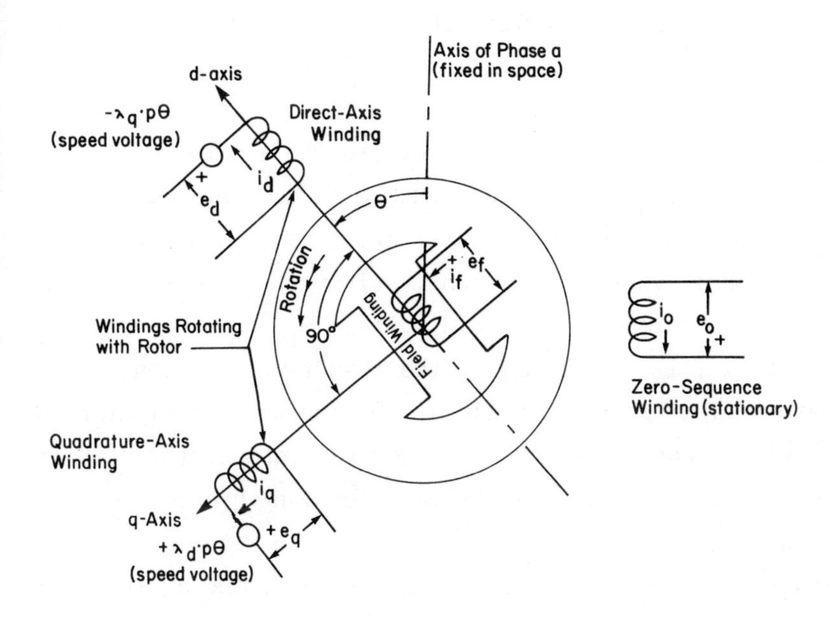

Figure 1.3.3. Representation of a Synchronous Machine in terms of Direct-Axis, Quadrature-Axis, and Zero-Sequence Windings.

The complete set of machine-performance equations consisting of flux-linkage relations and circuit-voltage equations have now been obtained. These are known as Park's equations, which may be realized to be linear differential equations with constant coefficients, if the rotor speed is a constant.

Another point needs to be made in this context. Looking at the flux-linkage equations given by Eqs. (1.3.41) to (1.3.43) and (1.3.47), (1.3.48), the reciprocity of the mutual-inductance coefficients, which is an essential condition for the existence of a static equivalent circuit, is not fulfilled. This difficulty is due to the transformation used for both current and flux linkage, with the choice of $K = 2/3$. It could have been avoided easily by other choices of transformation. However, the difficulty could be circumvented by suitably defining a per-unit system, which will be introduced in Section 1.4.

It is often desirable to rewrite the armature flux-linkage relations in a more suitable form. This can be done by substituting the rotor flux-linkage relations into the rotor-circuit voltage equations, solving those for the rotor currents in terms of the field voltage e_{fd} and the armature currents i_d, i_q, and plugging the resultant relations in the armature flux-linkage equations. During these manipulations, it would be easier to treat the derivative operator $p(=d/dt)$ algebraically, as it is legitimate for many problems since all the flux-linkage equations and all the rotor-circuit voltage relations are linear. As a result, one arrives at the equations of the following form:

$$\lambda_d = G(p)\, e_{fd} - L_d(p)\, i_d \tag{1.3.56}$$

$$\lambda_q = -L_q(p)\, i_q \tag{1.3.57}$$

$$\lambda_o = -L_o i_o \tag{1.3.58}$$

where $G(p)$, $L_d(p)$ and $L_q(p)$ are operators expressed as functions of the derivative operator p. $L_d(p)$, $L_q(p)$ and L_o are known as the operational inductances.

Power Output:

The instantaneous power output of a three-phase synchronous generator is given by

$$P = [e_a i_a + e_b i_b + e_c i_c] \tag{1.3.59}$$

In terms of the d, q, o components, it can be shown to be

$$P = \frac{3}{2} [e_d i_d + e_q i_q + 2e_o i_o] \tag{1.3.60}$$

For balanced operation, under which the zero-sequence quantities vanish, the power output may be expressed as

$$P = \frac{3}{2} [e_d i_d + e_q i_q] \tag{1.3.61}$$

Equation (1.3.60) may be rewritten by substituting the armature voltage relations given by Eqs. (1.3.53) to (1.3.55) for e_d, e_q and e_o:

$$P = \frac{3}{2} [i_d\, p\lambda_d + i_q\, p\lambda_q + 2i_o\, p\lambda_o] + \frac{3}{2} [i_q\, \lambda_d - i_d\, \lambda_q]\, p\theta + - \frac{3}{2} r\, [i_d^2 + i_q^2 + 2i_o^2] \tag{1.3.62}$$

The above expression for the power output may be interpreted further. The first term on the right-hand-side shows the rate of decrease of armature magnetic energy; the second term indicates the power transferred across air gap electromagnetically; and the third term is the armature resistance loss.

Electromagnetic Torque:

The electromagnetic torque is obtained by dividing the power transferred across air gap by the rotor speed, $p\theta$. Thus one has

$$T = \frac{3}{2} [\lambda_d\, i_q - \lambda_q\, i_d] \tag{1.3.63}$$

We have so far developed the volt-ampere relations, flux-linkage equations, expressions for power output and electromagnetic torque produced for the case of a synchronous machine.

1.4. Per-Unit Representation

A per-unit system is essentially a system of dimensionless parameters occuring in a set of wholly or partially dimesnionless equations. Systems of this kind are adopted extensively to simplify phenomena over a wide range of different physical problems. The dimensionless groups mostly have forms which are quite simple, nearly all being derived by a process of normalization in which one physical parameter is divided by another of the same dimension. The denominators are referred to as "base" quantities. These are chosen because of the ways in which they characterize particular features of the physical system. The magnitudes of some base quantities may be chosen freely and quite arbitrarily. The magnitudes of others follow by dependence through the laws governing the physical nature of the system. The choice of bases should be so made that the computational effort is minimized as much as possible, and evaluation as well as understanding of the main characteristics is made as simple and direct as possible. Normally for any device, the principal per-unit variables assume unit magnitude under rated conditions. The parameters of machines lie in a reasonably narrow numerical range when expressed in a per-unit system related to their rating. The per-unit system is also very useful in simulating machine systems on analog and digital computers for transient and dynamic behavior.

The rated volt-amperes of the machine and the rated frequency will be chosen as the stator 3-phase base volt-amperes (or power) and base frequency respectively. The stator base voltage is selected as the peak value of the rated line to neutral voltage, and the stator base current is the peak value of the rated phase current.

$$e_{s_{base}} = \sqrt{2}\, V_{ph} \tag{1.4.1}$$

and

$$i_{s_{base}} = \sqrt{2}\, I_{ph} \tag{1.4.2}$$

where V_{ph} is the rated rms line to neutral voltage, and I_{ph} is the rated rms phase current. Note that I_{ph} may simply be calculated as

$$I_{ph} = \frac{\text{Rated volt-amperes}}{3V_{ph}} \tag{1.4.3}$$

The stator base impedance works out as

$$Z_{s_{base}} = \frac{e_{s_{base}}}{i_{s_{base}}} \tag{1.4.4}$$

Then it follows for the stator inductance base as

$$L_{s_{base}} = \frac{Z_{s_{base}}}{\omega_{base}} \tag{1.4.5}$$

where

$$\omega_{base} = 2\pi f_{base} \tag{1.4.6}$$

The stator base flux linkage is given by

$$\lambda_{s_{base}} = \frac{e_{s_{base}}}{\omega_{base}} \tag{1.4.7}$$

and also, the volt-ampere rating of the machine may be expressed as

$$\text{stator 3-phase base volt-amperes (or power)} = \frac{3}{2}\, e_{s_{base}}\, i_{s_{base}} \tag{1.4.8}$$

The time t (seconds) can be made dimensionless by mutliplying it with ω_{base}, and the rotor speed $p\theta$ for steady-state synchronous-speed operation becomes unity in per unit notation. The per-unit reactance can be seen to be the same as per-unit inductance, and as such the terms can be used interchangeably.

The base systems for stator quantities is now completed and the armature voltage equations may now be expressed in per-unit notation. It turns out that the Eqs. (1.3.53) to (1.3.55) are unchanged, when all the quantities involved are expressed in per-unit system as described above.

Next we need to choose the base system for rotor quantities. In this context we may recall from the flux-linkage relations given by Eqs. (1.3.41) to (1.3.43) and (1.3.47), (1.3.48) that the reciprocity of the mutual inductance coefficients was not satisfied. So it would be desirable to select the rotor base system such that the reciprocal per-unit mutual inductances between the rotor and stator are maintained. With this aim in mind we shall proceed to utilize the degrees of freedom we have in choosing the bases for e_{fd}, i_{fd}, i_{kd}, i_{kq} and λ_{fd}.

Let us now consider the following flux linkage equations

given by our earlier Eqs. (1.3.41) and (1.3.47):

$$\lambda_{fd} = -\frac{3}{2} L_{afd} i_d + L_{ffd} i_{fd} + L_{f1d} i_{1d} + \cdots \tag{1.4.9}$$

$$\lambda_d = -L_d i_d + L_{afd} i_{fd} + L_{a1d} i_{1d} + \cdots \tag{1.4.10}$$

Using the basic relationships

$$\lambda_{s_{base}} = L_{s_{base}} i_{s_{base}} \tag{1.4.11}$$

and

$$\lambda_{fd_{base}} = L_{fd_{base}} i_{fd_{base}} \tag{1.4.12}$$

one can rewrite the Eqs. (1.4.9) and (1.4.10) as follows in per-unit notation:

$$\lambda_{fd} = -L_{fda} i_d + L_{ffd} i_{fd} + L_{f1d} i_{1d} + \cdots \tag{1.4.13}$$

$$\lambda_d = -L_d i_d + L_{afd} i_{fd} + L_{a1d} i_{1d} + \cdots \tag{1.4.14}$$

where

$$L_{fda} \text{ in per-unit} = \frac{3}{2} L_{afd} \frac{i_{s_{base}}}{\lambda_{fd_{base}}} \tag{1.4.15}$$

$$L_{f1d} \text{ in per-unit} = L_{f1d} \frac{i_{1d_{base}}}{\lambda_{fd_{base}}}, \text{ and so on.} \tag{1.4.16}$$

$$L_{afd} \text{ in per-unit} = L_{afd} \frac{i_{fd_{base}}}{\lambda_{s_{base}}} \tag{1.4.17}$$

and

$$L_{a1d} \text{ in per-unit} = L_{a1d} \frac{i_{1d_{base}}}{\lambda_{s_{base}}}, \text{ and so on.} \qquad (1.4.18)$$

In order to have equal values for L_{afd} and L_{fda} in per-unit so that reciprocity is maintained, it follows

$$\lambda_{fd_{base}} i_{fd_{base}} = \frac{3}{2} \lambda_{s_{base}} i_{s_{base}} \qquad (1.4.19)$$

Multiplying by ω_{base} on both sides of the above equation, one has

$$e_{fd_{base}} i_{fd_{base}} = \frac{3}{2} e_{s_{base}} i_{s_{base}} \qquad (1.4.20)$$

Thus we have to take the volt-ampere base of each rotor circuit to be equal to the three-phase stator volt-ampere base in order to have reciprocal per-unit mutual inductances between the rotor and stator.

Next let us attempt to make all per-unit mutual inductances equal between the rotor and stator circuits in each axis, as it is more convenient. Towards this end, let us consider the inductance L_d as made up of two parts:

$$L_d = L_{ad} + L_\ell \qquad (1.4.21)$$

where L_{ad} is the mutual or magnetizing inductance between the stator and rotor in d-axis, and L_ℓ is the leakage inductance due to flux that does not link any rotor circuit, such as air-gap leakage, slot-leakage, and end-turn leakage.

Similarly, let us consider

$$L_q + L_{aq} + L_\ell \tag{1.4.22}$$

where L_{aq} is the mutual or magnetizing inductance in the q-axis between the stator and rotor, and L_ℓ is the leakage, which is assumed to be the same in both the axes.

Then it follows

$$L_{ad} \text{ in per-unit} = \frac{L_{ad}}{L_{s_{base}}} \tag{1.4.23}$$

$$L_{aq} \text{ in per-unit} = \frac{L_{aq}}{L_{s_{base}}} \tag{1.4.24}$$

and

$$L_\ell \text{ in per-unit} = \frac{L_\ell}{L_{s_{base}}} \tag{1.4.25}$$

Now let us try to equalize all per-unit mutual inductances between the stator and rotor circuits in each axis. That is to say, in per-unit notation

$$L_{ad} = L_{afd} = L_{a1d} = \ldots \text{ in per-unit} \tag{1.4.26}$$

$$L_{aq} = L_{a1q} = \ldots \text{ in per-unit} \tag{1.4.27}$$

Comparing Eqs. (1.4.23), (1.4.17) and (1.4.18), one obtains

$$i_{fd_{base}} = \frac{L_{ad}}{L_{afd}} i_{s_{base}} \tag{1.4.28}$$

$$i_{1d_{base}} = \frac{L_{ad}}{L_{a1d}} i_{s_{base}}, \text{ and so on} \tag{1.4.29}$$

and

$$i_{1q_{base}} = \frac{L_{aq}}{L_{a1q}} i_{s_{base}}, \text{ and so on.} \tag{1.4.30}$$

Also, with the use of Eqs. (1.4.26) and (1.4.27), the flux-linkage relation given by Eq. (1.3.47), for example, can be simplified as follows in per-unit notation:

$$\lambda_d = L_{ad} (-i_d + i_{fd} + i_{1d} + \dots) - L_\ell i_d \tag{1.4.31}$$

all quantities expressed in per-unit system.

The effect of our chosen per-unit system is such that all Park's equations in per-unit notation will have the same form as before, except the factor of 3/2 will be lost in the flux-linkage equations. The reciprocity of the mutual inductance coefficients has been satisfied in the flux-linkage relations using the per-unit notation. Also all per-unit mutual inductances in each axis have been made equal.

The adopted per-unit system is also known as L_{ad}-base system, which inherently assumes a sinusoidal mutual flux-density distribution in the air gap. It can be seen that the base field current is that which establishes the same space fundamental air-gap flux as unit-peak three-phase armature currents. The most significant features of the L_{ad} base are that it is physically the most meaningful, that it permits the performance equations to be represented by simple equivalent circuits in the main axes, and that the reactances in its equivalent circuit correspond to the reactances normally calculated by the designer.

Before proceeding further, let us summarize the choice of rotor base quantities:

(a) $i_{fd_{base}} = \frac{L_{ad}}{L_{afd}} i_{s_{base}}$

(b) $e_{fd_{base}} = \frac{\text{Rated 3-phase volt-Amperes}}{i_{fd_{base}}}$

(c) $Z_{fd_{base}}$ and $L_{fd_{base}}$ may then be calculated.

(d) Similar procedure as above may be followed for 1d, 2d, ..., and 1q, 2q, ... circuits.

(e) It may also be seen that

$$L_{f1d_{base}} = L_{fd_{base}} \frac{i_{fd_{base}}}{i_{1d_{base}}} \quad \text{and so on.}$$

Next let us consider the effect of per-unit notation on the expressions developed earlier for power output given by Eq. (1.3.60), and for the electromagnetic torque given by Eq. (1.3.63). It may easily be seen that the factor of 3/2 disappears in both the equations, once all the quantities involved are expressed in per-unit system. Thus we have in per-unit notation:

$$P = e_d i_d + e_q i_q + 2e_o i_o \tag{1.4.32}$$

and

$$T = \lambda_d i_q - \lambda_q i_d \tag{1.4.33}$$

all quantities expressed in per-unit system.

A comment is probably necessary at this stage, as one may wonder why different notations have not been clearly adopted for quantities, when expressed in their normal physical units and in per-unit system. It is felt, since the only primary difference is the factor of 3/2 occuring at a few places, it may not be necessary to use two distinctly different notations. Further, we shall be using all performance equations in per-unit system only here afterwards.

Problems

1-1. Consider a motor-generator set consisting of two directly coupled three-phase synchronous machines. The motor is supplied from a 60Hz power supply system and the generator is producing electrical power at 25Hz. For such a system of operation, find

(a) the minimum number of poles the motor may have;
(b) the minimum number of poles the generator may have; and
(c) the speed at which the set specified with the number of poles in (a) and (b) operates.

1-2. Let us reconsider the example worked out in the text of a fractional-slot double-layer armature winding of a three-phase, 10-pole, 42-slot synchronous generator. Choose this time the coil pitch to be three slots, instead of four slots as in the example.

(a) Obtain the index number of the winding;
(b) Draw the layout (letter diagram) of the winding for all the poles in a repeatable group;
(c) Compare the results you have obtained with those of the solution of the example in the text, and

comment.

1-3. A 216-slot armature of a three-phase, 28-pole synchronous machine is equipped with a fractional-slot double-layer armature winding, with a coil pitch of seven slots. Determine

(a) the number of poles in a repeatable group;
(b) the index number of the winding;
(c) the distribution factor for the fundamental; and
(d) the pitch factor for the fundamental.

1-4. The field form of a salient-pole synchronous machine has been analyzed for the relative magnitudes of the harmonics. The ratios of the harmonics to the fundamental are given below:

$k_{f1} = 1.0000$; $k_{f3} = -0.061$; $k_{f5} = -.004$; $k_{f7} = 0.035$.

The double-layer star-connected armature winding of the machine has three slots per pole per phase, with a fractional pitch of 8/9. Determine the effect of the harmonics on the rms values of

(a) the line-to-neutral voltage; and
(b) the line-to-line voltage.

1-5. Following the procedure that is indicated in the text, obtain the flux-linkage relations given by Eqs. (1.3.47) to (1.3.49) starting from Eqs. (1.3.44) to (1.3.46).

1-6. Develop the armature voltage relations given by Eqs. (1.3.53) to (1.3.55) starting from Eqs. (1.3.2) to (1.3.4), indicating clearly the transformations you go through.

1-7. A three-phase, 13800-volt (rms line-to-line voltage), 60Hz, two-pole, star-connected synchronous machine operates at its rated speed. The maximum value of the mutual inductance between the field winding and anyone of the armature phase windings is 0.04 henry. Calculate the required field current for the machine to develop normal rated voltage on open circuit.

1-8. Suppose the negative sequence stator currents are flowing from a generator and have the form given below:

$$i_a = 1.0 \cos (t + \alpha)$$

$$i_b = 1.0 \cos (t + \alpha + 120°)$$

$$i_c = 1.0 \cos (t + \alpha - 120°)$$

Let the machine be operating at synchronous speed under steady-state conditions. Evaluate the corresponding i_d, i_q and i_o.

1-9. Consider a machine operating at synchronous speed under steady-state conditions, and let d.c. stator currents flow from the generator having the following magnitudes:

$$i_a = 1.0; \; i_b = i_c = -1/2$$

Find the corresponding i_d, i_q and i_o.

1-10. Flux-linkage and voltage relations given by Park's equations are given below:

$$\lambda_{fd} = -\frac{3}{2} L_{afd} i_d + L_{ffd} i_{fd} + L_{f1d} i_{1d} + \cdots$$

$$\lambda_{1d} = -\frac{3}{2} L_{a1d} i_d + L_{1fd} i_{fd} + L_{11d} i_{1d} + \cdots$$

$$\vdots$$

$$\lambda_{1q} = -\frac{3}{2} L_{a1q} i_q + L_{11q} i_{1q} + L_{12q} i_{2q} + \cdots$$

$$\vdots$$

$$\lambda_d = -L_d i_d + L_{afd} i_{fd} + L_{a1d} i_{1d} + \cdots$$

$$\lambda_q = -L_q i_q + L_{a1q} i_{1q} + L_{a2q} i_{2q} + \cdots$$

$$e_d = p\lambda_d - \lambda_q p\theta - ri_d$$

$$e_q = p\lambda_q + \lambda_d p\theta - riq$$

(a) For the case of the machine operating at steady-state synchronous speed, rewrite these equations making all possible simplifications.

(b) Repeat part (a) with the additional constraint that the armature is open circuited. If rated frequency is 60Hz and L_{afd} = 0.05 henry, calculate the line-to-line voltage of the machine corresponding to a field current of 1500 amperes.

(c) Repeat part (a) with the additional constraint of the armature being short-circuited. Neglect armature resistance. If the rated frequency is 60Hz, L_{afd} = 0.05 henry and L_d = 0.025 henry, calculate the line current for a star-connected generator corresponding to a field current of 1500 amperes.

1-11. A three-phase, 60Hz, 12.1KV, 20MVA, four-pole synchronous generator has the following inductances and resistances:

L_ℓ = 0.00214h

L_{ad} = 0.02145h

L_{aq} = 0.0191h

L_{afd} = 0.045h

L_{ffd} = 0.1538h

r_{fd} = 0.0208 ohm

L_{a1d} = 0.017h

L_{11d} = 0.0226h

L_{f1d} = 0.0541h

r_{11d} = 0.0315 ohm

L_{a1q} = 0.007h

r_{11q} = 0.0611 ohm

r_a = 0.0147 ohm

L_{11q} = 0.00446h

Convert all of the above in per-unit quantities.

1-12. A three-phase, 60Hz, 13KV, 25MVA, star-connected synchronous generator is given to have x_d = 1.2 p.u. and a time constant of 1000 radians. Find x_d in ohms and the time constant in seconds.

CHAPTER II

STEADY-STATE THEORY UNDER BALANCED OPERATION

2.1 Steady-State Analysis and Vector Diagrams

Let us now consider the balanced steady-state operation of a synchronous generator. The corresponding flux-linkage relations in per-unit notation can be written as follows based on our knowledge gained in the previous chapter:

$$\lambda_d = -L_d i_d + L_{ad} i_{fd} \tag{2.1.1}$$

$$\lambda_q = -L_q i_q \tag{2.1.2}$$

The voltage relations will be given by the following in per-unit system:

$$e_d = -\lambda_q - r i_d = +x_q i_q - r i_d \tag{2.1.3}$$

$$e_q = \lambda_d - r i_q = x_{ad} i_{fd} - x_d i_d - r i_q \tag{2.1.4}$$

Note that $p\lambda_d$ and $p\lambda_q$ vanish under steady-state conditions; all amortisseur currents will be zero; also $p\theta = 1.0$, and zero-sequence components do not come into the picture under balanced operation. Further, since per-unit inductance is the same as per-unit reactance, the symbols L and x may be used interchangeably.

On open-circuit condition ($i_d = i_q = 0$), one can easily see that

$$e_d = 0 \tag{2.1.5}$$

and

$$e_q = x_{ad} i_{fd} \tag{2.1.6}$$

which simply states that a speed-voltage is generated in the q-axis due to the field excitation in the d-axis. Let us define for convenience

$$E = x_{ad} i_{fd} \tag{2.1.7}$$

We may note here that the field circuit satisfies the voltage equation

$$e_{fd} = r_{fd} i_{fd} \tag{2.1.8}$$

and all the voltages as well as currents involved are only d.c. quantities. Now one may rewrite the voltage equations (2.1.3) and (2.1.4) as

$$e_d = x_q i_q - r i_d \tag{2.1.9}$$

and

$$e_q = E - x_d i_d - r i_q \tag{2.1.10}$$

As the generator begins to supply power, the magnitude and phase angle of the terminal voltage will change. Taking the bus or terminal voltage e as reference, let δ be the angle between the open-circuit voltage and the terminal voltage. Since the open-circuit voltage lies along the q-axis, it follows that

$$e_d = e \sin \delta \tag{2.1.11}$$

and

$$e_q = e \cos \delta \tag{2.1.12}$$

It is often desirable to calculate the excitation voltage E and the angle δ for assigned balanced terminal voltages and currents. Knowing the terminal conditions (i.e., e, i, and the power-factor angle ϕ), it is not apparent from the above equations as to how E and δ could be evaluated, although there should be a way to do it.

Expressing the real numbers along the d-axis and the imaginary numbers along the q-axis, one may express

$$e = e_d + je_q \tag{2.1.13}$$

and

$$i = i_d + ji_q \tag{2.1.14}$$

Let us define for convenience a new quantity

$$E_q = E - (x_d - x_q)\, i_d \tag{2.1.15}$$

While E_q may not have any simple physical interpretation, it is known to lie along the q-axis and is related to the excitation voltage E as per Eq. (2.1.15). Substituting Eq. (2.1.15) in Eq. (2.1.10), one obtains

$$e_q = E_q - x_q i_d - r i_q \tag{2.1.16}$$

From Eqs. (2.1.9) and (2.1.16), one can rewrite Eq. (2.1.13) as

$$e = jE_q - (r + jx_q)\, i \tag{2.1.17}$$

or

$$jE_q = e + (r + jx_q)\, i \tag{2.1.18}$$

It is now a simple matter to locate jE_q and hence the q-axis from the known terminal conditions, as can be seen from the above equation. The vector diagram corresponding to the steady-state analysis may now be constructed as in Figure 2.1.1 by following the steps given below:

(a) Draw OA equal to the terminal voltage e, taken as reference.

(b) Draw OB equal to the load current i, at the load power-factor angle ϕ.

(c) AC is drawn parallel to OB, equivalent to the ir drop.

(d) CD is drawn perpendicular to AC or OB, indicating jix_q.

(e) OD gives the location of the q-axis and OD stands for $jE_q = e + (r + jx_q)\, i$.

(f) Now the d-axis may easily be located, 90° lagging the q-axis.

(g) The voltage e may be resolved into its components along the axes: OH = e_d and OG = je_q.

(h) The current i may be resolved into its components

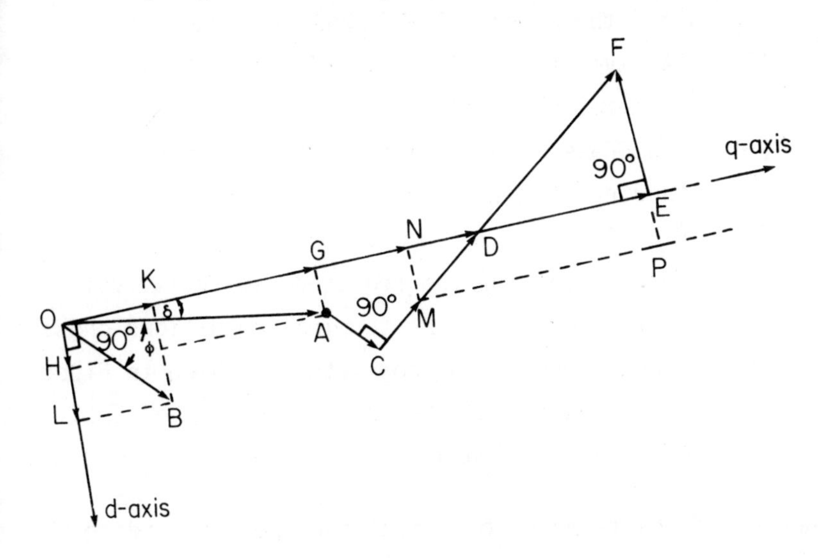

Figure 2.1.1. Steady-State Vector Diagram.

OA: Bus or Terminal Voltage e (taken as reference)

OB: Load current i

A$\hat{O}$B: Load power-factor angle ϕ (lagging case shown here)

AC: ir drop (drawn parallel to OB)

CD: jix_q (drawn perpendicular to AC or OB)

OD: $jE_q = e + (r + jx_q)\ i$

OG: je_q

OH: e_d

OK: ji_q

OL: i_d

D$\hat{O}$A: δ

DE: $j(x_d - x_q)\ i_d$

OE: $jE = jE_q + j(x_d - x_q)\ i_d$

DF: $j(x_d - x_q)\ i$

CF: $jx_d i$

EF: $-(x_d - x_q)\ i_q$

OF: $e + (r + jx_d)\ i$

along the axes: $OL = i_d$ and $OK = ji_q$.

(i) $D\hat{O}A$ gives the angle δ by which the q-axis leads the terminal voltage e.

(j) Draw DE as an extension of OD along the q-axis, standing for $j(x_d - x_q)\ i_d$.

(k) OE of Figure 2.1.1 now gives $jE = jE_q + j(x_d - x_q)id$, thereby yielding the required excitation voltage for the corresponding terminal conditions chosen.

(l) Triangle DFE may be completed as shown in Figure 2.1.1, noting that DF is an extension of CD (i.e., $\perp$r to AC or OB) and EF is perpendicular to DE.

A number of points need to be mentioned here regarding the steady-state vector diagrams:

i. The armature resistance may usually be neglected as it is very small, thereby simplifying the construction procedure.

ii. If $x_d = x_q$, as is the case for the round- or cylindrical-rotor synchronous machines, ΔDEF will vanish.

iii. Figure 2.1.1 is drawn for the case of a synchronous generator, delivering the load at a lagging power-factor. The leading power-factor case for the synchronous generator may easily be worked out by the student, following the same underlying general principles. For the synchronous-motor operation, the terminal current needs to be drawn in exactly the opposite direction (i.e., opposite to OB of Figure 2.1.1) to that of the generator terminal current, and similar construction procedure may then be followed to obtain the corresponding steady-state vector diagram. It may be expected that the terminal voltage would lead the excitation voltage for the synchronous-motor operation.

iv. Although the voltages and currents have been treated here for convenience in exactly the same way as is conventionally done with the complex-number representation of alternating voltages and currents, it should be emphasized that all the voltages and currents are constants (i.e., d.c. quantities). Any balanced external circuit may be simply added to the diagram in the conventional manner.

v. Phasor diagrams conventionally represent the coefficient of $e^{j\omega t}$ in the alternating complex voltages and currents. They do not represent the real voltages and currents, or even the complete complex quantities. In the case of Figure 2.1.1, we are representing the complete d-axis and q-axis voltages and currents. That is why it is called the vector diagram in contrast to our conventional phasor diagram and in fact it may be seen to be a true space vector diagram viewed from rotor.

Looking at Eq. (2.1.7) which gives the generated speed-voltage on open-circuit condition, it would be desirable to extend our per-unit system so as to find the base field current related to the terminal voltage on open-circuit. It is easy to see

when $i_{fd} = 1.0$ p.u., $E = x_{ad}$ in per-unit notation.

Thus one can utilize the above to find the base field current from the open-circuit saturation curve (OCC) as shown in Figure 2.1.2. In other words, it may be stated

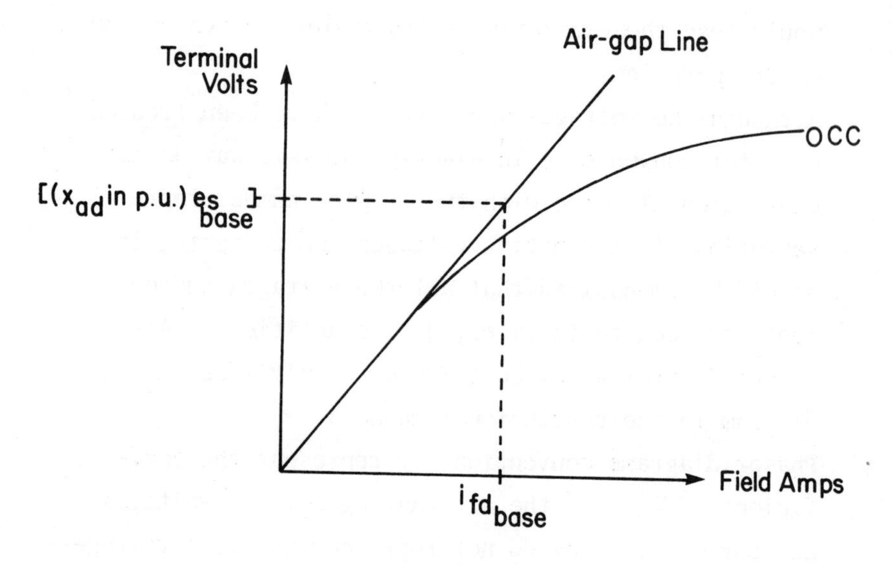

Figure 2.1.2.

that the x_{ad} in per-unit may now be determined with the use of stator quantities only. However, it needs to be pointed out that the air-gap line (i.e., no saturation) is considered and not the actual open-circuit saturation curve, as shown in Figure 2.1.2.

The direct-axis armature flux linkages in per-unit notation are given by

$$\lambda_d = x_{ad} i_{fd} \tag{2.1.19}$$

and

$$e_q = \lambda_d \tag{2.1.20}$$

under open circuit condition.

For normal armature terminal voltage $e_q = 1.0$ p.u., the required field current in per-unit is

$$i_{fd} \text{ in p.u.} = \frac{1}{x_{ad}} \tag{2.1.21}$$

where x_{ad} is expressed in per-unit. The required field voltage in per-unit is given by

$$e_{fd} \text{ in p.u.} = \frac{r_{fd}}{x_{ad}} \tag{2.1.22}$$

where r_{fd} and x_{ad} are expressed in per-unit. Now if the required field voltage and current are known corresponding to the no-load rated voltage in volts and amperes, neglecting saturation, the base field quantities may be calculated. That is to say, if the actual field current is i_{fo} amperes, the base field current in amperes is

$$i_{fd_{base}} = x_{ad} i_{fo} \tag{2.1.23}$$

where x_{ad} is in per-unit. Similarly, if the actual field voltage is e_{fo} volts, the base field voltage in volts is given by

$$e_{fd_{base}} = \frac{x_{ad}}{r_{fd}} e_{fo} \tag{2.1.24}$$

where x_{ad} and r_{fd} are expressed in per-unit. On the other hand, the per-unit machine impedances may be calculated as

$$x_{ad} \text{ in p.u.} = \frac{i_{fd_{base}} \text{ (in amperes)}}{i_{fo} \text{ (in amperes)}} \tag{2.1.25}$$

and

$$r_{fd} \text{ in p.u.} = \frac{i_{fd_{base}} e_{fo}}{i_{fo} e_{fd_{base}}} \tag{2.1.26}$$

Short-Circuit Ratio:

The short-circuit ratio (SCR) is the ratio of the field excitation current required to generate rated voltage at rated speed on open circuit to the field current required to produce rated armature current under a sustained three-phase short circuit. Saturation has the effect of increasing the short-circuit ratio. With the advent of recent improvements in excitation systems, the trend has been to design generators with smaller short-circuit ratios (such as 0.7) in order to lower the cost. Steam-turbine generators have SCR in the range of 0.5 to 1.1, while the water-wheel-driven generators usually have higher SCR up to 2.0, whereas synchronous condensers may have SCR as low as 0.4. The short-circuit characteristic (SCC) is a relationship (usually almost linear) between the field excitation current and the short-circuit armature current under a sustained three-phase short circuit applied at the armature terminals of a synchronous generator driven at rated speed.

Field-Flux Linkage:

Sometimes the phenomena of interest are such that the field-flux linkage tends to remain nearly constant rather than the field current. Neglecting damper-circuit currents, one has in per-unit notation

$$\lambda_d = -x_d i_d + x_{ad} i_{fd} \tag{2.1.27}$$

and

$$\lambda_{fd} = -x_{afd}i_d + x_{ffd}i_{fd} \tag{2.1.28}$$

Eliminating the field current i_{fd}, one obtains

$$\lambda_{fd} = \frac{x_{ffd}}{x_{afd}}\left[\lambda_d + \left(x_d - \frac{x_{afd}^2}{x_{ffd}}\right) i_d\right] \tag{2.1.29}$$

The quantity $[x_d - (x_{afd}^2/x_{ffd})]$ is a short-circuit reactance of the armature direct-axis circuit, indicating the demagnetizing effect of the armature current, and can be measured at the direct-axis armature terminals with zero field resistance. Here one may define for convenience

$$\left(x_d - \frac{x_{afd}^2}{x_{ffd}}\right) = x_d' \tag{2.1.30}$$

which is called the transient reactance, and

$$\frac{x_{afd}}{x_{ffd}} \lambda_{fd} = E_q' \tag{2.1.31}$$

which is the voltage back of transient reactance and is a quantity proportional to the field-flux linkage. Using the above new definitions, one gets from Eq. (2.1.29)

$$E_q' = \lambda_d + x_d' i_d \tag{2.1.32}$$

and since

$$e_q = \lambda_d - ri_q \tag{2.1.4}$$

it follows then

$$e_q = E'_q - x'_d i_d - r i_q \qquad (2.1.33)$$

From Eqs. (2.1.31), (2.1.28) and (2.1.30), it is possible to write

$$E'_q = -\frac{x_{afd}^2}{x_{ffd}} i_d + E = E - (x_d - x'_d)\, i_d \qquad (2.1.34)$$

Now it is fairly easy to see that the whole derivation of the vector diagram may be repeated simply by replacing x'_d for x_d, and E'_q for E everywhere. The details have been added on to Figure 2.1.1 as given below:

CM: jix'_d
ON: jE'_q
NE: $ji_d(x_d - x'_d)$
PF: $-i_q(x_d - x'_d)$

It may be remarked here that the quantity $(x'_d - x_q)$ is negative. We have now identified a quantity corresponding to the field-flux linkage on the vector diagram of Figure 2.1.1.

A Note Regarding the Per-Unit System:

From the viewpoint of steady-state operation, a per-unit system (even though nonreciprocal!) is sometimes employed such that for steady-state, open-circuit normal operation, field current of 1.0 p.u., field voltage of 1.0 p.u., and 1.0 p.u. field-flux linkages exist simultaneously producing a terminal voltage of 1.0 p.u. However, in this text throughout, the reciprocal L_{ad}-base per-unit system will be used.

2.2 An Alternate Approach to the Steady-State Analysis

Assuming balanced steady-state operation of a synchronous generator and neglecting the armature resistance, one has the voltage relations

$$e_d = -\lambda_q = x_q i_q = e \sin \delta \tag{2.2.1}$$

$$e_q = \lambda_d = x_{ad} i_{fd} - x_d i_d = E - x_d i_d = e \cos \delta \tag{2.2.2}$$

The notation is the same adopted in the previous section 2.1. The terminal current i may be expressed as

$$i = \frac{E - e \cos \delta}{x_d} e^{j(\delta-\pi/2)} + \frac{e \sin \delta}{x_q} e^{j\delta} \tag{2.2.3}$$

which may be rewritten as the following:

$$i = e \frac{x_d + x_q}{2x_d x_q} e^{j\pi/2} + e \frac{x_d - x_q}{2x_d x_q} e^{j(2\delta-\pi/2)} +$$

$$+ \frac{E}{x_d} e^{j(\delta-\pi/2)} \tag{2.2.4}$$

When connected to a normal bus, taking e = 1 p.u., one has

$$i = \frac{x_d + x_q}{2x_d x_q} e^{j\pi/2} + \frac{x_d - x_q}{2x_d x_q} e^{j(2\delta-\pi/2)} +$$

$$+ \frac{E}{x_d} e^{j(\delta-\pi/2)} \tag{2.2.5}$$

Thus there are three components of current the salient-pole synchronous machine delivers to the bus. The second term on the right-hand-side of Eq. (2.2.5) vanishes for the case of round-rotor synchronous machine as x_d equals x_q. Even for no excitation (i.e., E = 0), in the case of a salient-pole synchronous machine, it can be seen that the current has an active component. With sufficient excitation, the current may be made equal to zero; for example, at no-load, for $\delta = 0$, with E = e, i becomes zero. Further, it may be seen that, for E = 0,

$$i = i_{max} = \frac{e}{x_q} \quad , \qquad \text{for } \delta = \frac{\pi}{2} \tag{2.2.6}$$

and

$$i = i_{min} = \frac{e}{x_d} \quad , \qquad \text{for } \delta = 0 \tag{2.2.7}$$

The above may be called as excitation currents since they are displaced by $\pi/2$ or 90° from the voltage axis. The current wave is an amplitude modulated wave, provided δ is varied slowly.

The first term on the right-hand-side of Eq. (2.2.4) is dependent on the average amplitude of the exciting current, i.e., $(1/2)(1/x_d + 1/x_q)$ e. It is fixed by the machine design parameters and is clearly dependent on bus voltage only. Its phase does not depend on δ.

The second component depends on the amplitude of variations in exciting current, i.e., $(1/2)(1/x_q - 1/x_d)$ e. It is fixed by the machine design parameters and is dependent on the bus voltage only. However, its phase varies with 2δ.

The third term depends on the transform of field current

and may be varied by the excitation voltage E in magnitude. Its phase also varies with δ. With suitable manipulation, we could make the machine operate at unity power factor.

The current in general depends on both δ and ϕ, the external power-factor angle. δ would be seen later to correspond with the torque angle of the machine. The steady-state vector diagram, based on Eq. (2.2.4), may now be constructed as in Figure 2.2.1 by following the steps given below:

(a) Taking the bus voltage e as reference, draw the terminal current i, given by AC, at a power-factor angle of ϕ.

(b) AB is drawn perpendicular to the bus voltage vector such that $AB = e/x_q$.

(c) Locate the point 0 on AB such that

$$AO = \frac{x_d + x_q}{2x_d x_q} e \quad .$$

(d) With 0 as center and radius equal to

$$\frac{x_d - x_q}{2x_d x_q} e \quad ,$$

construct a circle. Let the circle intersect the line AB at B and F. It may be seen that $AF = e/x_d$.

(e) Join BC such that the line BC intersects the circle at D. Then $A\hat{B}C = \delta$, $A\hat{O}D = 2\delta$, and $DC = E/x_d$.

Thus for given terminal conditions of e, i and ϕ, and with known values of x_d and x_q, one can find E, the excitation voltage, and the torque angle, δ. Figure 2.2.1 has been

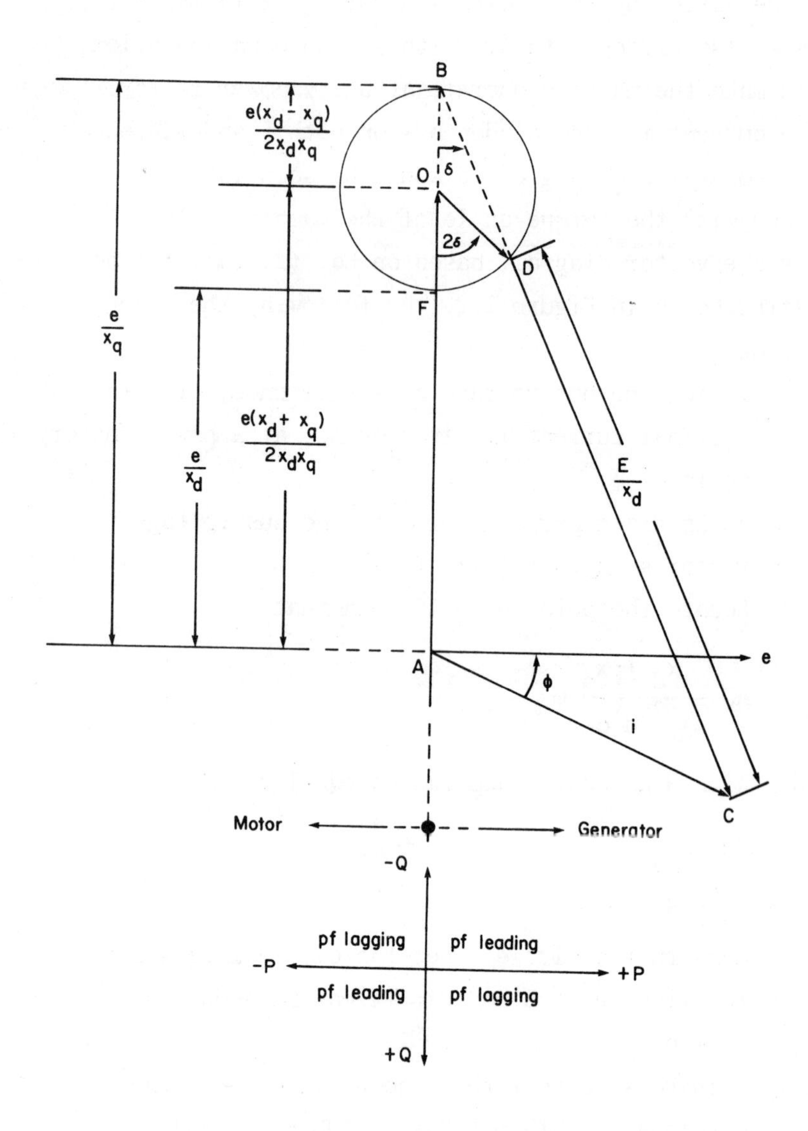

Figure 2.2.1. Steady State Vector Diagram of a Synchronous Machine.

drawn for the case of a generator delivering the load at a lagging power-factor. The diagram may appropriately be modified for the case of leading power-factor load and also for the synchronous motor operation. Further observation of additional details given in Figure 2.2.1 leads to the following statements:

i. The generator action corresponds to $\pi/2 > \delta > 0$.

ii. -Q means delivering capacitive vars for a generator action.

iii. -Q means receiving capacitive vars for a motor action.

iv. +Q means delivering inductive vars for a generator action.

v. +Q means receiving inductive vars for a motor action.

vi. If the real power P and the reactive power Q are of the same sign, power-factor will be lagging.

vii. If P and Q are of opposite signs, the power-factor will be leading.

Figure 2.2.2 shows the four possible cases of operation of a synchronous machine, in which resistance is neglected for simplicity and a round-rotor machine (with $x_d = x_q$) is considered.

It may be pointed out here that the current loci are limacons, and when

$$E = \frac{x_d - x_q}{x_q} \quad ,$$

the locus is a cordioid, and when E = 0, it is a circle, as shown in Figure 2.2.3. The points corresponding to the minimum current may be seen to be the stable points of operation.

We may now develop the expressions for the real and

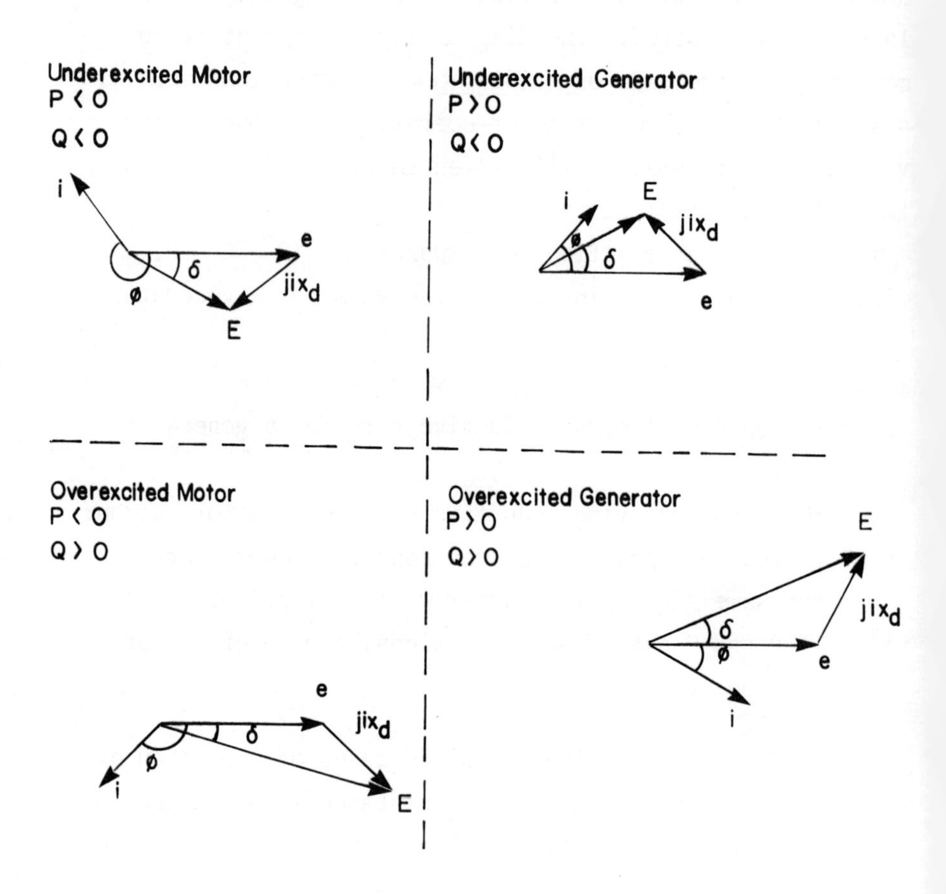

Figure 2.2.2. Four Possible Cases of Operation of a Synchronous Machine.
(For simplicity, resistance is neglected and a round-rotor machine with $x_d = x_q$ is considered).

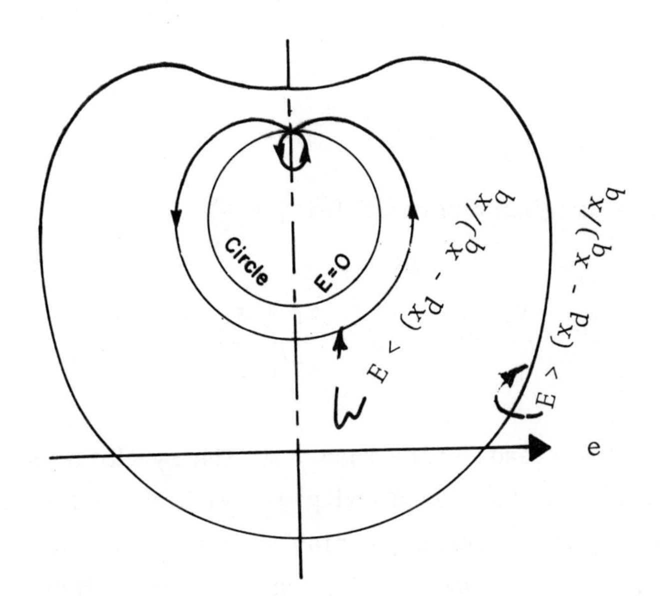

Figure 2.2.3. Current Loci

reactive power. In rectangular form, one can write for the conjugate of i from Eq. (2.2.4) as

$$i^* = \left[\frac{E}{x_d}\sin\delta + \frac{x_d - x_q}{2x_dx_q}e\sin 2\delta\right] +$$

$$+ j\left[\frac{E}{x_d}\cos\delta + \frac{x_d - x_q}{2x_dx_q}e\cos 2\delta - \frac{x_d + x_q}{2x_dx_q}e\right] \tag{2.2.8}$$

Then it follows that

$$P = \text{Real part-of } [ei^*] = \frac{Ee}{x_d} \sin \delta + \frac{x_d - x_q}{2x_d x_q} e^2 \sin 2\delta \tag{2.2.9}$$

and

$$Q = \text{Imaginary-part of } [ei^*] = \frac{Ee}{x_d} \cos \delta +$$

$$+ \frac{x_d - x_q}{2x_d x_q} e^2 \cos 2\delta - \frac{x_d + x_q}{2x_d x_q} e^2 \ . \tag{2.2.10}$$

In the steady-state theory of the synchronous machine, with known bus voltage e and given machine design constants x_d and x_q, the operating variables are six in number, given by P, Q, δ, ϕ, i, and E. One can only write four independent equations relating these six variables and any other equation relating them would be derivable from these four. So one can choose arbitrarily two quantities, the choice of which is to be consistent with the operating characteristics of the machine. Thus, the synchronous machine may be said to have two degrees of freedom. The selection of any two (such as ϕ and i, P and Q, δ and E) determines the operating point and establishes the other four quantities.

It may be added here that the variation of P with respect to ϕ, with E as a parameter, would be given by a Finger-print characteristic, the name of which may be justified by the appearance of the sketch shown in Figure 2.2.4. Some of the typical steady-state operating characteristics of synchronous machines are shown in Figures 2.2.5 through

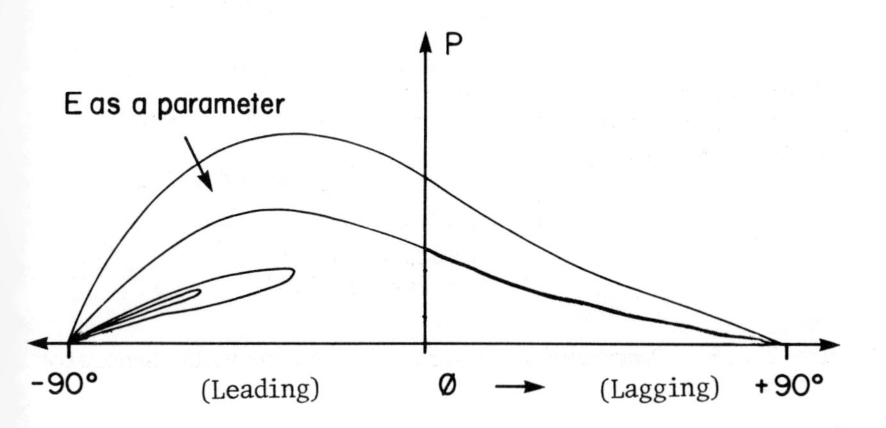

Figure 2.2.4. Finger-Print Characteristic.

2.2.8*.

Parallel Operation of Synchronous Generators

In order to assure the continuity of power supply within the prescribed limits of frequency and voltage at all the load points scattered over the service area, it becomes necessary in any modern power system to operate several alternators in parallel, interconnected by various transmission lines, in a well-coordinated and optimized manner such that the operation is most economical. A generator can be paralleled with an infinite bus (or another generator running at rated voltage and frequency supplying the load) by driving it at synchronous speed corresponding to the system-frequency and adjusting its field excitation such

*See Fitzgerald, A. E., et al, Electric Machinery, 3rd edition, Art. 6.4, McGraw-Hill Book Company, New York, 1971.
(The figures are adapted by permission of the publisher).

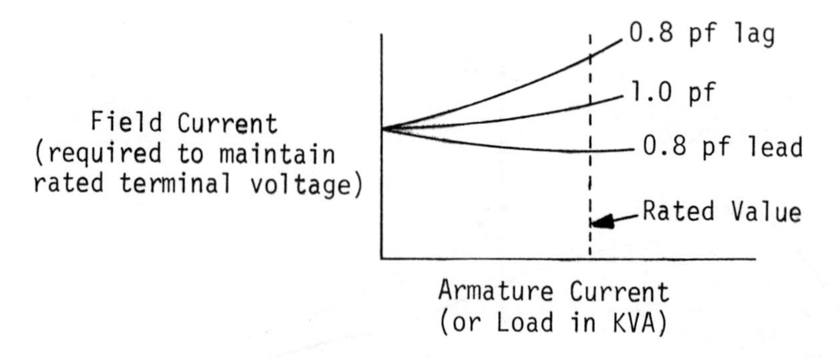

Figure 2.2.5. Compounding Curves of a Synchronous Generator

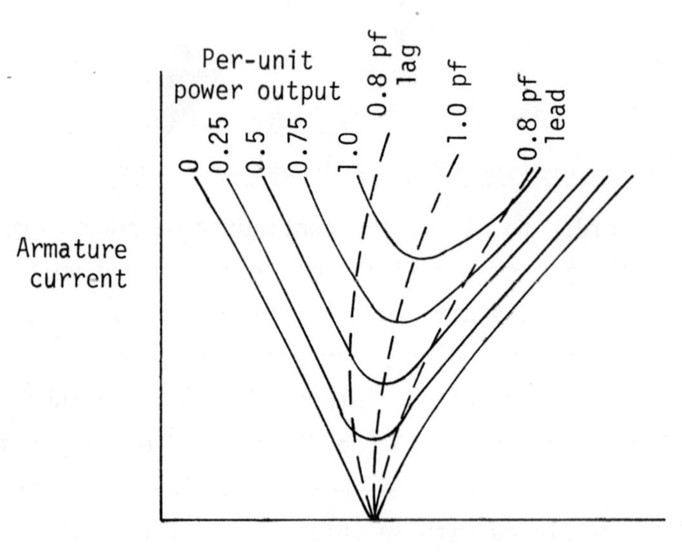

Figure 2.2.6. V-Curves of a Synchronous Motor

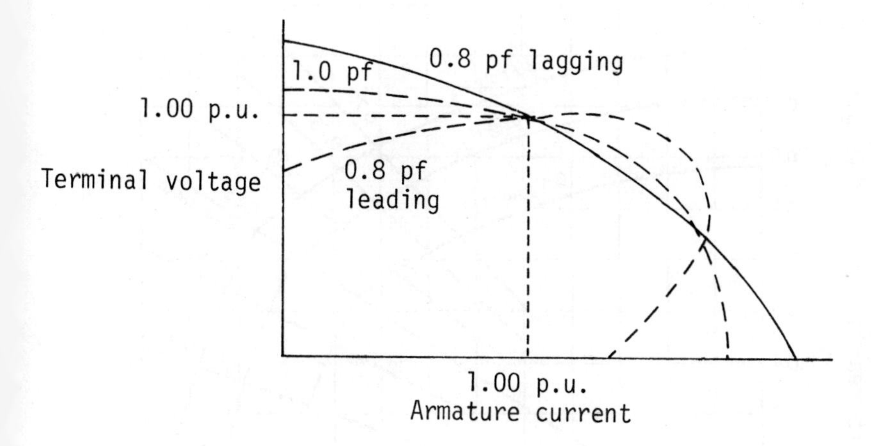

Figure 2.2.7. Generator Volt-Ampere Characteristics

(*Note:* Each curve is drawn for a different value of constant field current, which equals the value required to give rated terminal voltage at rated armature current.)

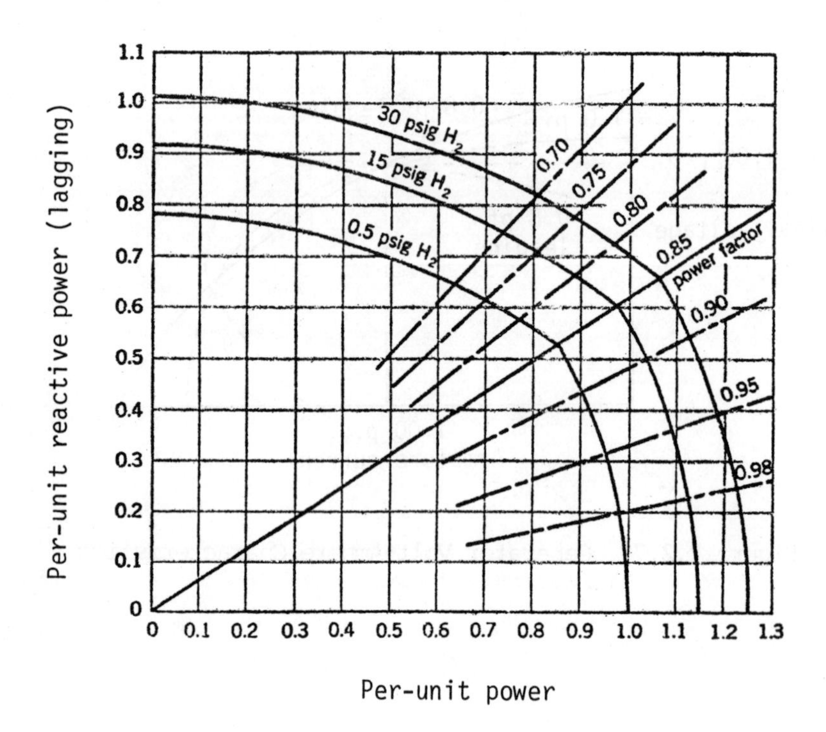

Figure 2.2.8. Typical Reactive-Capability Curves of Hydrogen-Cooled Turbine Generator (0.85 pf; 0.80 SCR; Base KVA is taken as rated KVA at 0.5 psig hydrogen; operation at rated voltage)

that its terminal voltage equals that of the bus. If the frequency of the incoming machine is not exactly equal to that of the system, the phase relation between its voltage and the bus voltage will vary at a frequency equal to the difference between the frequencies of the machine and bus voltages. In normal practice this difference can be made usually very small to a fraction of a hertz and the synchronizing switch is thrown in with the same phase sequence on either side when the two voltages are momentarily in phase and the voltage across the switch is zero. A synchroscope or a bright lamp synchronizing method or a dark lamp synchronizing method* is used for indicating the appropriate moment for synchronization. After the machine has been synchronized and is now a part of the system, it can be made to take its share of the active and reactive power by appropriate adjustments of its prime-mover throttle and field rheostat. The system frequency and the division of active power admidst the generators are controlled by means of prime-mover throttles regulated by governors and automatic frequency regulators, whereas the terminal voltage and reactive volt-ampere division amidst the generators are controlled by voltage regulators acting on the generator-field circuits and by transformers with automatic tap-changing devices.

2.3 Steady-State Power-Angle Characteristics

The expression for power output in terms of direct-, quadrature- and zero-axis quantities is given by Eq. (1.4.32):

---

*See for details any laboratory manual on Electrical Machine Experiments.

$$P = e_d i_d + e_q i_q + 2e_o i_o \tag{2.3.1}$$

For balanced steady-state operation, under which zero-sequence quantities vanish, the power output may be expressed as

$$P = e_d i_d + e_q i_q \tag{2.3.2}$$

The steady-state volt-ampere relations of a synchronous machine are given by the following:

$$e_d = x_q i_q - r i_d = e \sin \delta \tag{2.3.3}$$

$$e_q = E - x_d i_d - r i_q = e \cos \delta \tag{2.3.4}$$

The notation is the same adopted in the previous sections 2.1 and 2.2. The currents i_d and i_q may be solved for:

$$i_d = \frac{-re \sin \delta + x_q(E - e \cos \delta)}{x_d x_q + r^2} \tag{2.3.5}$$

$$i_q = \frac{x_d\ e \sin \delta + r(E - e \cos \delta)}{x_d x_q + r^2} \tag{2.3.6}$$

Substituting the above in Eq. (2.3.2), in terms of the bus voltage, the power output may be seen to be

$$P = \frac{Ee(x_q \sin \delta + r \cos \delta) - re^2 + \frac{e^2}{2}(x_d - x_q) \sin 2\delta}{x_d x_q + r^2} \tag{2.3.7}$$

The power input may be calculated by adding the armature copper losses, given by $[r(i_d^2 + i_q^2 + 2i_o^2)]$, to the power output. The per-unit power input is numerically equal to the per-unit torque, as the machine will be operating at synchronous speed.

If the armature resistance is neglected, Eq. (2.3.7) reduces to the following:

$$P = \frac{Ee}{x_d} \sin \delta + \frac{(x_d - x_q)\, e^2}{2x_d x_q} \sin 2\delta \tag{2.3.8}$$

Equation (2.3.8) gives the steady-state power-angle equation for zero armature resistance and fixed field excitation; it is sketched in Figure 2.3.1. It may be observed that the effect of saliency ($x_d \neq x_q$) is to introduce a second-harmonic term, which reaches its maximum at $\delta = 45°$ and is responsible for the reluctance torque. The maximum power occurs at an angle lying between $\pi/4$ and $\pi/2$, rather than right at $\pi/2$ as in a round-rotor machine (with $x_d = x_q$). It will be observed that the effect of the reluctance power is to steepen the curve in the stable region near the origin.

The angle at which maximum power occurs may be computed by setting $\frac{dP}{d\delta}$ equal to zero. Thus one has

$$\frac{dP}{d\delta} = \frac{Ee}{x_d} \cos \delta + \frac{(x_d - x_q)}{x_d x_q} e^2 \cos 2\delta = 0 \tag{2.3.9}$$

or

$$\cos^2 \delta + \frac{x_q E}{2(x_d - x_q)e} \cos \delta - \frac{1}{2} = 0 \tag{2.3.10}$$

which leads to

$$\cos\delta = -\frac{x_q E}{4(x_d - x_q)e} \pm \sqrt{\left[\frac{x_q E}{4(x_d - x_q)e}\right]^2 + \frac{1}{2}} \tag{2.3.11}$$

Only the positive value of the radical will correspond to maximum power, giving $\delta < \frac{\pi}{2}$, for normal values of the circuit parameters; the other root would be greater than unity.

Assuming the field-flux linkage to remain constant, one may derive a transient power angle characteristic in a similar manner in terms of the voltage E_q', introduced in Section 2.1. The derivation proceeds exactly as before except that E_q' replaces E, and x_d' replaces x_d. Then one obtains

$$P = \frac{E_q' e}{x_d'} \sin\delta + \frac{x_d' - x_q}{2x_d' x_q} e^2 \sin 2\delta \tag{2.3.12}$$

Noting that $(x_d' - x_q)$ is negative, the second-harmonic term (transient reluctance power) reaches its maximum at $\delta = 135°$ so that the maximum transient power occurs at an angle between $\pi/2$ and $3\pi/4$. Equation (2.3.12) is sketched in Figure 2.3.2. The condition for which maximum transient power occurs may be obtained, yielding the desired root as

$$\cos\delta = +\frac{x_q E_q'}{4(x_q - x_d')e} - \sqrt{\left[\frac{x_q E_q'}{4(x_q - x_d')e}\right]^2 + \frac{1}{2}} \tag{2.3.13}$$

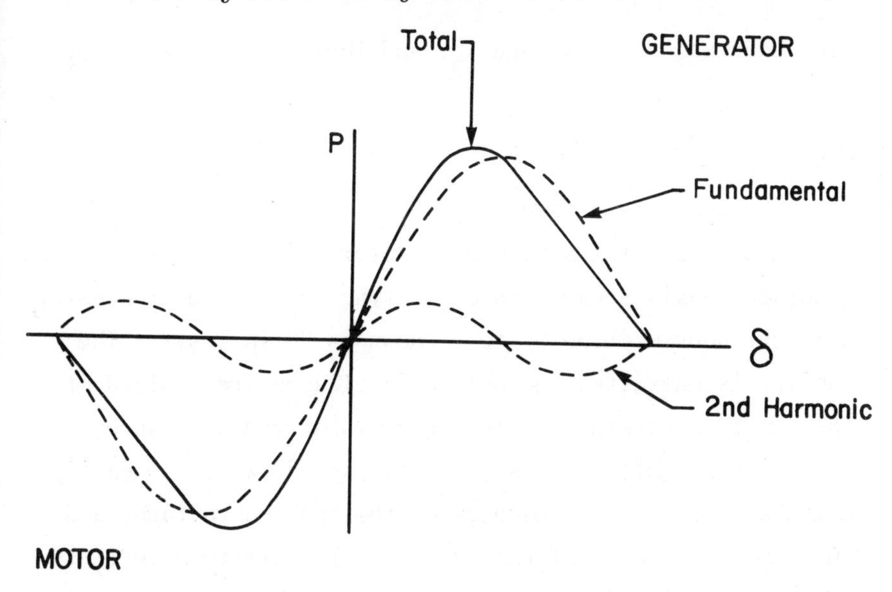

Figure 2.3.1. Steady-state Power-Angle Characteristic of a Salient-pole Synchronous Machine.

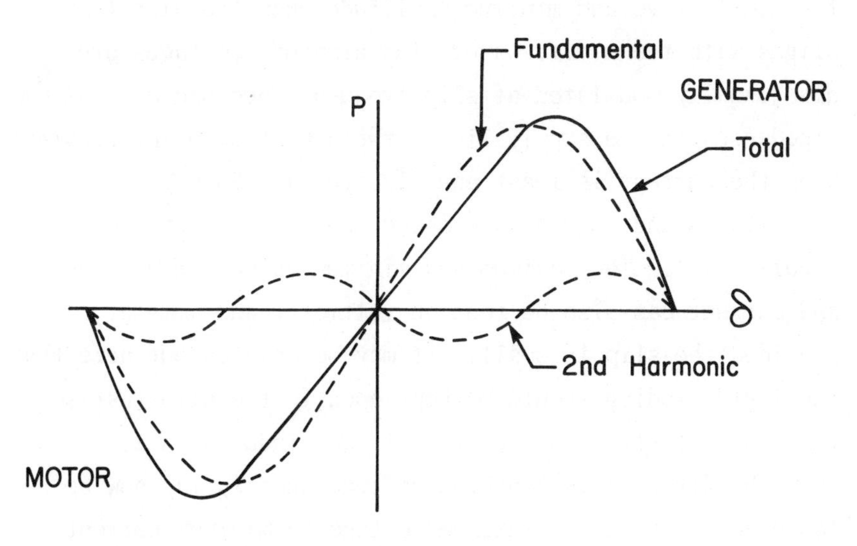

Figure 2.3.2. Transient Power-Angle Characteristic of a Salient-pole Synchronous Machine.

2.4 Reactances x_d, x_q and x_o, and Their Determination by Test Data.

Slip Test for x_d and x_q:

The analysis presented in Section 2.2 and Eqs. (2.2.6) and (2.2.7) suggest a method of determining the direct- and quadrature-axis steady-state, or synchronous, reactances of a synchronous machine by test known as "Slip Test". The machine is unexcited and balanced voltages are applied at the armature terminals. The rotor is driven at a speed differing slightly from synchronous speed, which is easily calculated from the frequency of the applied voltages and the number of poles of the machine. The armature currents are then modulated at slip frequency by the machine, having maximum amplitude when the quadrature axis is in line with the m.m.f. wave and minimum amplitude when the direct axis aligns with the m.m.f. wave. The armature voltages are also usually modulated at slip frequency because of impedances in the supply lines, the amplitude being greatest when the current is least and vice versa. Such variations of voltages and current are shown in an oscillogram of Figure 2.4.1. The maximum and minimum values of voltage and current can also be read on voltmeter and ammeter, provided the slip is small. It may be pointed out here that the field winding should be kept open in the slip test so that the slip-frequency current is not induced in it.

The direct-axis synchronous reactance x_d may now be found as the ratio of maximum voltage to minimum current. The quadrature-axis synchronous reactance x_q is the ratio of minimum voltage to maximum current. For the test to be successful, the slip must be sufficiently small.

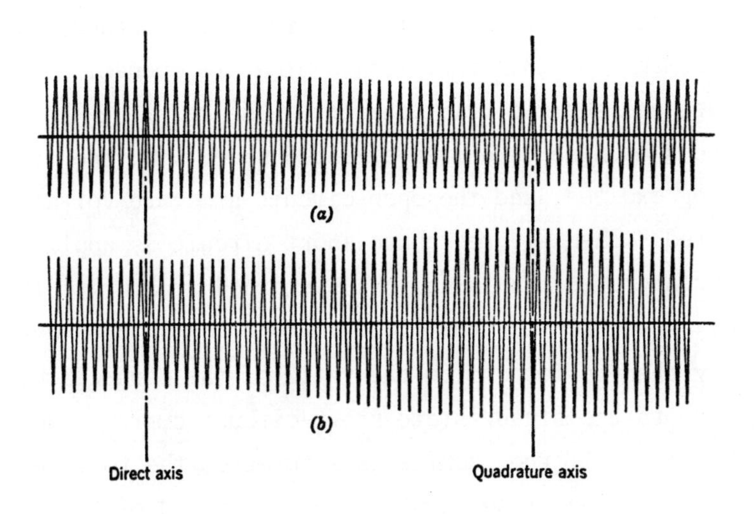

(a) Armature voltage variation

(b) Armature current variation

$$x_d = \frac{\text{Max. voltage}}{\text{Min. current}} \quad ; \quad x_q = \frac{\text{Min. voltage}}{\text{Max. current}}$$

Figure 2.4.1. Slip-test oscillogram.

Short-Circuit Test for x_d:

The analysis presented in Section 2.1 suggests another way of measuring the direct-axis armature reactance x_d from measurements of the steady-state armature open-circuit voltage and short-circuit current. The rotor of the synchronous machine is driven at rated speed with the field winding excited, and the open-circuit armature voltage measured. Then a three-phase short circuit is applied to the armature terminals and the sustained armature current is measured, keeping the field current the same as before. x_d may then be found from the ratio of the open-circuit voltage to the steady-state short-circuit current, neglecting saturation. If the machine is saturated, the voltage read from the air-gap line should be used instead of the open-circuit voltage read from the no-load saturation curve. The result is then the unsaturated value of x_d. It may be pointed out here that this method does not require a separate source of three-phase power, and the armature m.m.f. is automatically in line with the direct axis of the rotor.

In actual practice it may be found more convenient to calculate x_d as the ratio of the field current required to produce unit armature current on short circuit to the field current required to produce unit terminal voltage on open circuit. This ratio gives directly the per-unit direct-axis steady-state reactance.

Other Methods for Measuring x_d and x_q:

An alternate method of measuring x_d is by applying sustained positive-sequence currents to the armature, driving the rotor forward at synchronous speed such that its direct axis coincides with the peak of the space-fundamental rward-rotating m.m.f. wave, and measuring the sustained sitive-sequence armature voltage in quadrature with the rrent. The ratio of the voltage to the current under the ove conditions yields the direct-axis synchronous actance x_d. This method is not as convenient to carry t as the slip test or the short-circuit test described rlier.

A procedure similar to the above one may be suggested r the measurement of x_q. Positive-sequence currents uld be applied to the armature, and the rotor would be iven forward at synchronous speed with the quadrature is aligning with the peak of the rotating m.m.f. wave. e sustained positive-sequence armature voltage would then measured, and the ratio of the voltage to the current uld give x_q under the above conditions. However, this ocedure just described is rather difficult to carry out, the rotor would be in unstable equilibrium on account of luctance torque when the quadrature axis is in line with e crest of the m.m.f. wave.

Although the test methods discussed above are not nvenient to conduct, it may be observed that the pro- dures have a direct bearing on the fundamental concepts f the synchronous reactances and their definitions.

ero-Sequence Reactance x_o:

The zero-sequence reactance is easily measured by onnecting the three armature windings in series and passing ingle-phase current through them. The ratio of the terminal oltage of one-phase winding to the current is the zero- equence impedance, which is very nearly the same in magni- ude as the zero-sequence reactance. It may be mentioned ere that the zero-sequence reactance is the lowest of the synchronous-machine reactances.

The actual value of x_o varies over a wide range compared to that of other reactances and depends on the pitch of the armature coils. For the case of a two-thirds pitch the reactance would be the least, because each slot has then two coil sides carrying equal and opposite currents.

The general range of x_d is anywhere from 0.6 to 2.2 per-unit; x_q typically has the range of 0.4 to 1.4 per-unit and x_o may vary from 0.01 to 0.25 per-unit.

Armature Leakage Reactance x_ℓ:

Armature leakage reactance is a calculated quantity and cannot be measured. As mentioned already and indicated earlier in Section 1.4, it forms a part of the direct-axis and quadrature-axis synchronous reactances.

x_ℓ is the reactance caused by the difference between the total flux produced by the armature currents acting alone and the space fundamental of the flux in the air gap. It is sometimes treated in terms of components due to slot-leakage flux, end-leakage flux and differential leakage flux, which consists of the space harmonics of the air-gap flux that induce fundamental-frequency armature voltage.

2.5 Equivalent Circuits

Based on the analysis presented in Section 2.1, the cylindrical-rotor synchronous machine, for which $x_d = x_q$, may be very simply modeled as given in Figure 2.5.1 while neglecting the armature resistance.

For transient stability studies, neglecting transient saliency and assuming constant field-flux linkages, the simple equivalent circuit in Figure 2.5.2 is used often.

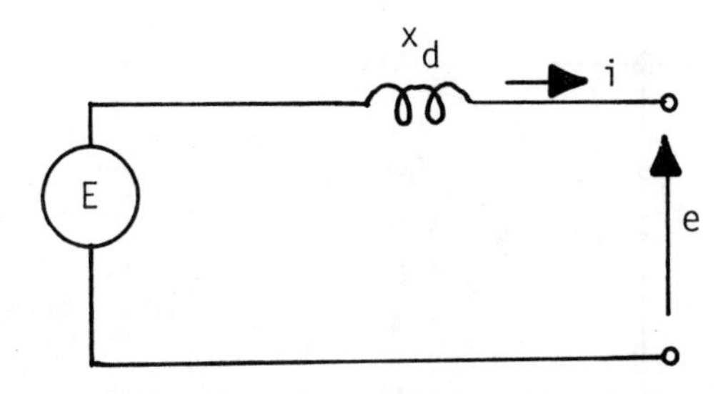

Figure 2.5.1.

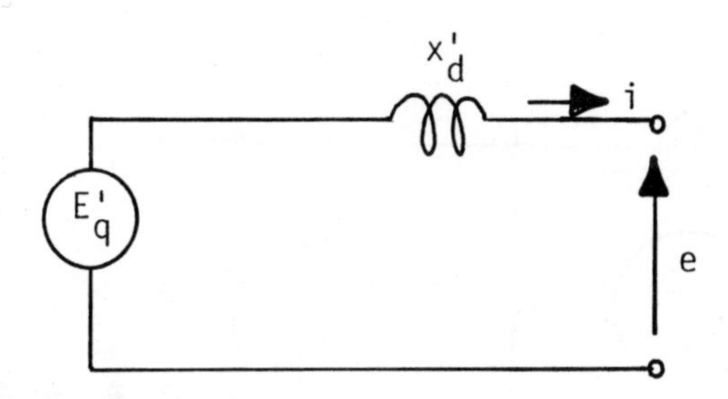

Figure 2.5.2.

The notation is the same as that employed in Section 2.1.

The equivalent direct-axis and quadrature-axis circuits based on the flux-linkage equations may be represented as shown in Figures 2.5.3 and 2.5.4.

A few comments are in order here. These simplified equivalent circuits are drawn with the introduction of some new inductances:

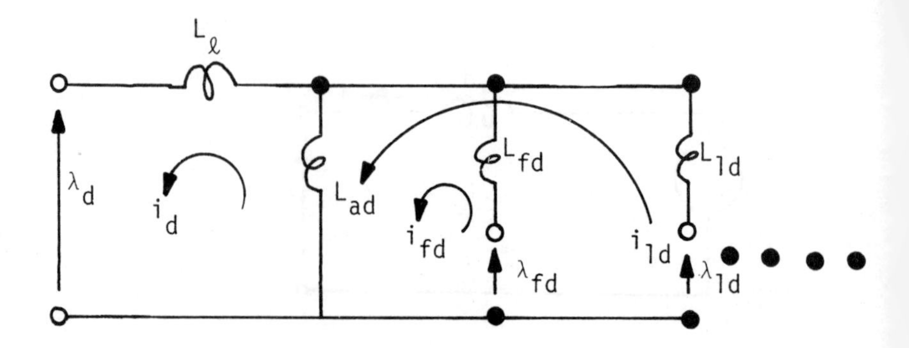

Figure 2.5.3. D-Axis Equivalent Circuit

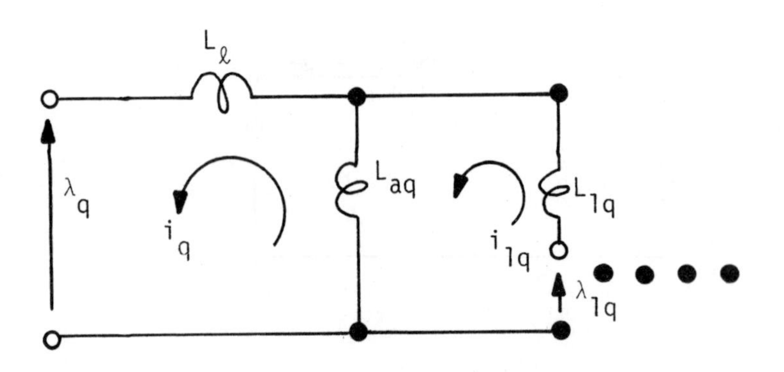

Figure 2.5.4. Q-Axis Equivalent Circuit

$L_{fd} = L_{ffd} - L_{ad}$, field-leakage inductance

$L_{1d} = L_{11d} - L_{ad}$, direct-axis amortisseur leakage inductance

⋮

$L_{1q} = L_{11q} - L_{aq}$, quadrature-axis amortisseur leakage inductance

⋮

Also L_{f1d}, L_{f2d}, etc. are taken to be equal to L_{ad}, thereby neglecting the quantities such as $(L_{f1d} - L_{ad})$, called the eripheral leakage, which is generally a very small quantity ompared to other terms involved and which can usually be eglected.

Problems

-1. Consider the steady-state operation of a synchronous machine and three-phase balanced set of voltages given below in per-unit:

$$e_a = e \cos t$$

$$e_b = e \cos \left(t - \frac{2\pi}{3}\right)$$

$$e_c = e \cos \left(t + \frac{2\pi}{3}\right)$$

(a) Find the corresponding e_d, e_q and e_o, assuming $\theta = \theta_o + t$ where θ is the angle between the direct axis and the axis of phase a of the machine.

(b) Comment on the results you obtain.

2-2. A salient-pole synchronous generator rated 20-MVA, 2.3-KV is operating under balanced steady-state conditions at 14 MVA, rated voltage, and at a power factor of 0.8 lagging. The reactances of the machine are given below in per-unit:

$$x_d = 1.40$$

$$x_q = 0.90$$

$$x_{ad} = 1.25$$

(a) Determine the required field current i_{fd} and draw

the corresponding steady-state vector diagram.

(b) Repeat the problem at unity power factor and 0.8 power factor leading.

2-3. A salient-pole synchronous generator is operating under balanced steady-state conditions:

$e = 1.0; \quad i = 1.0; \quad \phi = 15°$ lagging

The reactances are given as $x_d = 1.2$ and $x_q = 0.8$.

(a) Compute the excitation voltage E and the angle δ.

(b) Using E as found in part (a), to what value could P, the real power output, be increased without loss of synchronism?

2-4. A 3-phase, 60-HZ, synchronous condenser rated 200-MVAR, 20-KV, is to be operated at rated voltage and MVAR output. Neglecting the losses and taking the reactances to be

$x_d = x_q = 1.4$, and $x_{ad} = 1.25$

compute the required field current in per-unit.

2-5. A manufacturing plant has 2000-KVA of 0.8pf induction motors and 1000-KW of lighting load. It is desired to install a 1500-HP synchronous motor (rated pf = 0.8; efficiency = 80.5%) to correct the power factor on the line to the power plant to unity. If the motor operates at rated voltage, what must its field current be in per-unit? The reactances of the synchronous machine are given as $x_d = 1.5$, $x_q = 1.1$, and $x_{ad} = 1.35$.

2-6. A 3-phase, 60-HZ, synchronous motor rated 500-HP, 2300-volts is to be operated at rated load and voltage, and unity power factor. Calculate the required field current in per-unit, given the following:

Efficiency = 90%; rated power factor = 0.9

$x_d = x_q = 1.4$; $x_{ad} = 1.25$

2-7. Starting from Eqs. (2.2.1) and (2.2.2), work out the intermediate steps leading to Eqs. (2.2.4), (2.2.9) and (2.2.10).

2-8. Starting from the fundamental flux-linkage equations, justify the equivalent circuits shown in Figures 2.5.3 and 2.5.4 along with the newly introduced leakage inductances in Section 2.5.

[illegible] phase [illegible] is [illegible] from [illegible]

[illegible] the [illegible] (not [illegible]

[illegible] and [illegible] in [illegible]

[illegible] in per unit [illegible]

[illegible] Section 2.9 [illegible]

[illegible] 1.23 [illegible]

[illegible]

intermediate steps leading to Eqs. [illegible] 1), [illegible]

and [illegible] 0)

Starting from the fundamental flux linkage equations, [illegible]

[illegible] equations [illegible] shown [illegible]

[illegible] along [illegible] introduced [illegible]

[illegible] in Section [illegible]

CHAPTER III

TRANSIENT PERFORMANCE - I

3.1 Introduction to Transient Reactances and Time Constants

One would be interested during transients in the changes that occur in the values of variables, just after ($t = 0_+$) and just before ($t = 0_-$) the disturbance. By letting such differences as

$$\Delta\lambda_d = \lambda_d\Big|_{t=0_+} - \lambda_d\Big|_{t=0_-} \tag{3.1.1}$$

$$\Delta i_d = i_d\Big|_{t=0_+} - i_d\Big|_{t=0_-} \tag{3.1.2}$$

etc. .....

one could easily verify that the flux-linkage relations in terms of the change-variables are of the same form as the original equations, and hence the equivalent circuits of the same form as before can be drawn for the change-variables:

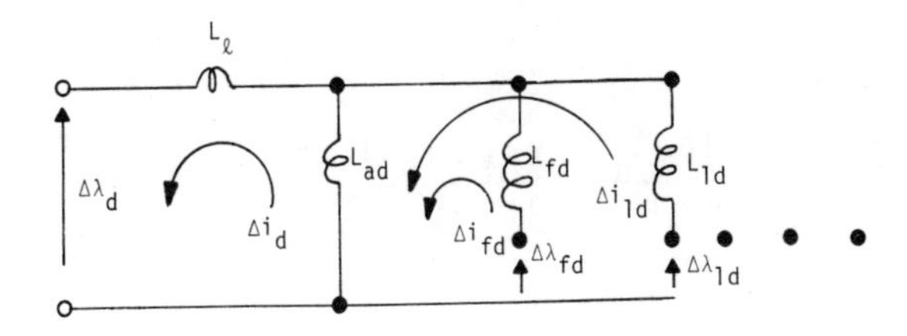

Figure 3.1.1. D-axis equivalent circuit.

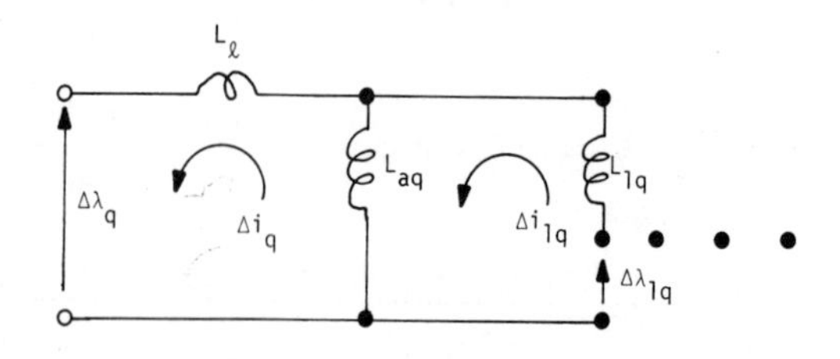

Figure 3.1.2. Q-axis equivalent circuit.

One can also write the rotor volt-ampere equations as follows:

$$\Delta e_{fd} = p(\Delta\lambda_{fd}) + r_{fd}\,\Delta i_{fd} \tag{3.1.3}$$

$$0 = p(\Delta\lambda_{1d}) + r_{1d}\,\Delta i_{1d} \tag{3.1.4}$$

$$\vdots$$

$$0 = p(\Delta\lambda_{1q}) + r_{1q}\,\Delta i_{1q} \tag{3.1.5}$$

$$\vdots$$

solving for

$\Delta\lambda_{fd}$, $\Delta\lambda_{1d}$, ..., $\Delta\lambda_{1q}$, ... etc., one obtains

$$\Delta\lambda_{fd} = \int_{t=0_-}^{t=0_+} [\Delta e_{fd} - r_{fd}\,\Delta i_{fd}]\,dt \tag{3.1.6}$$

$$\Delta\lambda_{1d} = \int_{t=0_-}^{t=0_+} [-r_{1d}\, i_{1d}]\, dt \tag{3.1.7}$$

$$\vdots$$

$$\Delta\lambda_{1q} = \int_{t=0_-}^{t=0_+} [-r_{1q}\, i_{1q}]\, dt \tag{3.1.8}$$

$$\vdots$$

From the well-known theorem of constant flux linkage, it follows that the flux linkage of any closed circuit of finite resistance and e.m.f. cannot change instantly, and the flux linkage of any closed circuit having no resistance or e.m.f. remains constant. If the changes occur in a time that is short compared with the time constants of the circuit, the flux linkages will remain substantially constant during the change. Thus one can redraw the equivalent circuits as follows, that hold good for the short period of disturbance.

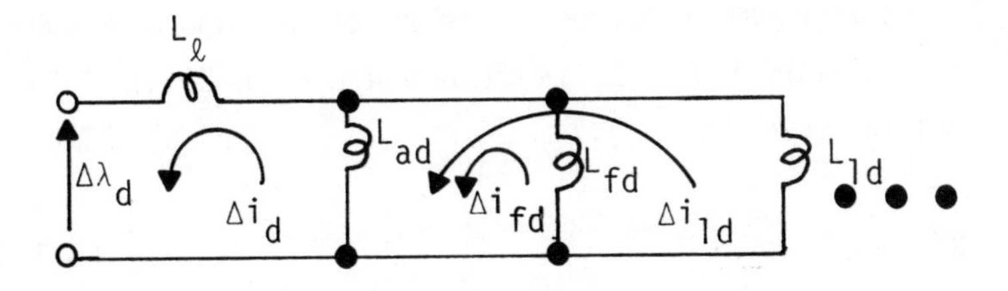

Figure 3.1.3. D-axis equivalent circuit for the sub-transient period.

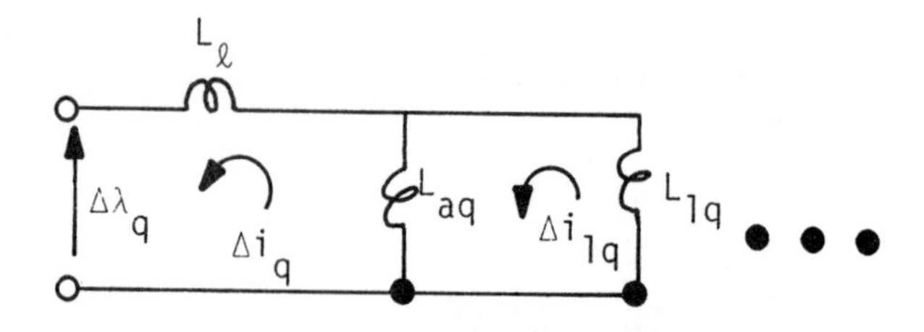

Figure 3.1.4. Q-axis equivalent circuit for the subtransient period.

When rotor circuits in addition to the main field winding, such as amortisseur windings, are considered, the armature flux linkage per armature ampere is defined as the direct-axis subtransient inductance L''_d. The decrement of the currents in the additional rotor circuits is very rapid compared to that of the field current so that all rotor currents except the field current will become negligibly small after a few cycles. The effective armature reactance will then increase from the subtransient to the transient value. The idea of a transient of very short time duration is conveyed through the term "subtransient".

The subtransient direct-axis reactance and the subtransient quadrature-axis reactance are then defined in per-unit notation as

$$x''_d = \frac{\Delta\lambda_d}{\Delta i_d} \tag{3.1.9}$$

and

$$x''_q = \frac{\Delta\lambda_q}{\Delta i_q} \tag{3.1.10}$$

and it follows from the equivalent circuits of Figures 3.1.3 and 3.1.4 that

$$x''_d = L_\ell + \frac{1}{\frac{1}{L_{ad}} + \frac{1}{L_{fd}} + \frac{1}{L_{1d}} + \dots} \quad (3.1.11)$$

$$x''_q = L_\ell + \frac{1}{\frac{1}{L_{aq}} + \frac{1}{L_{1q}} + \dots} \quad (3.1.12)$$

Neglecting the amortisseur winding circuits, one can go through similar procedure and define transient reactances:

Transient Direct-Axis Reactance $x'_d = L_\ell + \frac{1}{\frac{1}{L_{ad}} + \frac{1}{L_{fd}}}$

(3.1.13)

and

Transient Quadrature-Axis Reactance $x'_q = L_\ell + \frac{1}{\frac{1}{L_{aq}}} = L_\ell + L_{aq} = x_q$ (3.1.14)

For salient-pole machines, x'_q is equal to x_q; x'_d is less than x'_q; x''_q is much less than x'_q; x''_q is usually slightly greater than x''_d. For salient-pole machines without amortisseur windings, x''_q is equal to x'_q, which is equal to x_q.

For solid round-rotor machines, the changing reactance due to quadrature-axis eddy currents may be represented as the sum of two exponentials, one of which is rapid and the other slow. The slower transient yields x'_q when extrapolated back to zero time and the initial rapid transient gives x''_q. In the case of such machines, x''_q is slightly smaller than

x'_q, and x'_q lies between the values of x'_d and x_q.

Negative Sequence Reactance x_2:

If the unexcited field structure is rotated forward at synchronous speed with all rotor circuits closed and negative-sequence currents are applied to the armature terminals, a backward m.m.f. rotating at synchronous speed with respect to the armature would be set up. This m.m.f. would be rotating backward at twice the synchronous speed with respect to the rotor and as such currents of twice the rated frequency would be induced in all rotor circuits, keeping the flux linkages of those circuits almost constant at zero value. The armature flux linkage per armature ampere under these conditions is defined as the negative-sequence inductance L_2. The value of x_2 lies between x''_d and x''_q, and it is usually taken as the average value given by

$$x_2 = \frac{x''_d + x''_q}{2} \tag{3.1.15}$$

Synchronous-Machine Resistances:

The positive-sequence resistance of a synchronous machine is its armature a.c. resistance and it is somewhat greater than the d.c. armature resistance. It is nearly always neglected in power-system studies under normal operation and also in system-stability studies. The only exception is the case of a three-phase short circuit on or near the generator terminals, as neglecting resistance in this case leads to pessimistic conclusions regarding stability.

The negative-sequence resistance of a synchronous machine is greater than the positive-sequence resistance, since the currents are induced in all rotor circuits under negative-sequence conditions. Its value depends greatly on the resistance of the damper windings.

The zero-sequence resistance of a synchronous machine is equal to or somewhat greater than the positive-sequence resistance, as it depends in a way on the rotor copper bars due to rotor currents induced by zero sequence armature currents. It is usually neglected. However, external zero-sequence resistance may not be negligible, as three times the resistance of the resistor should be added to the zero-sequence resistance of the machine, if the neutral of a star-connected armature winding is grounded through a resistor.

Synchronous-Machine Time Constants:

The direct-axis transient open-circuit time constant T'_{do}, which is also known as the field open-circuit time constant in per-unit, is given by

$$T'_{do} = \frac{x_{ffd}}{r_{fd}} \tag{3.1.16}$$

When the armature is open-circuited and there is no amortisseur winding, the change of field current in response to the sudden application, removal, or change of e.m.f. in the field circuit is governed by this time constant. The open-circuit a.c. terminal voltage of an unsaturated machine is directly proportional to the field current, and therefore changes with the same time constant. This time constant is of the order of 750 to 4000 radians (2 to 11 seconds), and is greater than any of the other time constants discussed below.

The direct-axis transient short-circuit time constant T'_d is defined by

$$T'_d = \frac{x'_d}{x_d} T'_{do} \tag{3.1.17}$$

because the ratio of the short-circuit inductance of the field winding (when the armature is short-circuited) to its open-circuit inductance (when the armature is open-circuited) is x'_d/x_d and its resistance is unchanged. This time constant is about one-fourth as large as the open-circuit time constant.

For a machine with damper windings, direct-axis subtransient open-circuit time constant T''_{do} and short-circuit time constant T''_d may be introduced. The subtransient time constant is shorter than the transient time constant, and the short-circuit value is related to the open-circuit value as

$$T''_d = \frac{x''_d}{x'_d} T''_{do} \tag{3.1.18}$$

The open-circuit time constant with one amortisseur circuit may be defined as

$$T''_{do} = \frac{x_{11d} - \dfrac{x^2_{f1d}}{x_{ffd}}}{r_{1d}} \tag{3.1.19}$$

noting that the amortisseur winding and field winding are coupled, and a mutual reactance exists between the two.

The quadrature-axis subtransient open-circuit time

constant T''_{qo} may now be introduced for the machine with one additional rotor circuit as the ratio given by

$$T''_{qo} = \frac{x_{11q}}{r_{1q}} \tag{3.1.20}$$

and the short-circuit time constant T''_q is related to T''_{qo} as

$$T''_q = \frac{x''_q}{x_q} T''_{qo} \tag{3.1.21}$$

Usually T''_q is very nearly equal to T''_d.

Next the quadrature-axis transient open-circuit time constant T'_{qo} and short-circuit time constant T'_q may be introduced. For the case of a salient-pole machine, these time constants are meaningless; however, for a machine with solid round rotor, these constants may be applicable; the value of T'_q is about one-half of that of T'_d.

The armature short-circuit time constant T_a applies to the direct current in the armature windings and to the induced alternating currents in the field and damper windings, when both are closed. It is equal to the ratio of the average armature short-circuit inductance to the armature resistance under the stated conditions. It is approximately given by

$$T_a = \frac{2x''_d\, x''_q}{r(x''_d + x''_q)} \tag{3.1.22}$$

The justification for this approximation comes from actual short-circuit tests and also by the mathematical analyses of several cases. The armature time constant is also sometimes taken to be

$$T_a = \frac{x_2}{r} = \frac{(x''_d + x''_q)}{2r} \tag{3.1.23}$$

for simplicity and convenience.

The currents and voltages of a synchronous machine under transient conditions have components whose magnitudes change in accordance with one or more of the above time constants discussed. Investigation of this aspect would be carried out in detail in the next few Sections.

Table 3.1.1 gives typical average values of synchronous-machine constants* for various types of machines in per-unit notation, while Table 3.1.2 gives a summary of reactances and time-constants.

3.2 Determination of Transient Reactances and Time Constants from Sudden Three-Phase Short-Circuit Test Data

Reactances and time constants of a synchronous machine are of great assistance for predicting the short-circuit currents. Conversely, a short-circuit current oscillogram may be utilized to evaluate some of the reactances and time constants. Before we go into the details of computing the constants from the test data, let us try to get a physical picture of the happenings under short-circuit conditions.

Consider a three-phase synchronous generator operating at synchronous speed with constant excitation initially unloaded. Let a three-phase short-circuit be suddenly applied at the armature terminals. We shall now attempt to explore the nature of the currents of the three armature phases and the field circuit.

The flux that is produced by the field circuit links

---

*See also Appendix C for typical constants of three-phase synchronous machines.

Table 3.1.1

Typical Average Values of Synchronous-Machine Constants

(Adapted from Kimbark, E. W., Power System Stability: Synchronous Machines, Ch. XII, Dover Publications, Inc., New York, 1956 (1968).

Machine Constant	Turbo-generator (solid rotor)	Water-wheel generator (with dampers)	Synchronous Condenser	Synchronous Motor
x_d	1.1	1.15	1.80	1.20
x_q	1.08	0.75	1.15	0.90
x'_d	0.23	0.37	0.40	0.35
x'_q	0.23	0.75	1.15	0.90
x''_d	0.12	0.24	0.25	0.30
x''_q	0.15	0.34	0.30	0.40
x_2	0.13	0.29	0.27	0.35
x_0	0.05	0.11	0.09	0.16
r (d.c.)	0.003	0.012	0.008	0.01
r (a.c.)	0.005	0.012	0.008	0.01
r_2	0.035	0.10	0.05	0.06
T'_{do}	5.6x377	5.6x377	9.0x377	6.0x377
T'_d	1.1x377	1.8x377	2.0x377	1.4x377
$T''_d=T''_q$	0.035x377	0.035x377	0.035x377	0.035x377
T_a	0.16x377	0.15x377	0.17x377	0.15x377

Table 3.1.2

Summary of Reactances and Time-Constants

Reactances

Synchronous: d-axis $x_d = x_{ad} + x_\ell$

q-axis $x_q = x_{aq} + x_\ell$

Transient: d-axis $x'_d = \dfrac{x_{ad}\, x_{fd}}{x_{ad} + x_{fd}} + x_\ell$

Subtransient: d-axis $x''_d = \dfrac{x_{ad}\, x_{fd}\, x_{1d}}{x_{ad}\, x_{fd} + x_{fd}\, x_{1d} + x_{1d}\, x_{ad}} + x_\ell$

q-axis $x''_q = \dfrac{x_{aq}\, x_{1q}}{x_{aq} + x_{1q}} + x_\ell$

Time-Constants

Open-circuit transient: d-axis $T'_{do} = \dfrac{1}{r_{fd}} [x_{ad} + x_{fd}]$

Table 3.1.2 (Cont'd)

Open-circuit subtransient:	d-axis	$T''_{do} = \frac{1}{r_{1d}}\left[\frac{x_{ad}\, x_{fd}}{x_{ad} + x_{fd}} + x_{1d}\right]$
	q-axis	$T''_{qo} = \frac{1}{r_{1q}}\left[x_{aq} + x_{1q}\right]$
Short-circuit transient:	d-axis	$T'_{d} = \frac{1}{r_{fd}}\left[\frac{x_{ad}\, x_{\ell}}{x_{ad} + x_{\ell}} + x_{fd}\right] = \frac{x'_d}{x_d} T'_{do}$
Short-circuit subtransient:	d-axis	$T''_{d} = \frac{1}{r_{1d}}\left[\frac{x_{ad}\, x_{fd}\, x_{\ell}}{x_{ad}\, x_{fd} + x_{fd}\, x_{\ell} + x_{ad}\, x_{\ell}} + x_{1d}\right] = \frac{x''_d}{x'_d} T''_{do}$
	q-axis	$T''_{q} = \frac{1}{r_{1q}}\left[\frac{x_{aq}\, x_{\ell}}{x_{aq} + x_{\ell}} + x_{1q}\right] = \frac{x''_q}{x_q} T''_{qo}$
Short-circuit armature (d.c.)		$T_a = \frac{1}{r}\left[\frac{2x''_d\, x''_q}{x''_d + x''_q}\right]$

the armature circuits. When the three-phase short-circuit fault occurs at $t = 0$, the armature flux linkages that are trapped are given by

$$\lambda_a \propto \cos \alpha \tag{3.2.1}$$

$$\lambda_b \propto \cos (\alpha - 2\pi/3) \tag{3.2.2}$$

$$\lambda_c \propto \cos (\alpha + 2\pi/3) \tag{3.2.3}$$

where α is the angle at $t = 0$ between the phase-a axis and the d-axis. Thus it is seen that the flux linking each phase is different. As the field moves away after $t = 0$, since the flux cannot change immediately, the d.c. current of approproate magnitude appears in each phase to preserve the flux. Since the flux is different for all the phases depending on the angle α, the d.c. currents which appear will also be of different magnitudes, depending on α. These currents will damp out eventually with the armature time constant T_a.

If the magnitudes of all the d.c. currents appearing in the armature phases were to be the same, there would not be any net resultant flux. But as they happen to be all unequal, there will be a resultant flux which produces a damped fundamental current in the field circuit because of the relative motion between field circuit and armature circuits, the damping being dictated by the armature time constant T_a.

To an observer on rotor, the above-mentioned fundamental component produces a flux which is a uniaxial pulsating one. This can be resolved into two rotating flux waves of same magnitude traveling in opposite directions at synchronous speed with respect to an observer on rotor. The one traveling in the positive direction with respect to the rotor travels at twice the synchronous speed with respect to the stator, and the other one would be stationary with respect to the stator. So the latter cannot induce anything; but the former produces the double-frequency component currents in the armature phases which are all of the same magnitude. These second harmonic currents give rise to zero net resultant flux. Hence the reflection across the air-gap ceases. These second harmonic currents in the armature phases will damp out eventually, the damping being affected by the armature time constant T_a.

Thus we see that the unequal d.c. currents in the armature phases give rise to the fundamental component in the field circuit which in turn is responsible to produce equal second-harmonic components in the armature phases. Since the process started in the armature, all these will be damped out with the armature time constant, T_a.

The constant flux from the sustained d.c. current in the field circuit induces sustained fundamental three-phase armature currents. These currents produce an m.m.f. rotating forward at synchronous speed with respect to the stator, but stationary with respect to the field and centered on the direct axis of the field. This armature m.m.f. opposes the field m.m.f. and tends to reduce the field flux linkages and damper-winding flux linkages. In order to prevent such flux linkage changes, increased field current as well as amortisseur currents are induced. Thus there exist the transient and subtransient d.c. components in the rotor windings, damped by the direct-axis transient short-circuit time constant T'_d, and the direct-axis subtransient short-circuit time constant T''_d, respectively. Correspondingly, there will be the a.c. components in the armature windings; the transient and subtransient fundamental components damped by time constants T'_d and T''_d respectively.

Thus we have the following components in the field current: (i) sustained d.c., (ii) damped d.c. with time constant T'_d, (iii) damped d.c. with time constant T''_d, and (iv) damped fundamental a.c. component with time constant T_a.

The armature phase windings will be seen to have the following component currents: (i) sustained fundamental a.c., (ii) damped fundamental a.c. with time constant T'_d, (iii) damped fundamental a.c. with time constant T''_d, (iv) damped d.c. components with time constant T_a, which depend on the instant the fault occurs, and (v) damped second-harmonic a.c., with time constnat T_a. Thus it will be observed that each current wave in general consists of two kinds of components: (a) alternating current components, and (b) direct current components, the former of which are equal in all three phases and the latter of which are dependent upon the particular point on the cycle at which the short circuit occurs.

It may be noted that the short-circuit currents would not decay if the flux linkages were to remain absolutely constant. Actually they have decrements controlled by different time constants of the synchronous machine, as explained above.

Figure 3.2.1 shows an enlarged view of the alternating-components of a symmetrical short-circuit armature current in a synchronous machine. The envelope of the current for the subtransient, transient, and steady-state periods may be observed. The subtransient period lasts only for the first few cycles, during which the decrement of the current

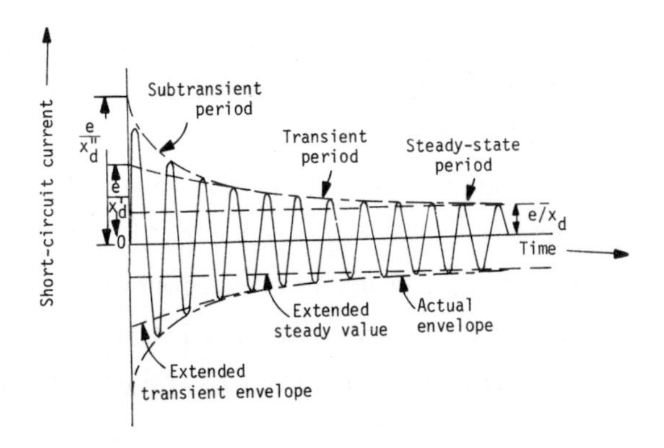

Figure 3.2.1. Alternating component of a symmetrical short-circuit armature current in a synchronous machine.

is very rapid; the transient period covers a relatively longer time during which the decrement of the current is more moderate; and finally the steady-state is attained during which the current has a sustained value. From the open-circuit prefault armature voltage and the initial values of different armature current components, the direct-axis reactances x_d, x_d' and x_d'' can be computed. The direct-axis short-circuit time constants T_d' and T_d'' can be evaluated from the logarithmic plots of the transient and subtransient components.

For a prolonged short circuit the armature current finally attains the sustained value (during the steady-state period), the magnitude of which is given by e/x_d, where e is the open-circuit voltage of an unsaturated machine, or the voltage read from the air-gap line of a saturated machine. The initial value of the alternating component of the armature current at the beginning of the transient period is given by e/x_d', which may be seen by extrapolation of the transient envelope in Figure 3.2.1. The initial value of the alternating component of the armature current at the commencement of the subtransient period is given by e/x''_d as shown in Figure 3.2.1.

Figure 3.2.2 shows the envelope of the symmetrical short-circuit current. The difference $\Delta i'$ indicated in Figure 3.2.2 between the transient envelope (including the extended part) and the steady-state amplitude is plotted to a logarithmic scale as a function of time in Figure 3.2.3. Similarly, the difference $\Delta i''$ between the subtransient and extrapolated transient envelope is plotted to a logarithmic scale as a function of time in Figure 3.2.4. It would be seen that both the plots closely approximate to straight lines, thereby illustrating the essentially exponential nature of the current decrement. From Figures 3.2.3 and 3.2.4, one can evaluate the time constants T'_d and T''_d, by reading the times corresponding to 0.368 of their initial current values.

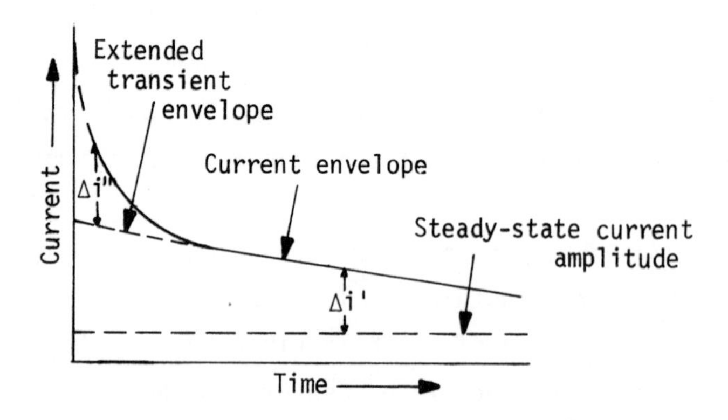

Figure 3.2.2. Envelope of the symmetrical short-circuit current.

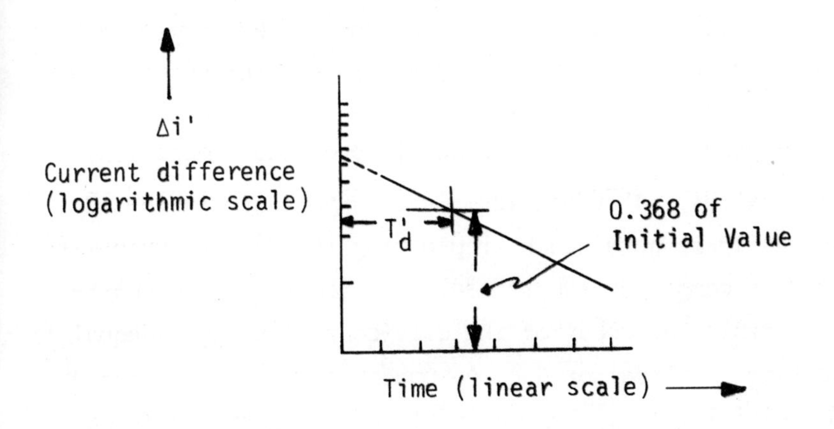

Figure 3.2.3. Semilog plot of $\Delta i'$ as a function of time.

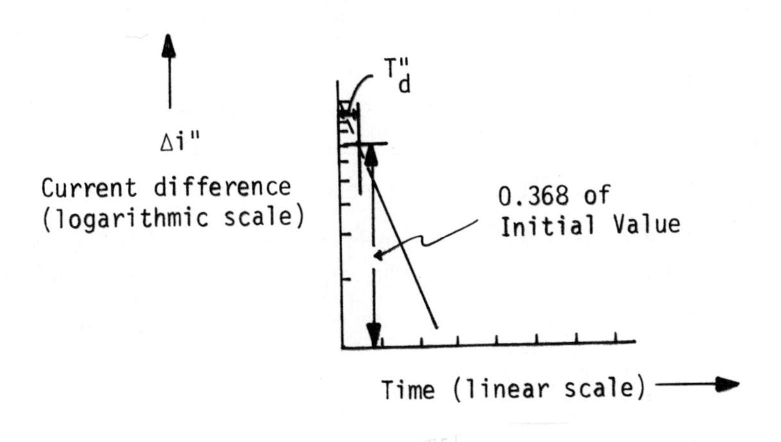

Figure 3.2.4. Semilog plot of $\Delta i''$ as a function of time.

A typical short-circuit oscillogram in Figure 3.2.5 shows the three-phase armature current waves as well as the field current. The traces of the armature phase currents are not symmetrical about the zero-current axis, and definitely exhibit d.c. components which are responsible for offset waves. A symmetrical wave as that of Figure 3.2.1 may easily be obtained by replotting the offset waves with the d.c. component subtracted. The armature time constant T_a controls the decay of the d.c. component and is equal to the time required for d.c. component to decay to 0.368 of its initial value. Semilog plots of d.c. components of armature currents shown in Figure 3.2.5 and a.c. component of field current may be utilized for determining the time constant T_a. From the semilog plot of excess d.c. component of field current over sustained value, time constants T'_d and T''_d may also be evaluated.

It may be pointed out here that the induced second harmonic armature currents are rather small if the machine has damper windings.

As for the experimental determination of quadrature-axis transient and subtransient reactances, negative-sequence resistance and reactance, the reader is referred to the technical article by S. H. Wright (listed in Bibliography), the book by C. F. Wagner and R. D. Evans (quoted in Bibliography), and the IEEE Test Code on synchronous machines (cited in Appendix B).

3.3 Mathematical Analysis of a Three-Phase Short-Circuit of a Synchronous Machine

We shall consider a three-phase fault of a synchronous machine, the fault having taken place after the field circuit has reached its steady state. Let the short circuit occur at

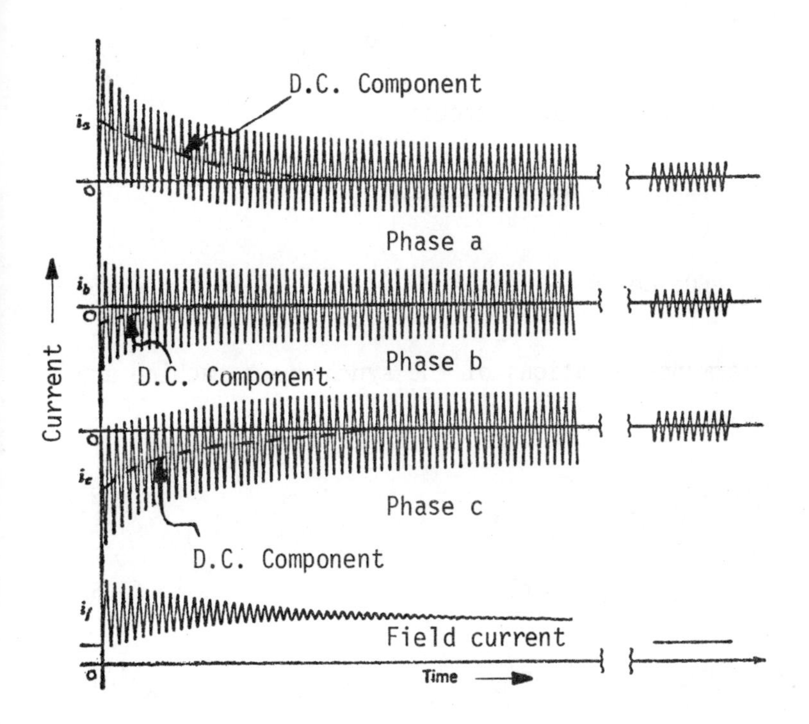

Figure 3.2.5. Short-circuit three-phase armature current and field current waves.

[*Adapted by permission of the author:* Kimbark, E. W., Power System Stability: Synchronous Machines, p. 41, Dover Publications, Inc., New York, 1956/1968.]

the armature terminals of the machine initially unloaded and normally excited. We shall consider the armature resistance of the machine to be small and assume no amortisseur windings to start with, for the sake of mathematical simplicity. Thus we have the following in per-unit notation:

$$\text{At } t = 0_-,\ e_d = 0;\ e_q = 1,\ e_{fd} = \frac{r_{fd}}{x_{ad}}, \tag{3.3.1}$$

just prior to short-circuit

$$\text{At } t = 0_+,\ e_d = 0;\ e_q = 0;\ e_{fd} = \frac{r_{fd}}{x_{ad}}\ , \tag{3.3.2}$$

just after short-circuit

The differences are then obtained as

$$\Delta e_d = 0;\ \Delta e_q = -1;\ \Delta e_f = 0 \tag{3.3.3}$$

The performance equations of the synchronous machine are given below:

$$e_d = p\lambda_d - \lambda_q p\theta - ri_d \tag{3.3.4}$$

$$e_q = p\lambda_q + \lambda_d p\theta - ri_q \tag{3.3.5}$$

$$e_o = p\lambda_o \qquad\qquad - ri_o \tag{3.3.6}$$

$$e_{fd} = p\lambda_{fd} + r_{fd}\, i_{fd} \tag{3.3.7}$$

where

$$\lambda_d = -L_d i_d + L_{ad}\, i_{fd} = G(p)\, e_{fd} - x_d(p)\, i_d \tag{3.3.8}$$

$$\lambda_q = -L_q i_q = -x_q(p)\, i_q \tag{3.3.9}$$

$$\lambda_o = -L_o i_o \tag{3.3.10}$$

$$\lambda_{fd} = -L_{ad} i_d + L_{ffd}\, i_{fd} \tag{3.3.11}$$

For balanced cases, such as the balanced three-phase short-circuit case, the zero-sequence components do not come into the picture as they do not exist. It should be noted that

$$p\theta = 1.0 \tag{3.3.12}$$

corresponds to the synchronous speed in per-unit notation, and no amortisseur windings have been assumed to exist. Applying the above equations of performance to the difference voltages obtained in Eq. (3.3.3), and noting that Δe_f is equal to zero in the case under consideration, one has

$$\Delta e_d = o = -[px_d(p) + r]\ \Delta i_d + x_q(p)\ \Delta i_q \tag{3.3.13}$$

$$\Delta e_q = -1 = -x_d(p)\ \Delta i_d - [px_q(p) + r]\ \Delta i_q \tag{3.3.14}$$

where

$$x_d(p) = x_d - \frac{px_{ad}^2}{px_{ffd} + r_{fd}} \tag{3.3.15}$$

Equation (3.3.15) can be obtained from Eqs. (3.3.7), (3.3.8) and (3.3.11). It is convenient to switch from the Heaviside operator p to the Laplace operator s. Noting that the Laplace transform* of the unit function is 1/s, the solution for $\Delta i_d(s)$ may be expressed as follows in terms of determinants:

$$\Delta i_d(s) = \frac{\begin{vmatrix} 0 & x_q(s) \\ -\frac{1}{s} & -[sx_q(s) + r] \end{vmatrix}}{\begin{vmatrix} -[sx_d(s) + r] & x_q(s) \\ -x_d(s) & -[sx_q(s) + r] \end{vmatrix}} \tag{3.3.16}$$

*See Appendix D for details on Laplace transforms.

The above may be simplified as

$$\Delta i_d(s) = \frac{\frac{1}{s}\, x_q(s)}{(s^2 + 1)\; x_d(s)\; x_q(s) + rs[x_d(s) + x_q(s)] + r^2}$$

$$= \frac{}{sx_d(s)\left[(s^2+1) + rs(\frac{1}{x_q(s)}) + \frac{1}{x_d(s)}) + \frac{r^2}{x_d(s)x_q(s)}\right]}$$

(3.3.17)

Neglecting the term

$$\frac{r^2}{x_d(s)\; x_q(s)} \quad ,$$

and making the following approximation in the coefficient of rs:

$$\frac{1}{x_d(s)} \simeq \frac{1}{x_d'} \tag{3.3.18}$$

and noting that

$$\frac{1}{x_q(s)} = \frac{1}{x_q} \tag{3.3.19}$$

one obtains

$$\Delta i_d(s) = \frac{1}{sx_d(s)[s^2 + 2as + 1]} \tag{3.3.20}$$

where

$$2a = r\,\frac{x_d' + x_q}{x_d'\; x_q} \tag{3.3.21}$$

In terms of the time constants, Eq. (3.3.15) may be rewritten as

$$x_d(s) = x_d - \frac{s x_{ad}^2}{s x_{ffd} + r_{fd}} = x_d \frac{sT'_d + 1}{sT'_{do} + 1} = x'_d \frac{s + \frac{1}{T'_d}}{s + \frac{1}{T'_{do}}} \tag{3.3.22}$$

It may be further noted that the armature time constant for this case with no damper windings is given by

$$T_a = \frac{1}{a} \tag{3.3.23}$$

Then one can rewrite Eq. (3.3.20) as follows:

$$\Delta i_d(s) = \frac{1}{x'_d} \frac{s + \frac{1}{T'_{do}}}{s\left(s + \frac{1}{T'_d}\right)(s^2 + 2as + 1)} \tag{3.3.24}$$

which may also be expressed in terms of the partial fractions* as

$$\Delta i_d(s) = \frac{1}{x'_d}\left[\frac{A}{s} + \frac{B}{\left(s + \frac{1}{T'_d}\right)} + \frac{Cs + D}{(s^2 + 2as + 1)}\right] \tag{3.3.25}$$

where A, B, C and D are the constants. The evaluation of these constants* may be done either by the method of comparing the coefficients of similar terms or by any other method, and for our case these will be seen to be

---

*See Appendix D for details on Partial-Fraction Expansion.

$$A = \frac{x'_d}{x_d} \tag{3.3.26}$$

$$B = 1 - \frac{x'_d}{x_d} \tag{3.3.27}$$

$$C = -1 \tag{3.3.28}$$

and $D = a$ (3.3.29)

when reasonable approximations are made. The details of evaluation are left as an exercise to the student. Thus we have the solution for $\Delta i_d(s)$ as

$$\Delta i_d(s) = \frac{\left(\frac{1}{x_d}\right)}{s} + \frac{\left(\frac{1}{x'_d} - \frac{1}{x_d}\right)}{\left(s + \frac{1}{T'_d}\right)} - \frac{1}{x'_d}\frac{s + a}{(s + a)^2 + 1 - a^2} \tag{3.3.30}$$

Taking the inverse Laplace transform*, the expression for $\Delta i_d(t)$ is obtained as

$$\Delta i_d(t) = \frac{1}{x_d} + \left(\frac{1}{x'_d} - \frac{1}{x_d}\right) e^{-(t/T'_d)} - \frac{1}{x'_d} \cos t \; e^{-(t/T_a)} \tag{3.3.31}$$

The steady-state value may also be checked by taking

$$\left\{ \lim_{s \quad 0} [s \; \Delta i_d(s)] \right\}$$

and noting that

*See Appendix D for Table of Laplace Transforms.

$$\frac{T'_d}{T'_{do}} = \frac{x'_d}{x_d}$$

Since there is no $i_d(t)$ prior to the fault, the total solution for the direct-axis armature current is given by

$$i_d(t) = \frac{1}{x_d} + \left(\frac{1}{x'_d} - \frac{1}{x_d}\right) e^{-(t/T'_d)} - \frac{1}{x'_d} \cos t \; e^{-(t/T_a)} \tag{3.3.32}$$

Based on the physical picture of the happenings under short-circuit conditions discussed in Section 3.2, it should be fairly easy to see that the mathematical analysis including the amortisseur windings would lead to the following total solution for the direct-axis armature current under sudden three-phase short-circuit conditions:

$$i_d(t) = \left(\frac{1}{x''_d} - \frac{1}{x'_d}\right) e^{-(t/T''_d)} + \left(\frac{1}{x'_d} - \frac{1}{x_d}\right) e^{-(t/T'_d)} +$$

$$+ \frac{1}{x_d} - \frac{1}{x''_d} \cos t \; e^{-(t/T_a)} \tag{3.3.33}$$

where

$$T_a = \frac{2x''_d \, x''_q}{r(x''_d + x''_q)} \tag{3.3.34}$$

The solution for the quadrature-axis armature current under sudden three-phase short-circuit conditions could be obtained by following a similar procedure. From Eqs. (3.3.13) and (3.3.14), one can obtain

$$\Delta i_q(s) = \frac{1}{sx_q(s)\left[s + \dfrac{r}{x_q(s)} + \dfrac{1}{s + \dfrac{r}{x_d(s)}}\right]} \tag{3.3.35}$$

which could be rewritten as

$$= \frac{s + \dfrac{r}{x_d(s)}}{sx_q(s)\left[s^2 + rs\left(\dfrac{1}{x_d(s)} + \dfrac{1}{x_q(s)}\right) + \dfrac{r^2}{x_d(s)x_q(s)} + 1\right]} \tag{3.3.36}$$

Neglecting the terms

$$\frac{r}{x_d(s)} \quad \text{and} \quad \frac{r^2}{x_d(s)\, x_q(s)} \quad ,$$

and making the following approximation in the coefficient of rs:

$$\frac{1}{x_d(s)} = \frac{1}{x''_d} \tag{3.3.37}$$

and

$$\frac{1}{x_q(s)} = \frac{1}{x''_q} \tag{3.3.38}$$

we have

$$\Delta i_q(s) = \frac{1}{x_q(s)\left[s^2 + r\,\dfrac{x''_d + x''_q}{x''_d\, x''_q}\, s + 1\right]} \tag{3.3.39}$$

Now recognizing that the inclusion of the damper-winding

circuits would lead to

$$x_q(s) = x_q \frac{(sT''_q + 1)}{(sT''_{qo} + 1)} \quad (3.3.40)$$

one can rewrite

$$\Delta i_q(s) = \frac{(sT''_{qo} + 1)}{x_q(sT''_q + 1)\left[s^2 + \frac{r(x''_d + x''_q)}{x''_d\, x''_q} s + 1\right]} \quad (3.3.41)$$

or

$$= \frac{1}{x''_q} \frac{(s + \frac{1}{T''_{qo}})}{(s + \frac{1}{T''_q})(s^2 + 2as + 1)} \quad (3.3.42)$$

where

$$2a = \frac{r(x''_d + x''_q)}{x''_d\, x''_q} \quad (3.3.43)$$

It may be seen that the armature time constant is given by

$$T_a = \frac{1}{a} \quad (3.3.44)$$

and also

$$\frac{T''_{qo}}{T''_q} = \frac{x_q}{x''_q} \quad (3.3.45)$$

In Eq. (3.3.42), the ratio

$$\frac{(s + \frac{1}{T''_{qo}})}{(s + \frac{1}{T''_q})}$$

may be approximated to be unity in which case one has

$$\Delta i_q(s) = \frac{1}{x''_q} \frac{1}{s^2 + 2as + 1} \tag{3.3.46}$$

or

$$= \frac{1}{x''_q} \frac{1}{(s + a)^2 + 1 - a^2} \tag{3.3.47}$$

or

$$\simeq \frac{1}{x''_q} \frac{1}{(s + a)^2 + 1} \tag{3.3.48}$$

as a^2 may be neglected compared to 1. Taking the inverse Laplace transform, we get

$$\Delta i_q(t) = \frac{1}{x''_q} \sin t \, e^{-(t/T_a)} \tag{3.3.49}$$

Approximations made are to be justified from experience, judgment, physical picture of the happenings, typical values of the various constants involved, and finally the comparison with test data. Since there is no $i_q(t)$ prior to the fault, the total solution for the quadrature axis armature current is given by

$$i_q(t) = \frac{1}{x''_q} \sin t \, e^{-(t/T_a)} \tag{3.3.50}$$

From the expressions we have for $i_d(t)$ and $i_q(t)$, one may now obtain the solution for the armature phase currents as follows through the appropriate transformation:

$$i_a(t) = \left[\left(\frac{1}{x''_d} - \frac{1}{x'_d}\right) e^{-(t/T''_d)} + \left(\frac{1}{x'_d} - \frac{1}{x_d}\right) e^{-(t/T'_d)} + \frac{1}{x_d}\right] \cdot$$

$$\cdot \cos(t + \alpha) - \left[\frac{1}{2}\left(\frac{1}{x''_d} + \frac{1}{x''_q}\right) \cos \alpha + \right.$$

$$\left. + \frac{1}{2}\left(\frac{1}{x''_d} - \frac{1}{x''_q}\right) \cos(2t + \alpha)\right] e^{-(t/T_a)} \qquad (3.3.51)$$

The other two phase currents i_b and i_c would be given by similar expressions with α replaced by $(\alpha - 2\pi/3)$ and $(\alpha + 2\pi/3)$ respectively. The angle α is the angle at $t = 0$ between the phase-a axis and the d-axis. It may now be verified that the expression obtained in Eq. (3.3.51) for the armature phase current contains all the components as expected from the physical picture of the happenings under sudden three-phase short-circuit conditions (Section 3.2). In case the amortisseur windings are neglected, x''_d would be replaced by x'_d, and x''_q by x_q.

Next let us consider the same conditions of three-phase short-circuit fault and try to obtain the expression for the field current. From Eq. (3.3.8) one can write

$$i_{fd} = \frac{\lambda_d + L_d i_d}{L_{ad}} \qquad (3.3.52)$$

Utilizing Eq. (3.3.15) and recognizing that Δe_f is equal to zero in the case under consideration, one has

$$\Delta i_{fd}(s) = \frac{s x_{ad}}{s x_{ffd} + r_{fd}} \Delta i_d(s) \qquad (3.3.53)$$

or

$$\frac{x_{ad}}{x_{ffd}} \cdot \frac{s}{s + \frac{1}{T'_{do}}} \cdot \Delta i_d(s) \tag{3.3.54}$$

Substituting for $\Delta i_d(s)$ from Eq. (3.3.24) which is derived with no amortisseur windings assumed, one gets

$$\Delta i_{fd}(s) = \frac{x_{ad}}{x_{ffd}\, x'_d} \frac{1}{(s + \frac{1}{T'_d})(s^2 + 2as + 1)} \tag{3.3.55}$$

where

$$2a = \frac{r(x'_d + x_q)}{x'_d\, x_q} \tag{3.3.56}$$

Equation (3.3.55) may be expressed as

$$\Delta i_{fd}(s) = \frac{x_{ad}}{x_{ffd}\, x'_d} \left[\frac{A}{(s + \frac{1}{T'_d})} + \frac{Bs + C}{(s^2 + 2as + 1)} \right] \tag{3.3.57}$$

Evaluation of the constant coefficients in the partial-fraction expansion would lead to

$$A \simeq 1; \quad B \simeq -1; \quad C = -a \text{ (which is very small)} \tag{3.3.58}$$

when appropriate approximations are made.

Thus we have

$$\Delta i_{fd}(s) = \frac{x_{ad}}{x_{ffd}\, x'_d} \left[\frac{1}{(s + \frac{1}{T'_d})} - \frac{s + a}{(s + a)^2 + 1 - a^2} \right] \tag{3.3.59}$$

Neglecting a^2 compared to unity and taking the inverse

Laplace transform, one obtains

$$\Delta i_{fd}(t) = \frac{x_{ad}}{x_{ffd}\, x'_d}\left[e^{-(t/T'_d)} - \cos t\, e^{-(t/T_a)}\right] \quad (3.3.60)$$

where

$$T_a = \frac{1}{a} \quad (3.3.61)$$

Onto the above solution is to be superimposed the value of $i_f(t)$ prior to the fault, which is $(1/x_{ad})$ per-unit, in order to get the total solution. Thus we have

$$i_{fd}(t) = \frac{1}{x_{ad}} + \frac{x_{ad}}{x_{ffd}\, x'_d}\, e^{-(t/T'_d)} - \frac{x_{ad}}{x_{ffd}\, x'_d} \cos t\, e^{-(t/T_a)}$$

$$(3.3.62)$$

The above expression may suitably be modified for the case in which amortisseur windings are considered. It may now be verified that Eq. (3.3.62) giving the solution for the field current does contain all the components as expected from the physical picture of the happenings under sudden three-phase short-circuit conditions (Section 3.2).

We have now completed the mathematical analysis of a particular case of three-phase short-circuit of a synchronous machine. Other cases of symmetrical faults may similarly be analyzed mathematically. A proper understanding of the nature of the short-circuit currents of synchronous machines, particularly of large machines, is necessary to evaluate short-circuit forces, stresses and torques, and to design proper protective relaying and switchgear. In practice, however, one needs to consider an integrated power system as a whole, of which the synchronous machine is just one of the components.

3.4 Unbalanced Short Circuits of a Synchronous Machine

Unbalanced short circuits include single-phase cases such as line-to-line short circuit and line-to-neutral short circuit, as well as double-line-to-ground short circuit and sequential faults. In what follows, the mathematical analysis of the line-to-line short circuit will be presented to give a feel to the reader regarding the method of attack with various approximations and the complexity of the problem in spite of the reasonable assumptions made. An indication of how to analyze the case of line-to-neutral short circuit will be given. Later, the use of symmetrical components based upon the representation of the fundamental frequency components is presented for analyzing the unbalanced short-circuit currents. We shall restrict ourselves to the case of small resistance and shall therefore be able to make use of the powerful concept of constant flux linkages, without which the solution procedure would be more complex. Also constant rotor speed will be assumed.

Line-to-Line Short Circuit:

Let us consider a line-to-line fault of a synchronous machine which is unloaded and normally excited before fault. We shall assume no amortisseur windings to start with for the sake of simplicity. Then we have the following, based upon Figure 3.4.1:

$$i_a = 0 \qquad (3.4.1)$$

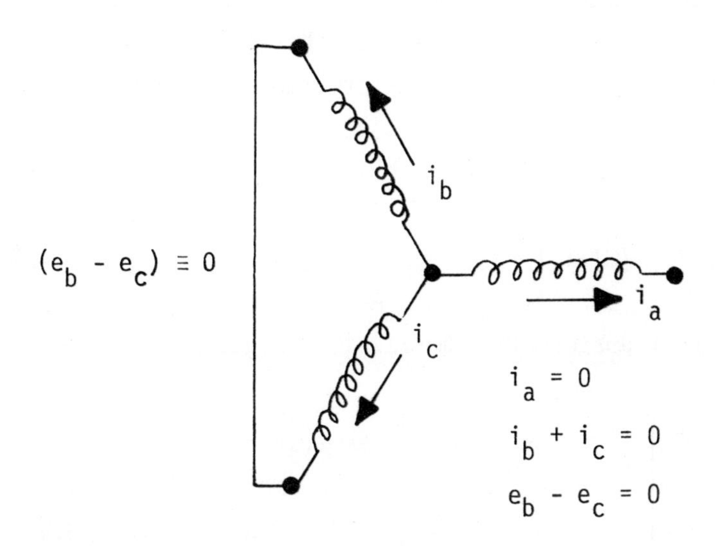

Figure 3.4.1. Line-to-line short circuit on a synchronous machine initially unloaded and normally excited before fault.

$$i_b + i_c = 0 \tag{3.4.2}$$

$$e_b - e_c = 0 \tag{3.4.3}$$

When the above are expressed in terms of direct- and quadrature-axis quantities, one obtains

$$i_o = 0 \tag{3.4.4}$$

$$i_d \cos \theta - i_q \sin \theta = 0 \tag{3.4.5}$$

$$e_d \sin \theta + e_q \cos \theta = 0 \tag{3.4.6}$$

The form of the above equations suggests that it may be

advisable to introduce a new set of variables, known as α, β and o Clarke components, for the study of line-to-line short-circuit case. The transformations relating the phase (a,b,c) components, Blondel (d,q,o) components and Clarke (α,β,o) components are given below:

$$[i_C] = [C][i_p] \tag{3.4.7}$$

where the transformation matrix [C] is given by

$$[C] = \begin{bmatrix} \frac{2}{3} & -\frac{1}{3} & -\frac{1}{3} \\ 0 & \frac{1}{\sqrt{3}} & -\frac{1}{\sqrt{3}} \\ \frac{1}{3} & \frac{1}{3} & \frac{1}{3} \end{bmatrix} \tag{3.4.8}$$

Since we had already that

$$[i_p] = [A]^{-1}[i_B] \tag{3.4.9}$$

one can write

$$[i_C] = [C][A]^{-1}[i_B] \tag{3.4.10}$$

where

$$[C][A]^{-1} = \begin{bmatrix} \frac{2}{3} & -\frac{1}{3} & -\frac{1}{3} \\ 0 & \frac{1}{\sqrt{3}} & -\frac{1}{\sqrt{3}} \\ \frac{1}{3} & \frac{1}{3} & \frac{1}{3} \end{bmatrix} \begin{bmatrix} \cos\theta & -\sin\theta & 1 \\ \cos(\theta - \frac{2\pi}{3}) & -\sin(\theta - \frac{2\pi}{3}) & 1 \\ \cos(\theta + \frac{2\pi}{3}) & -\sin(\theta + \frac{2\pi}{3}) & 1 \end{bmatrix}$$

$$= \begin{bmatrix} \cos\theta & -\sin\theta & 0 \\ \sin\theta & \cos\theta & 0 \\ 0 & 0 & 1 \end{bmatrix} \tag{3.4.11}$$

which is simply an orthogonal rotation matrix, whose inverse may be seen to be the transpose of itself.

The a,b,c, phase components may be obtained from the α,β,o components through the following relation:

$$[i_p] = [C]^{-1}[i_C] \tag{3.4.12}$$

where

$$[C]^{-1} = \begin{bmatrix} 1 & 0 & 1 \\ -\frac{1}{2} & \frac{\sqrt{3}}{2} & 1 \\ -\frac{1}{2} & -\frac{\sqrt{3}}{2} & 1 \end{bmatrix} \tag{3.4.13}$$

Same transformation matrices can be applied to the flux linkages as well as the voltages, just as they have been applied to the currents. In terms of the new components introduced, Eqs. (3.4.1) to (3.4.3) may be rewritten as

$$i_o = 0 \tag{3.4.14}$$

$$i_\alpha = 0 \tag{3.4.15}$$

$$e_\beta = 0 \tag{3.4.16}$$

Further, if the armature resistance is neglected the total flux linkages of phases b and c will be maintained constant at their initial values as determined by λ_{bo} and λ_{co} such that

$$\lambda_b - \lambda_c = \lambda_{bo} - \lambda_{co} \tag{3.4.17}$$

which can be expressed as

$$\lambda_\beta = \lambda_{\beta 0} \tag{3.4.18}$$

The α,β quantities may be seen to be referred to the stator of the synchronous machine, their axes being fixed in the armature; i_α and i_β may be regarded as equivalent two-phase currents producing the same air-gap flux as the original three-phase currents. The location of the α,β axes with rexpect to the a,b,c phase axes and d,q axes is shown in Figure 3.4.2. The boundary conditions may be seen to have been simplified in terms of α,β,o quantities, and the β phase of the equivalent two-phase machine may be seen to be short-circuited, with open α phase. The flux-linkage relations in terms of d,q,o components are given by

$$\lambda_d = x_{ad}\, i_{fd} - x_d\, i_d = E - x_d(p)\, i_d \tag{3.4.19}$$

$$\lambda_q = -x_q(p)\, i_q \tag{3.4.20}$$

$$\lambda_o = -x_o\, i_o \tag{3.4.21}$$

where

$$x_d(p) = x'_d \text{ and } x_q(p) = x_q \quad , \tag{3.4.22}$$

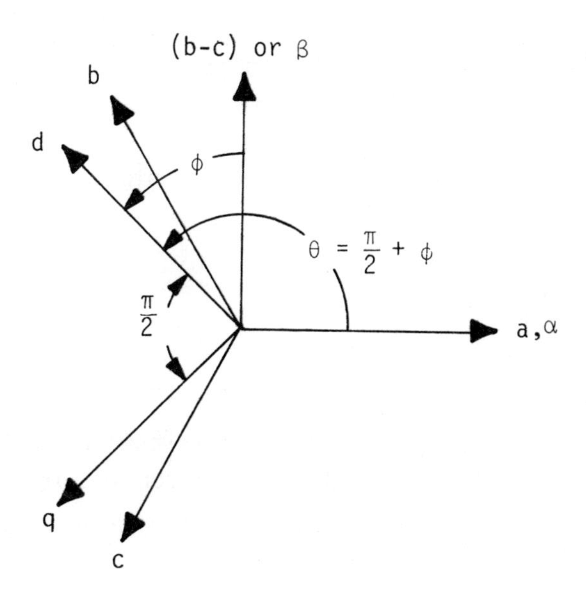

Figure 3.4.2. Location of α,β axes with respect to a,b,c phase axes and d.q axes.

if all rotor resistances are neglected and no amortisseur windings are assumed. The above equations may now be transformed into the new α,β,o variables and the constraints of Eqs. (3.4.14) and (3.4.15) be imposed. One obtains then:

$$\lambda_\alpha = E \cos\theta + \frac{1}{2}(x_q - x_d') \sin 2\theta \; i_\beta \tag{3.4.23}$$

$$\lambda_\beta = E \sin\theta - \frac{1}{2}[x_q + x_d' + (x_q - x_d') \cos 2\theta] \; i_\beta \tag{3.4.24}$$

Neglecting armature resistance, the armature current i_β may be evaluated by making use of Eq. (3.4.18).

$$i_\beta = \frac{2(E \sin\theta - \lambda_{\beta o})}{x_q + x_d' + (x_q - x_d')\cos 2\theta} \tag{3.4.25}$$

For the case of a short circuit on an unloaded machine, one has

$$\lambda_{\beta o} = \lambda_{do} \sin\alpha = E \sin\alpha \tag{3.4.26}$$

where α is the angle at $t = 0$ between the phase-a axis and the d-axis. From Eqs. (3.4.25) and (3.4.26), noting that i_α and i_o are zero for the case under consideration, one can obtain the following in terms of the phase components:

$$i_b = -i_c = \frac{\sqrt{3}\, E(\sin\theta - \sin\alpha)}{(x_q + x_d') + (x_q - x_d')\cos 2\theta} \tag{3.4.27}$$

which may also be written as

$$i_b = -i_c = \frac{\sqrt{3}\, E(\cos\phi - \cos\phi_o)}{(x_q + x_d') - (x_q - x_d')\cos 2\phi} \tag{3.4.28}$$

where ϕ is the angle between the β-axis and the d-axis; and ϕ_o is the measure of ϕ at $t = 0$. Thus we have obtained the expression for the short-circuit current corresponding to the line-to-line fault. If the amortisseur windings were to be taken into account and the same approximations are made, x_d' and x_q would be replaced by x_d'' and x_q'' respectively.

From Eq. (3.4.27), one can observe that only odd time harmonics appear in the current arising from $\sin\theta$ term, when $\alpha = 0$ and the initial trapped flux linkages in the short-circuited phases are zero. The even-harmonic series appearing in the current arising from the constant sin α term would be a maximum for $\alpha = \pm \pi/2$, when the initial trapped flux linkages are a maximum. (For details see Chapter 5 and Appendix A of the book by Concordia quoted in the Bibliography.)

Next let us take a look at the open-phase voltage or the voltage across the unfaulted line for the case of the line-to-line fault just considered.

$$e_a = p\lambda_a \tag{3.4.29}$$

where

$$\begin{aligned} \lambda_a &= \lambda_d \cos(\pi/2 + \phi) + \lambda_q \cos(\pi + \phi) \\ &= -\lambda_d \sin\phi - \lambda_q \cos\phi \\ &= -(E - x_d' i_d) \sin\phi + x_q i_q \cos\phi \end{aligned} \tag{3.4.30}$$

where again

$$i_d = \frac{2}{\sqrt{3}} i_b \cos\phi \tag{3.4.31}$$

and

$$i_q = -\frac{2}{\sqrt{3}} i_b \sin\phi \tag{3.4.32}$$

corresponding to the case under consideration. Thus the flux linkages will be given by

$$\lambda_a = -E \sin \phi - (x_q - x_d') E \sin 2\phi \cdot$$

$$\cdot \frac{\cos \phi - \cos \phi_o}{(x_q + x_d') - (x_q - x_d') \cos 2\phi} \qquad (3.4.33)$$

after the substitution of the result from Eq. (3.4.28). The voltage on the unfaulted phase is simply obtained by differentiating λ_a with respect to time and noting that $\phi = t + \phi_o$:

$$e_a = \frac{d\lambda_a}{dt} = -E \cos \phi - E(x_q - x_d') \cdot$$

$$\cdot \frac{N}{[(x_q + x_d') - (x_q - x_d') \cos 2\phi]^2} \qquad (3.4.34)$$

where

$$N = [(x_q + x_d') - (x_q - x_d') \cos 2\phi] \cdot$$

$$\cdot [\sin 2\phi(-\sin \phi) + (\cos \phi - \cos \phi_o) 2\cos 2\phi] +$$

$$- \sin 2\phi(\cos \phi - \cos \phi_o)[(x_q - x_d') 2\sin 2\phi] \qquad (3.4.35)$$

Usually the peak value of the voltage that occurs in the first cycle is of interest. We may reasonably expect by observation and physical reasoning that the maximum value would occur when $\phi_o = 0$ (i.e., when the trapped flux linkages are a maximum) and $\phi = \pi$ (i.e., half-cycle after closing, which may be seen from the nature of the denominator in the expression for e_a). The corresponding maximum comes out as

$$e_a\Big|_{max} = E\left(2\,\frac{x_q}{x'_d} - 1\right) \tag{3.4.36}$$

from which it may be seen that the maximum voltage is a function of the saliency ratio x_q/x'_d. By using the proper amortisseur windings one can cut down the saliency ratio, which will then be equal to x''_q/x''_d, and hence reduce the maximum possible voltage on the unfaulted phase. It should be observed that the peak value of the voltage does depend on the instant of fault (i.e., the angle ϕ_o). The value calculated by Eq. (3.4.36) yields the maximum of the possible maximum values that it can have.

In order to observe the harmonic components of the current, one may rewrite Eqs. (3.4.25) and (3.4.28) as given below: (The steps in between are left as an exercise to the reader.)

$$i_\beta = \frac{2E}{x'_d + x_2}\,[0] - \frac{E\cos\phi_o}{x_2} - \frac{2E\cos\phi_o}{x_2}\,a\,[E] \tag{3.4.37}$$

$$i_b = \frac{\sqrt{3}\,E}{x'_d + x_2}\,[0] - \frac{\sqrt{3}\,E\cos\phi_o}{2x_2} - \frac{\sqrt{3}\,E\cos\phi_o}{x_2}\,a\,[E] \tag{3.4.38}$$

where

$$x_2 = \sqrt{x'_d\,x_q} \tag{3.4.39}$$

$$[0] = \cos\phi + a\cos 3\phi + \ldots,\ \text{odd harmonic series} \tag{3.4.40}$$

$$[E] = \cos 2\phi + a \cos 4\phi + \ldots, \text{ even harmonic series} \tag{3.4.41}$$

and

$$a = \frac{\sqrt{x_q} - \sqrt{x'_d}}{\sqrt{x_q} + \sqrt{x'_d}} \tag{3.4.42}$$

Next let us investigate the field current for the line-to-line fault case under consideration, in which the machine is initially unloaded. Neglecting damping, one can write from the concept of constant flux linkages

$$\lambda_{fdo} = L_{ffd}\, i_{fdo} = \lambda_{fd} = L_{ffd}\, i_{fd} - L_{ad}\, i_d \tag{3.4.43}$$

Then one has

$$i_{fd} = \frac{L_{ffd}\, i_{fdo} + L_{ad}\, i_d}{L_{ffd}} = i_{fdo} + \frac{L_{ad}}{L_{ffd}}\, i_d =$$

$$= i_{fdo} + \frac{L_{ad}}{L_{ffd}}\, i_\beta \cos \phi \tag{3.4.44}$$

For normal excitation

$$L_{ad}\, i_{fdo} = \lambda_{do} = i_{qo} = E \tag{3.4.45}$$

which may be taken as unity and designating per-unit field current as

$$I = L_{ad}\, i_{fd} \tag{3.4.46}$$

so that it too will be unity for normal excitation, one can write the following in per-unit notation:

$$I = I_o + \frac{L_{ad}^2}{L_{ffd}} i_\beta \cos\phi = 1 + (x_d - x_d') i_\beta \cos\phi \tag{3.4.47}$$

in which I_o may be taken as unity as the machine is normally excited before fault. Recognizing that

$$\cos\phi\, i_\beta = \frac{1}{x_d' + x_2} + \frac{2}{x_d' + x_2} \frac{\sqrt{x_q}}{\sqrt{x_q} + \sqrt{x_d'}} [E] + $$

$$- \frac{\cos\phi_o}{x_2} (1 + a)[0] \tag{3.4.48}$$

one obtains for the field current

$$I = \frac{x_d + x_2}{x_d' + x_2} + \frac{2\sqrt{x_q}\,(x_d - x_d')}{(\sqrt{x_q} + \sqrt{x_d'})(x_d' + x_2)} [E] + $$

$$- \frac{2(x_d - x_d')\sqrt{x_q}}{\sqrt{x_q x_d'}\,(\sqrt{x_q} + \sqrt{x_d'})} [0] \cos\phi_o \tag{3.4.49}$$

where [E] and [0] are as defined by Eqs. (3.4.41) and (3.4.40) respectively. Further, the field current may be expressed as

$$I = \frac{x_d + x_2}{x_d' + x_2} + \frac{2\sqrt{x_q}\,(x_d - x_d')}{(\sqrt{x_q} + \sqrt{x_d'})(x_d' + x_2)} [E] + $$

$$- \frac{2(x_d - x_d')}{x_d' + x_2} [0] \cos \phi_o \tag{3.4.50}$$

where

$$\frac{x_d + x_2}{x_d' + x_2}$$

represents the undamped d.c. component. Thus one has for the field current just before and after fault:

$$I\Big|_{t = 0_-} = 1 \quad \text{and} \quad I\Big|_{t = 0_+} = \frac{x_d + x_2}{x_d' + x_2} \tag{3.4.51}$$

the latter of which is greater than one as x_d' is less than x_d. The transient part is then given by

$$I_{tr} = I\Big|_{t = 0_+} - I\Big|_{t = 0_-} = \frac{x_d - x_d'}{x_d' + x_2} =$$

$$= I\Big|_{t = 0_+} \frac{x_d - x_d'}{x_d + x_2} \tag{3.4.52}$$

The steady-state part of I may be expressed as

$$I_{ss} = I\Big|_{t = 0_+} - I_{tr} = 1 = I\Big|_{t = 0_+} \frac{x_d' + x_2}{x_d + x_2} \tag{3.4.53}$$

Since the field flux linkages are conservative, one can write

$$L_{ffd}\, I\Big|_{t = 0_-} = L_{ffd}'\, I\Big|_{t = 0_+} \tag{3.4.54}$$

from which it follows that

$$L'_{ffd} = L_{ffd} \frac{x'_d + x_2}{x_d + x_2}$$

or

$$T'_d = T'_{do} \frac{x'_d + x_2}{x_d + x_2} \tag{3.4.55}$$

Also the flux linkages $(\lambda_b - \lambda_c)$ are conservative and it therefore follows that

$$(\lambda_b - \lambda_c)\Big|_{t = 0_+} = \sqrt{3} \cos \phi_o \tag{3.4.56}$$

Since the d.c. component of

$$i_b\Big|_{t = 0_+}$$

is given by

$$\frac{\sqrt{3} \cos \phi_o}{2x_2} ,$$

one can write that

$$L'_a = 2x_2 \quad \text{and} \quad T'_a = \frac{x_2}{r} \tag{3.4.57}$$

noting r to be the armature resistance per phase and 2r needs to be considered here. One may then reasonably express the b-phase current and the field current as a function of time in the following manner:

$$i_b = \left[\frac{\sqrt{3}}{x_d + x_2} + \frac{\sqrt{3}\,(x_d - x_d')}{(x_d + x_2)(x_d' + x_2)}\, e^{-(t/T_d')}\right] [0] +$$

$$- \left[\frac{\sqrt{3}\cos\phi_o}{2x_2} + \frac{\sqrt{3}\cos\phi_o}{x_2}\, a[E]\right] e^{-(t/T_a')} \qquad (3.4.58)$$

and

$$I = 1 + \frac{x_d - x_d'}{x_d' + x_2}\, e^{-(t/T_d')} +$$

$$+ \frac{2\sqrt{x_q}\,(x_d - x_d')}{(\sqrt{x_q} + \sqrt{x_d'})(x_d + x_2)} \left[1 + \frac{x_d - x_d'}{x_d' + x_2}\, e^{-(t/T_d')}\right] [E] +$$

$$- \frac{2(x_d - x_d')\cos\phi_o}{x_d' + x_2}\, e^{-(t/T_a')} [0] \qquad (3.4.59)$$

It should be observed here that average time constants have been estimated, as the damping factors are in fact amplitude modulated and the differential equations would not have constant coefficients, in which case they could not be easily solved. The maximum value of the sustained steady-state armature current is given by

$$\frac{\sqrt{3}}{x_d + x_2} \cdot \frac{1}{1 - a}$$

where $1/(1 - a)$ is due to the summation of the odd series in geometric progression in Eq. (3.4.58). The rms or effective value of the sustained armature current is given by

$$\frac{\sqrt{3}}{x_d + x_2} \cdot \sqrt{\frac{1 + a^2 + a^4 + \ldots}{2}}$$

which may be rewritten as

$$\frac{\sqrt{3}}{x_d + x_2} \sqrt{\frac{1}{2(1 - a^2)}}$$

It may be seen that the effect of harmonics, due to which a and a^2 are appearing in the maximum value and the rms value of the sustained armature current respectively, is less significant on the rms value than on the maximum value. The value of a given by Eq. (3.4.42) is typically around 0.25. Neglecting the effect of harmonics, the fundamental component of the sustained armature current is given by

$$\frac{\sqrt{3}}{x_d + x_2} \cos \phi$$

which may also be obtained quite easily from the symmetrical component theory which will be introduced a little later on in this chapter.

The functional forms of the armature current as well as the field current given by Eqs. (3.4.58) and (3.4.59) can also be justified from the physical picture of the happenings in the synchronous machine initially unloaded, normally excited, and subjected to line-to-line short circuit. The chain of events may be described as follows: The d.c. component in armature (which would appear to preserve the flux and would eventually be damped with time constant T_a') would induce a fundamental component in the field circuit, which in turn would induce double-frequency component in the armature current. Thus the odd harmonics appear in the field current and the even harmonics in the armature-phase current, which will all be damped out with time constant T_a'. The sustained d.c. in the field gives rise to the sustained fundamental in the armature current. The d.c. component in the field appearing to preserve the flux (which would be damped with time constant T_d') will induce a fundamental component in the armature circuit, which in turn will induce a double-frequency component in the field circuit. Thus odd harmonics result in the armature current and even harmonics in the field current, all of which will be damped out with time constant T_d'. The effect of reflection from rotor to stator and vice versa does not cease as in the case of symmetrical three-phase short circuit and hence odd as well as even harmonic series will result.

Having completed the mathematical analysis of line-to-line short circuit of a synchronous machine, let us proceed to other cases of unbalanced faults.

Line-to-Neutral Short Circuit:

This case will work out to be very similar to the line-to-line short circuit case except for the addition of the zero-phase sequence components. Similar assumptions and approximations as in the previous case may be made, and the details of analysis are left out as desirable exercise for the reader. If the zero-sequence reactance x_o is taken as zero, it will be seen that the magnitude of the short-circuit current for the line-to-neutral fault is $\sqrt{3}$ times that for the line-to-line fault.

Symmetrical Component Theory for Analyzing Unbalanced Faults:

It is always possible to decompose a set of three unbalanced fundamental-frequency currents (i_a, i_b, i_c) or voltages (which may be unbalanced either in magnitude or angle or both) into three symmetrical-component sets: the positive-phase-sequence set consisting of balanced three-phase components of fundamental frequency and normal phase rotation (i_{a1}, i_{b1}, i_{c1}), the negative-phase-sequence set comprised of balanced three-phase components of fundamental frequency and opposite phase rotation (i_{a2}, i_{b2}, i_{c2}), and the zero-phase-sequence set consisting of three equal components of fundamental frequency which are all in phase with each other (i_{ao}, i_{bo}, i_{co}). This is shown in Figure 3.4.3 in terms of the corresponding phasor sets. The transformation relating the symmetrical-component phasors and the a, b, c phasors is given below:

$$\begin{bmatrix} I_1 \\ I_2 \\ I_o \end{bmatrix} = \frac{1}{3}\begin{bmatrix} 1 & a & a^2 \\ 1 & a^2 & a \\ 1 & 1 & 1 \end{bmatrix}\begin{bmatrix} I_a \\ I_b \\ I_c \end{bmatrix} \tag{3.4.60}$$

where the operator a (causing rotation by $\frac{2\pi}{3}$ radians) is defined as

$$a = e^{j\,2\pi/3} = -\frac{1}{2} + j\,\frac{\sqrt{3}}{2} \tag{3.4.61}$$

$$a^2 = e^{j\,4\pi/3} = -\frac{1}{2} - j\,\frac{\sqrt{3}}{2} \tag{3.4.62}$$

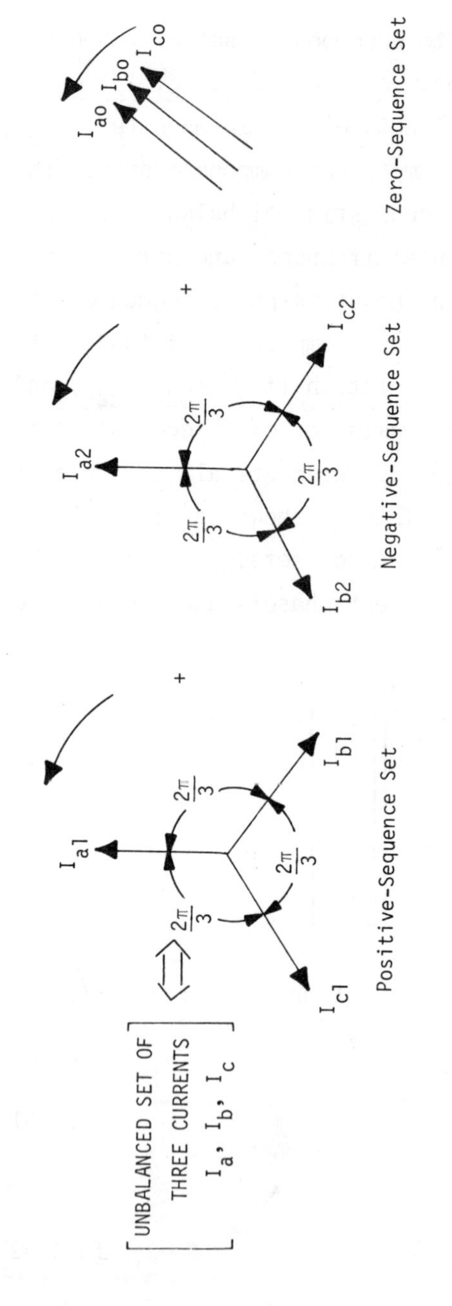

$$I_a = I_{a1} + I_{a2} + I_{ao}$$

$$I_b = I_{b1} + I_{b2} + I_{bo} = a^2 I_{a1} + aI_{a2} + I_{ao}$$

$$I_c = I_{c1} + I_{c2} + I_{co} = aI_{a1} + a^2 I_{a2} + I_{ao}$$

Figure 3.4.3. Symmetrical-component phasor sets

$$a^3 = e^{j\,6\pi/3} = 1.0 \qquad (3.4.63)$$

The transformations relating the symmetrical-component phasors and α, β, o-component phasors are given below, and may be verified easily by the reader:

$$\begin{bmatrix} I_1 \\ I_2 \\ I_o \end{bmatrix} = \begin{bmatrix} 1/2 & j/2 & 0 \\ 1/2 & -j/2 & 0 \\ 0 & 0 & 0 \end{bmatrix} \begin{bmatrix} I_\alpha \\ I_\beta \\ I_o \end{bmatrix} \qquad (3.4.64)$$

$$\begin{bmatrix} I_\alpha \\ I_\beta \\ I_o \end{bmatrix} = \begin{bmatrix} 1 & 1 & 0 \\ -j & +j & 0 \\ 0 & 0 & 1 \end{bmatrix} \begin{bmatrix} I_1 \\ I_2 \\ I_o \end{bmatrix} \qquad (3.4.65)$$

The time-functional form of any current or voltage may be obtained by taking the real part of its phasor after being multiplied by $e^{j\omega t}$, as is conventionally done. It is a common practice to refer to symmetrical-component phasors as the symmetrical components themselves.

We shall now proceed to utilize the symmetrical components based upon the representation of the fundamental frequency components for analyzing the unbalanced short circuits of a synchronous machine. The impedances to the positive, negative and zero sequence currents are given by z_1, z_2 and z_o respectively, which may be taken as x_d, x_2 and x_o while neglecting the resistances. Let us first analyze the line-to-line fault at the terminals of a symmetrical star-connected alternator which is initially unloaded and normally excited before fault. This case is chosen here first so that we may compare the results obtained by the use of symmetrical components with those by the mathematical analysis performed earlier in this section. The boundary conditions as well as the equivalent diagram showing the interconnection of sequence networks in order to satisfy the terminal onditions are shown in Figure 3.4.4. It follows then for this case that

$$I_{ao} = 0; \quad I_{a1} = \frac{-a + a^2}{3} I_a; \quad I_{a2} = -I_{a1} \tag{3.4.66}$$

and

$$E_{ao} = 0; \quad E_{a1} = E_{a2} \tag{3.4.67}$$

The zero sequence network is eliminated and the positive as well as negative sequence networks are connected as in Figure 3.4.4 satisfying the above conditions. One may now obtain

$$I_1 = \frac{E}{x_d + x_2}; \quad I_2 = -\frac{E}{x_d + x_2}; \quad I_o = 0 \tag{3.4.68}$$

from which it follows that

$$I_a = 0; \; I_b = (a^2 - a)\frac{E}{x_d + x_2} = -j\frac{\sqrt{3}\,E}{x_d + x_2}; \; I_c = -I_b \tag{3.4.69}$$

where E is the speed-voltage generated on open-circuit condition. The voltages E_a, E_b, E_c may also be obtained in terms of E:

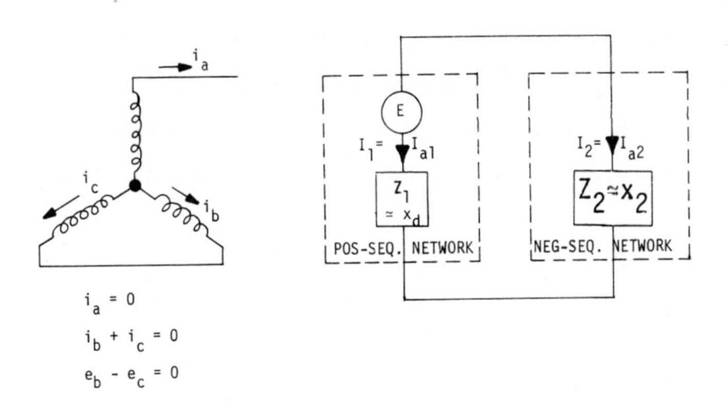

Figure 3.4.4. Line-to-line short circuit on an alternator.

$$E_a = \frac{2x_2}{x_d + x_2} E; \quad E_b = - \frac{x_2}{x_d + x_2} E; \quad E_c = E_b \qquad (3.4.70)$$

The armature b-phase current is given by

$$i_b = \text{real part of } [I_b\, e^{j\theta}] \qquad (3.4.71)$$

where θ is given by $(t + \alpha)$; it follows then that

$$i_b = \frac{\sqrt{3}\, E}{x_d + x_2} \sin\theta = \frac{\sqrt{3}\, E}{x_d + x_2} \cos\phi \qquad (3.4.72)$$

since θ is equal to $(\frac{\pi}{2} + \phi)$ as shown in Figure 3.4.2.

It may be observed that the fundamental component of the sustained armature current given above is the same as that obtained from the mathematical analysis [see Eq. (3.4.58)].

The analysis of single line-to-neutral short circuit with the use of symmetrical components proceeds as follows.

Assuming again the fault to occur at the terminals of a symmetrical star-connected alternator which is initially unloaded and normally excited before fault, the terminal conditions and the corresponding interconnection of sequence networks are shown in Figure 3.4.5. It follows then that

$$I_{a1} = I_{a2} = I_{ao} = \frac{I_a}{3} \tag{3.4.73}$$

$$E_a = E_{a1} + E_{a2} + E_{ao} = 0 \tag{3.4.74}$$

and the three networks need be connected in series. One can then obtain:

$$I_1 = I_2 = I_o = \frac{E}{x_d + x_2 + x_o} \tag{3.4.75}$$

The phase currents are obtained as

$$I_a = \frac{3E}{x_d + x_2 + x_o} \; ; \quad I_b = I_c = 0 \tag{3.4.76}$$

The phase voltages may be seen to be

$$E_a = 0; \quad E_b = \frac{E}{x_d + x_2 + x_o} \left[(a^2 - a)\, x_2 + (a^2 - 1)\, x_o \right];$$

$$E_c = \frac{E}{x_d + x_2 + x_o} \left[(a - a^2)\, x_2 + (a - 1)\, x_o \right] \tag{3.4.77}$$

One may observe here from Eqs. (3.4.69) and (3.4.76) that the magnitude of the short-circuit current for the line-to-neutral fault is $\sqrt{3}$ times that for the line-to-line fault, if the zero sequence reactance x_o is taken as zero.

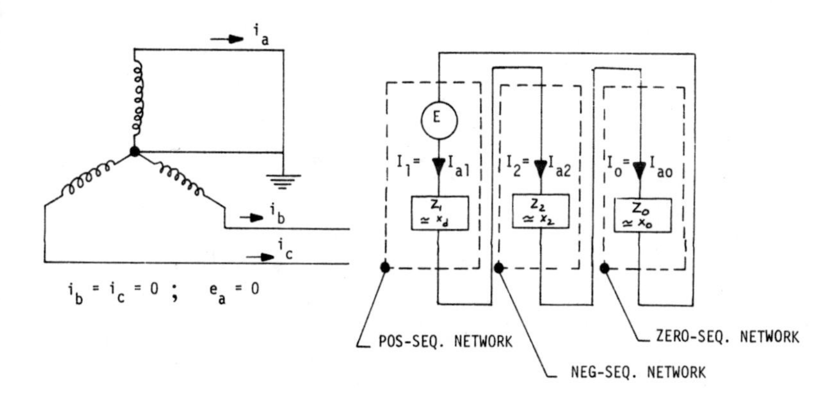

Figure 3.4.5. Line-to-neutral short circuit on an alternator.

Next, let us consider double line-to-ground fault with the use of symmetrical components. The terminal conditions and the corresponding interconnection of sequence networks are given in Figure 3.4.6. We have then in this case

$$E_{a1} = E_{a2} = E_{ao} = \frac{E_a}{3} \tag{3.4.78}$$

$$I_a = I_{a1} + I_{a2} + I_{ao} = 0 \tag{3.4.79}$$

and the three networks need be connected in parallel. The solution of the equivalent network leads to the following:

$$I_1 = \frac{(x_2 + x_o)}{(x_d x_2 + x_2 x_o + x_o x_1)} E \; ; \tag{3.4.80a}$$

$$I_2 = - \frac{x_o}{(x_d x_2 + x_2 x_o + x_o x_1)} E \; ; \tag{3.4.80b}$$

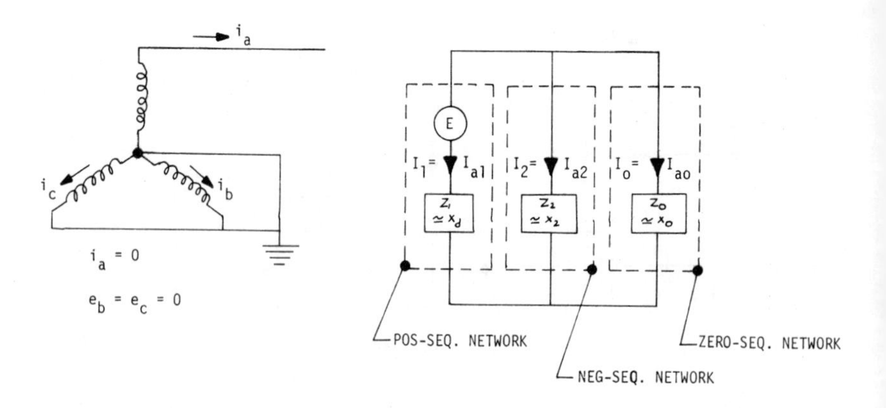

Figure 3.4.6. Double line-to-ground short circuit on an alternator.

$$I_o = - \frac{x_2}{(x_d x_2 + x_2 x_o + x_o x_1)} E \qquad (3.4.80c)$$

The corresponding phase currents and voltages are given below:

$$I_a = 0; \quad I_b = \frac{E}{(x_d x_2 + x_2 x_o + X_o x_1)} \cdot$$

$$\cdot \ [(a^2 - 1)\ x_2 + (a^2 - a)\ x_o];$$

$$I_c = \frac{E}{(x_d x_2 + x_2 x_o + x_o x_1)} [(a - 1)\ x_2 + (a - a^2)\ x_o] \qquad (3.4.81)$$

$$E_a = \frac{3 x_2 x_o}{(x_d x_2 + x_2 x_o + x_o x_1)} E; \quad E_b = E_c = 0 \qquad (3.4.82)$$

Regarding the application of symmetrical components for

analyzing unbalanced faults, the following remarks may be added at this point:

1. In case of simple line-to-ground and double-line-to-ground faults, if there is an external fault impedance z_n in the machine neutral, $3z_n$ needs to be added in series to the zero sequence impedance z_o in the zero sequence network.

2. The equivalent circuits corresponding to initial sub-transient and transient conditions may be obtained by simply replacing x_d by x''_d and x'_d respectively; also the corresponding subtransient or transient voltage source needs to be used in place of E.

3. Instead of mathematically analyzing with a number of approximations and assumptions without which the solution would be more difficult, one can go about using the symmetrical components to evaluate the fundamental frequency components of voltages and currents; the d.c. components may be calculated separately and the harmonics that may exist may be estimated; the average decrement factors may also be estimated. The solution in the final analysis should be justified from the physical picture of the happenings inside the machine under the particular fault condition.

It may be further pointed out that the sequential faults are not considered here in this text and the reader is referred to the article by Roeschlaub and the book by Concordia cited in the Bibliography. The principal effect of the application of sequential faults is on the d.c. component of short-circuit current.

3.5 Short-Circuit Torques

The instantaneous torque is in general given by

$$T = \lambda_d i_q - \lambda_q i_d \tag{3.5.1}$$

If the currents and the flux linkages could be calculated exactly for any balanced- or unbalanced-fault case, the torque may then be exactly evaluated by using Eq. (3.5.1). However, the procedure in calculating the currents may be very complex as we have seen in Section 3.4. Therefore we shall first attempt to make use of the approximate results obtained in Sections 3.3 and 3.4 in order to calculate short-circuit torques, and then analyze the deficiencies which may be corrected by adding additional components of torque.

We obtained expressions for $i_d(t)$, $i_q(t)$ and $i_{fd}(t)$ in Section 3.3 corresponding to the case of a three-phase short circuit of a synchronous machine with a small armature resistance and with no amortisseur circuits, the fault having taken place on the machine initially unloaded and normally excited.

$$i_d(t) = \frac{1}{x_d} + \left(\frac{1}{x'_d} - \frac{1}{x_d} \right) e^{-(t/T'_d)} - \frac{1}{x'_d} \cos t \; e^{-(t/T_a)} \tag{3.5.2.}$$

$$i_q(t) = \frac{1}{x_q} \sin t \; e^{-(t/T_a)} \tag{3.5.3}$$

$$i_{fd}(t) = \frac{1}{x_{ad}} + \frac{x_{ad}}{x_{ffd}\, x'_d} e^{-(t/T'_d)} - \frac{x_{ad}}{x_{ffd}\, x'_d} \cos t \; e^{-(t/T_a)} \tag{3.5.4}$$

Let us attempt to plot the first-half cycle of the torque following three-phase short circuit. Neglecting damping with due justification in the first-half cycle, we have

$$i_d(t) = \frac{1}{x'_d}(1 - \cos t) \tag{3.5.5}$$

$$i_q(t) = \frac{1}{x_q}\sin t \tag{3.5.6}$$

$$i_{fd}(t) = \frac{1}{x_{ad}} + \frac{x_{ad}}{x_{ffd}\, x'_d}(1 - \cos t) \tag{3.5.7}$$

The flux linkages are given by the following:

$$\lambda_d = -x_d i_d + x_{ad} i_{fd} = \cos t \tag{3.5.8}$$

$$\lambda_q = -x_q i_q = -\sin t \tag{3.5.9}$$

[Note that

$$x'_{\alpha} = x_d - \frac{x_{ad}^2}{x_{ffd}}$$

has been used in obtaining Eq. (3.5.8).]

Now the torque is obtained from Eq. (3.5.1) as

$$T = \frac{1}{x'_d}\sin t + \left(\frac{1}{2x_q} - \frac{1}{2x'_d}\right)\sin 2t \tag{3.5.10}$$

after substituting the results of Eqs. (3.5.5) through (3.5.9). Giving some typical per-unit values to x'_d and x_q as 0.2909 and 0.8 respectively, one has

$$T = 3.44 \sin t - 1.095 \sin 2t \tag{3.5.11}$$

which may be seen to contain a fundamental component as well as a double-frequency component with a negative sign. The maximum torque and the position where it occurs can be found by noting that the angle ϕ is in general given by $(t + \delta)$ and in our particular case $\phi = t$ (at $t = 0$, $\phi = 0$, taken as reference). Thus we have

$$T = 3.44 \sin \phi - 1.095 \sin 2\phi \tag{3.4.12}$$

from which one can obtain (by setting $\partial T/\partial \phi = 0$) the angle corresponding to maximum torque as

$$\phi]_{T_{max}} = 114°31' \text{ and } T_{max} = 3.96 \text{ p.u.} \tag{3.5.13}$$

It is seen that the maximum torque is nearly 4 times the full-load torque, following the three-phase short circuit. Further,

$$\phi|_{T_{max}}$$

occurs in the second quadrant, between $\pi/2$ and π radians. The alternating components as well as torque are plotted in Figure 3.5.1 as a function of the angle for the first-half cycle, neglecting damping. It may be observed that the shape of the torque-angle characteristic resembles that of the steady-state power-angle characteristic (see Figure 2.3.1), while the peak in the latter case occurs in the first quadrant, between $\pi/4$ and $\pi/2$ radians.

The expression for torque given by Eq. (3.5.10) may be easily modified as shown below, if the amortisseur

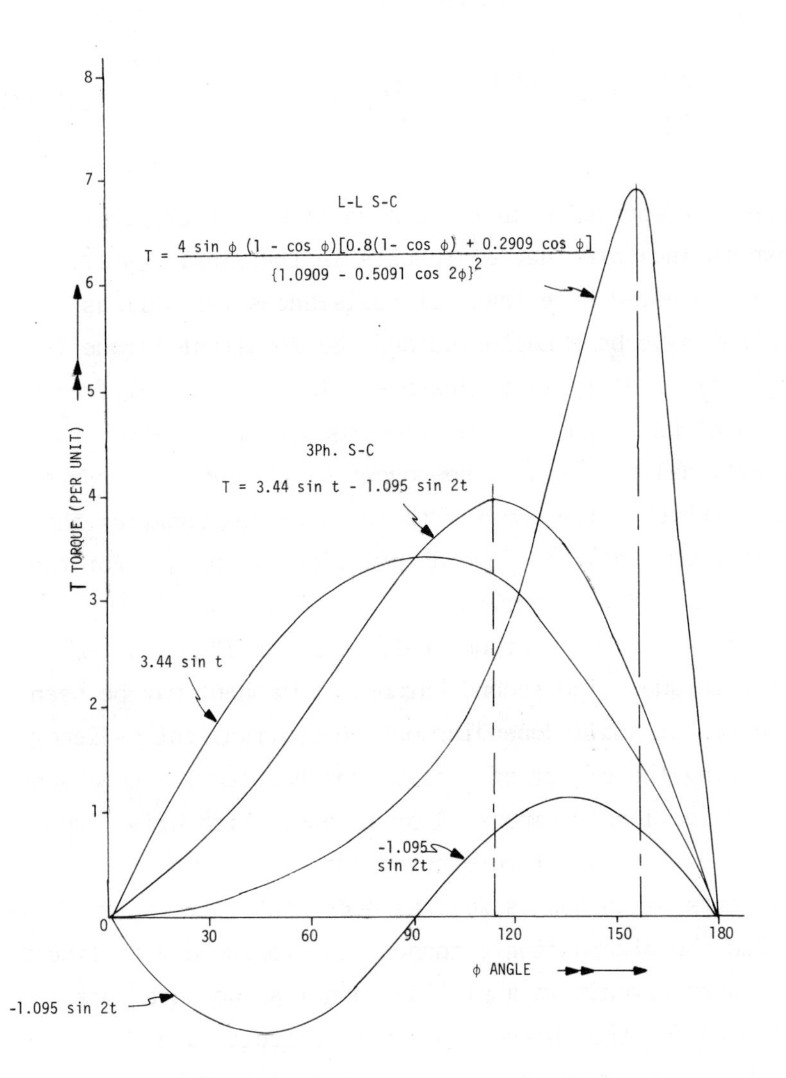

Figure 3.5.1. Alternating components of short-circuit torque (for the first-half cycle, neglecting damping).

windings are also considered:

$$T = \frac{e^2}{x''_d} \sin t - \frac{e^2}{2}\left(\frac{1}{x''_d} - \frac{1}{x''_q}\right) \sin 2t \qquad (3.5.14)$$

where the dependence on the bus voltage e is explicitly shown to indicate that e^2 results in the final result. It should be noted here that all resistances (as well as damping) have been neglected and the resultant torque is wholly oscillatory in its nature. The fundamental-frequency component has resulted from the product of i_d (which is proportional to the a.c. component of the armature current) and λ_q (which corresponds with the d.c. component of the armature current); it depends principally on x''_d. For the case of a turbine generator having x''_d of about 0.10 p.u., the initial maximum torque would be about 10 times the normal torque. The second harmonic component may be seen to be rather small depending on the subtransient saliency. Neglecting the effect of resistances has led to the absence of unidirectional components of torque, which arise from i^2r losses. We shall return to consider the non-alternating components of torque as well as decrement factors, after we analyze the alternating components of torque due to line-to-line short circuit in a similar manner as we have just completed for the three-phase short-circuit case.

Let us attempt to plot the first-half cycle of the torque following line-to-line short circuit, assuming no damping, no amortisseur windings, ϕ_o is equal to zero (for notation see Figure 3.4.2), and the fault having taken place on a synchronous machine initially unloaded and normally excited. We shall also assume, as before, that the constant speed is maintained during the period of our interest and that the effect of resistances is neglected in evaluating the alternating components of torque. Under the conditions stated above, we had the expressions in Section 3.4 for the short-circuit phase current i_b as

$$i_b = -i_c = \frac{\sqrt{3}\,(\cos\phi - \cos\phi_o)}{(x_q + x_d') - (x_q - x_d')\cos 2\phi} \tag{3.5.15}$$

Noting that $\cos\phi_o$ is unity in our case under consideration, θ is the same as $[(\pi/2) + \phi]$ as shown in Figure 3.4.2, the corresponding d,q,o components of current may be obtained through the transformation relating the Blondel components and the phase components [see Eq. (1.3.37)]. One has then the following:

$$i_d = \frac{2}{\sqrt{3}}\, i_b \cos\phi \tag{3.5.16}$$

$$i_q = -\frac{2}{\sqrt{3}}\, i_b \sin\phi \tag{3.5.17}$$

Next let us develop the corresponding flux-linkage expressions for λ_d and λ_q. As for the field, one can write from Eq. (3.4.47)

$$I = 1 + (x_d - x_d')\, i_\beta \cos\phi \tag{3.5.18}$$

Using Eq. (3.4.7) and noting that $i_a = 0$ and $i_b = -i_c$ for our case under consideration, one can obtain

$$i_\beta = \frac{2}{\sqrt{3}}\, i_b \tag{3.5.19}$$

The flux linkages are then given by: (with no amortisseurs)

$$\lambda_d = x_{ad}\, i_{fd} - x_d\, i_d = I - x_d\, i_d \tag{3.5.20}$$

$$\lambda_q = -x_q\, i_q \tag{3.5.21}$$

From Eqs. (3.5.16) through (3.5.21), we have the following:

$$\lambda_d = 1 + \frac{2}{\sqrt{3}} (x_d - x_d') \cos\phi \frac{\sqrt{3}\,(\cos\phi - 1)}{[(x_q + x_d') - (x_q - x_d')\cos 2\phi]} +$$

$$- x_d \frac{2}{\sqrt{3}} \cos\phi \frac{\sqrt{3}\,(\cos\phi - 1)}{[(x_q + x_d') - (x_q - x_d')\cos 2\phi]} \tag{3.5.22}$$

$$\lambda_q = \frac{2}{\sqrt{3}} x_q \sin\phi \frac{\sqrt{3}\,(\cos\phi - 1)}{[(x_q + x_d') - (x_q - x_d')\cos 2\phi]} \tag{3.5.23}$$

Having obtained all the relevant expressions, we are now in a position to arrive at an expression for the torque, which may be written as given below after some simplification:

$$T = \lambda_d i_q - \lambda_q i_d =$$

$$= \frac{4\sin\phi\,(1 - \cos\phi)[x_q(1 - \cos\phi) + x_d'\cos\phi]}{[(x_q + x_d') - (x_q - x_d')\cos 2\phi]^2} \tag{3.5.24}$$

Substituting some typical values for x_q as 0.8 and x_d' as 0.2909, we get

$$T = \frac{4\sin\phi\,(1 - \cos\phi)[0.8(1 - \cos\phi) + 0.2909\cos\phi]}{[1.0909 - 0.5091\cos 2\phi]^2} \qquad (3.5.25)$$

The first-half cycle of the torque angle characteristic given by the above equation is plotted in Figure 3.5.1 and it may be seen that the maximum torque is 5.86 per-unit occuring at an angle of 157°. The maximum torque following line-to-line short circuit is about seven times the full-load torque and roughly twice that of the maximum torque following three-phase short circuit. The effect of amortisseur windings may be included by replacing x'_d by x''_d, and x_q by x''_q. The expression obtained so far for the torque neglected the effect of resistances and damping, thereby ignoring the unidirectional components of the torque as well as the decrement factors. It may be observed that the expression for torque may be rewritten in terms of its harmonic components, and as may be reasonably expected, it would be seen to contain even as well as odd harmonic components. The d,q,o components have been again successfully utilized for the evaluation of the alternating components of torque.

Unidirectional Components of Torque:

The effect of neglecting all resistances is to miss entirely the unidirectional components of torque, which arise from the i^2r losses. It would be seen to be easier to utilize the symmetrical components for the evaluation of the i^2r losses caused by the fundamental a.c. component of armature current and thereby calculate unidirectional average torque, particularly for the case of unbalanced faults. The effect of harmonics contained in the armature currents may be neglected as their contribution to torque would be rather small. The contribution due to the d.c. component of the armature current would be calculated separately and added on later. The d.c. component of the armature current acts as d.c. excitation on the armature and generates a fundamental-frequency current in the rotor, which in turn is responsible for causing the rotor i^2r loss. The fundamental component of armature current causes a stator i^2r as well as a rotor i^2r loss, the latter of which is due to the negative-sequence component of the armature current. As a first step in evaluating the unidirectional average torque caused by the fundamental component of armature current, an expression for such a torque would be obtained in terms of symmetrical components.

The expression for the instantaneous torque in terms of the d,q,o components is given by

$$T = \lambda_d i_q - \lambda_q i_d \tag{3.5.26}$$

It may also be written in terms of α,β,o quantities as

$$T = \lambda_\alpha i_\beta - \lambda_\beta i_\alpha \tag{3.5.27}$$

The expression for the average torque may be obtained in terms of α,β,o component phasors as

$$T_{avg} = \frac{1}{2} (I_\beta \cdot \Lambda_\alpha - I_\alpha \cdot \Lambda_\beta) \tag{3.5.28}$$

or

$$T_{avg} = \text{real part of} \left[\frac{1}{2} (I_\beta^* \Lambda_\alpha - I_\alpha \Lambda_\beta^*)\right] \tag{3.5.29}$$

where I_α, I_β, Λ_α, and Λ_β are the corresponding phasor quantities of i_α, i_β, λ_α, and λ_β respectively; the dot product represents the sum of the products of the in-phase components of the two quantities I and Λ; I_β^* and Λ_β^* stand for the complex conjugate of the complex phasors I_β and Λ_β respectively and we are concerned only with the sinusoidally varying fundamental components of currents and flux linkages. In terms of the symmetrical-component phasors, the average torque may be expressed as

$$T_{avg} = I_1 \cdot (j\Lambda_1) - I_2 \cdot (j\Lambda_2) \tag{3.5.30}$$

Further, it can be shown (the details of which are left as an exercise to the reader) that the average torque is equal to the sum of the positive- and negative-sequence torques which may be calculated separately.

$$T_{avg} = (T_{avg})_1 + (T_{avg})_2 \tag{3.5.31}$$

The positive-sequence average torque is given by

$$(T_{avg})_1 = E \cdot I_1 \tag{3.5.32}$$

in terms of the notation of Section 3.4, and may also be expressed as the total i^2r loss in the equivalent circuit corresponding to the fault under consideration:

$$(T_{avg})_1 = |I_1|^2 R_1 + |I_2|^2 R_2 + |I_o|^2 R_o \tag{3.5.33}$$

where R_1, R_2 and R_o are the real parts of the corresponding positive-, negative- and zero-sequence impedances Z_1, Z_2 and Z_o respectively.

Next let us obtain an expression for the negative-sequence average torque by considering the negative-sequence currents given by

$$i_a = I_2 \cos (t + \theta) \tag{3.5.34}$$

$$i_b = I_2 \cos (t + \theta + \frac{2\pi}{3}) \tag{3.5.35}$$

$$i_c = I_2 \cos (t + \theta - \frac{2\pi}{3}) \tag{3.5.36}$$

from which it follows that

$$i_d = I_2 \cos (2t + \theta) \tag{3.5.37}$$

$$i_q = -I_2 \sin (2t + \theta) \tag{3.5.38}$$

$$I_o = 0 \tag{3.5.39}$$

By defining appropriately that

$$I_d = I_2 e^{j(2t+\theta)} \tag{3.5.40a}$$

and

$$I_q = jI_d \tag{3.5.40b}$$

and recognizing that

$$p\Lambda_d = j2\Lambda_d = -Z_d I_d \tag{3.5.41}$$

and

$$p\Lambda_q = j2\Lambda_q = -Z_q I_q \tag{3.5.42}$$

(Z_d and Z_q are the direct- and quadrature-axis impedances respectively), one can obtain the expression for the negative-sequence average torque as:

$$(T_{avg})_2 = \text{real part of} \left[\frac{1}{2}\,(I_q^* \Lambda_d - I_d^* \Lambda_q)\right] \tag{3.5.43}$$

which may be rewritten after making use of Eqs. (3.5.40) through (3.5.42) as

$$(T_{avg})_2 = I_2^2 \left(\text{real part of } \frac{Z_d + Z_q}{4}\right) \tag{3.5.44}$$

It can be shown that

$$\text{real part of } \frac{Z_d + Z_q}{4} = R_2 - R_1 \tag{3.5.45}$$

from which it follows that

$$(T_{avg})_2 = I_2^2\,(R_2 - R_1) \tag{3.5.46}$$

where R_1 and R_2 are the positive- and negative-sequence resistances respectively.

The total average torque due to the fundamental-frequency components of armature currents is then given by

$$T_{avg} = (T_{avg})_1 + (T_{avg})_2$$

$$= |I_1|^2 R_1 + |I_2|^2 R_2 + |I_o|^2 R_o + I_2^2\,(R_2 - R_1) \tag{3.5.47}$$

The unidirectional component of torque for the case of

three-phase fault would be given by

$$T_{avg} = \frac{E^2 R_1}{|Z_1|^2} \tag{3.5.48}$$

For the case of line-to-line fault, since

$$I_1 = -I_2 = \frac{E}{Z_1 + Z_2} \quad \text{and} \quad I_o = 0 \tag{3.5.49}$$

the average torque based on Eq. (3.5.47) comes out as

$$T_{avg} = \frac{E^2 \, 2R_2}{|Z_1 + Z_2|^2} \tag{3.5.50}$$

For the line-to-neutral short circuit, using the results of Section 3.4, it works out as

$$T_{avg} = \frac{E^2(2R_2 + R_o)}{|Z_1 + Z_2 + Z_o|^2} \tag{3.5.51}$$

As for the double line-to-ground fault, the result may be seen to be

$$T_{avg} = \frac{E^2[|Z_2 + Z_o|^2 R_1 + |Z_o|^2 (2R_2 - R_1) + |Z_2|^2 R_o]}{|Z_1 Z_2 + Z_2 Z_o + Z_o Z_1|^2} \tag{3.5.52}$$

Thus it is seen that symmetrical components have been very helpful in calculating the unidirectional components of torque caused by the fundamental component of armature current. However, it should be observed that such an analysis based on symmetrical components does not bring out the decrement factors.

Next let us consider the unidirectional component of torque due to the d.c. component of armature current, which causes the rotor i^2r loss because of the fundamental-frequency current in the rotor and the fundamental-frequency resistance of the rotor. Recognizing that the d.c. components of armature current are nearly the same as the a.c. fundamental components of i_d and i_q, one can express the average torque approximately as the average of the losses in the direct- and quadrature-axis rotor circuits given by

$$T_{avg} = \frac{1}{2} \left(|i_d|^2 R_d + |i_q|^2 R_q \right) \qquad (3.5.53)$$

where R_d and R_q are the effective rotor resistances as viewed from the armature; $|i_d|$ and $|i_q|$ are the magnitudes of the a.c. components of i_d and i_q respectively. The fundamental-frequency components of i_d and i_q may be obtained from the even harmonic components of i_α and i_β in the cases of unbalanced short circuits.

Concluding Remarks:

Approximate expressions for short-circuit torques have been obtained in this section under various assumptions including constant rotor speed. It is believed that the cases discussed here (short circuit occuring from no-load, open-circuit condition) are sufficient to illustrate the main points involved and to show the possible methods of analysis. The average decrement factors may be estimated from the analysis of the physical happenings in the machine. It is a normal practice to neglect the subtransient saliency (as a consequence of which $x''_q = x''_d = x_2$; $R_d = R_q$) in calculating the approximate short-circuit torques. The reader who is interested in pursuing further the analysis of short-circuit torques is referred to various relevant articles by a number of authors and the book by Concordia quoted in the Bibliography.

Problems

3-1. Consider a two-winding transformer operating under steady-state with voltage $e = E_m \sin \omega t$ applied to the primary terminals and with open secondary terminals. Let the transformer windings have self inductances L_1 and L_2, and a mutual inductance M. Let the secondary terminals be suddenly short circuited at time $t = t_1$. Neglecting resistance and assuming the primary voltage to be unaffected by the short circuit, obtain the expressions for the primary and secondary currents.

3-2. A 4-pole, 3-phase turbogenerator rated at 10 MVA, 11·8 KV, 60 HZ, is subjected to a sudden three-phase short circuit from an unloaded condition when the open-circuit voltage is 5.9 KV. An oscillogram of one of the armature currents is taken and the measurements made on the ordinates of the envelope of the oscillogram are given below in a Table. The scale factor of the oscillogram based on instantaneous value is 1930 amperes per centimeter. Compute the direct-axis reactances x_d, x'_d, x''_d in per-unit, and the time constants T'_d, T''_d and T_a in seconds.

Table for Problem 3-2

Time in Cycles	Ordinates of Envelopes in CMS		Time in Cycles	Ordinates of Envelopes in CMS	
	Upper	Lower		Upper	Lower
0	2.63	-0.63	15	0.86	-0.62
1	2.24	-0.50	20	0.75	-0.63
2	1.96	-0.44	25	0.67	-0.61
3	1.73	-0.41	30	0.63	-0.59
4	1.56	-0.42	40	0.54	-0.54
5	1.44	-0.44	50	0.48	-0.48
6	1.33	-0.45	60	0.43	-0.43
7	1.24	-0.48	90	0.32	-0.32
8	1.16	-0.50	120	0.25	-0.25
10	1.05	-0.55	∞	0.15	-0.15

3-3. Consider the 3-phase short-circuit analysis presented in the text. Which of the following are affected by the time at which the short-circuiting switch is closed?

(a) d.c. component in phase a
(b) d.c. component in direct-axis
(c) d.c. component in field
(d) torque

3-4. If $I_d(s)$ is given to be

$$i_d(s) = \frac{1}{x'_d} \frac{s + 1/T'_{do}}{s(s + 1/T'_d)(s^2 + 2as + 1)}$$

find the sustained (steady-state) value of $i_d(t)$.

3-5. Determine the field breaker duty (interrupting current and voltage) for a three-phase fault at the terminals of a synchronous machine whose data is given below. The machine is initially operating on open circuit at rated voltage. Assume a field discharge resistance equal to the cold (25°C) resistance of the generator field. Assume also that the field breaker trips 0.10 seconds after the fault occurs and that the excitation system output voltage has increased to 550 volts by this time. Neglect armature resistance.

Synchronous Machine Data:

2-pole, 3600 rpm, 625 MVA, 22 KV, 0.85 pf

$x_d = x_q = 1.73$; $x'_d = 0.28$

$T'_d = 0.61$ sec; $r_{fd} = 0.074$ ohms at 25°C

Field amperes corresponding to rated voltage at

no load = 1930 amps.

-6. A synchronous machine is initially operating at unity power factor, rated KVA, and rated terminal voltage. The machine parameters are given below:

$$x_d = x_q = 1.0;\ x_d' = 0.15;\ x_d'' = 0.10;$$

$$x_q'' = 0.12;\ r = 0$$

Corresponding to a 3-phase terminal short circuit, determine:

(a) the initial subtransient fundamental frequency fault current.

(b) the maximum d.c. offset that can occur in any phase current. Comment on the base of which this is expressed as per-unit.

(c) the initial transient fundamental frequency fault current (using voltage behind transient reactance) and the d.c. field current corresponding to this value (use e_q').

(d) the steady-state fault current.

3-7. Consider a synchronous machine with no amortisseur windings. Let a sudden three-phase short circuit occur at the machine terminals at the same instant when the field circuit is closed, so that

at $t = 0_-$, $e_d = 0$, $e_q = 0$; $e_{fd} = 0$ just before closing field circuit.

at $t = 0_+$, $e_d = 0$; $e_q = 0$; $e_{fd} = 0.0011$ just after closing field circuit.

Determine the functional form of the field current

$i_f(t)$

(a) when the machine is run at synchronous speed ($p\theta = 1$).

and

(b) when the machine is at standstill ($p\theta = 0$).

The synchronous machine data given below may be used:

$x_d = 1.2$; $x_q = 0.8$, $x_d' = 0.2909$; $r = 0.005$;

$r_{fd} = 0.0011$; $x_{ad} = 1.0$; $T_{do}' = 1000$ radians;

$T_d' = 242$ radians.

3-8. Consider a synchronous machine with damper windings at standstill. Let the field be short-circuited with no excitation and let a single phase voltage of rated frequency be applied as shown in the Figure below, with phase a open. Neglect all resistances. Evaluate the ratio of the voltage to the current, when

(a) $\alpha = 0$ and (b) $\alpha = \pi/2$

where α is the angle between the d-axis and a-phase axis at time $t = 0$.

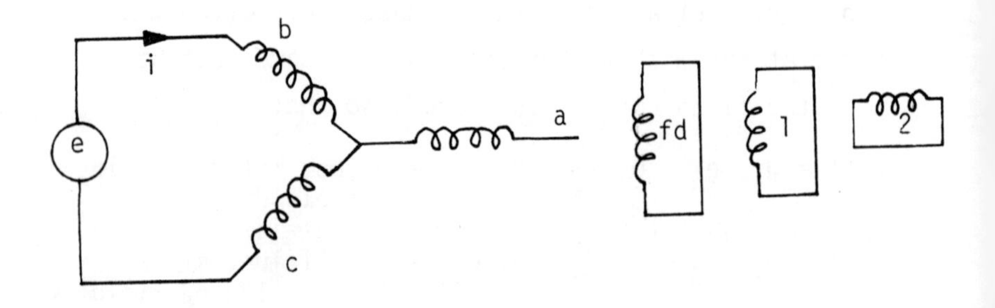

3-9. Given $i_a = \sin \omega t$, $i_b = \sin (\omega t - 2\pi/3)$, and $i_c = \sin (\omega t + 2\pi/3)$, evaluate i_d and i_β.

3-10. A synchronous machine is given to have the following data:

$$x_d = 1{\cdot}2;\ x_q = 0{\cdot}8;\ x_d' = 0.2909;\ r = 0.005;$$

$$r_{fd} = 0.0011;\ T_{do}' = 1000 \text{ radians};$$

$$T_d' = 242 \text{ radians}.$$

The machine is initially operating at no-load and rated voltage. Calculate and plot the following:

(a) first cycle of phase 'a' transient short-circuit current for the symmetrical three-phase fault, which occurs at the machine terminals at $\alpha = 0$ (where α is the angle between the d-axis and a-phase axis at time $t = 0$).

(b) first cycle of phase 'b' current for the case of line-to-line short circuit, which occurs between phases b and c at $\phi_o = 0$ (where ϕ_o is the angle between the d-axis and β-phase axis at time $t = 0$).

(c) steady-state components of short-circuit currents corresponding to cases (a) and (b) mentioned above in this problem.

(d) first cycle of open-a phase voltage for the case of line-to-line short circuit.

Neglect all amortisseurs, and neglect all damping when plotting first cycle after application of short circuits. Note that all decremented terms would reach zero value for steady-state conditions (i.e., for $t = \infty$).

3-11. Consider a line-to-neutral (phase 'a' to neutral) short circuit of a synchronous machine which is unloaded and normally excited before fault. Following the procedure of analysis for line-to-line short circuit, obtain an expression for the short-circuit current, assuming no amortisseur windings. Then find the fundamental component of i_a, which should be the symmetrical component solution

$$i_a = \frac{3 \cos \theta}{x_d' + x_2 + x_o} .$$

where

$$x_2 = x_2(x_d', x_q, x_o).$$

3-12. A synchronous machine with the data given below is initially operating at full-load, rated voltage and 0.8 pf lag. The machine is subjected to a three-phase short circuit at its terminals. Obtain the expressions for $i_d(s)$ and $i_d(t)$.

$x_d = 1.2$; $x_q = 0.8$; $x_d' = 0.2909$, $r = 0.005$;

$r_{fd} = 0.0011$; $T_{do}' = 1000$ radians;

$T_d' = 242$ radians.

Neglect amortisseur windings.

3-13. Consider the line-to-line short-circuit analysis presented in the text. Express Eq. (3.4.28) into even and odd harmonic series. Starting from Eqs. (3.4.31) and (3.4.32), evaluate the fundamental components of i_d and i_q.

3-14. A line-to-neutral (phase 'a' to neutral) short circuit is suddenly applied to a synchronous generator at its terminals. The generator is initially operating at no-load and rated voltage. The machine has its neutral solidly grounded and has the following data:

$$x_d = 1.0;\ x_q = 0.7;\ x''_d = 0.15;\ x'_d = 0.25,$$

$$x''_q = 0.17;\ x_o = 0.10,\ r = 0;\ r_2 = 0$$

Calculate the generator subtransient and transient phase 'a' fundamental-frequency component currents, using the symmetrical component theory and neglecting subtransient and transient saliency.

3-15. Starting from Eq. (3.5.26), show that Eq. (3.5.31) is true.

3-16. A 3-phase synchronous machine, operating initially at no-load and rated voltage, is subjected to a line-to-neutral fault on phase 'a'. Calculate the steady-state fault current and average torque, assuming that the voltage of phase 'b' is maintained at 1.0 per unit.

$$r = 0.002;\ x_\ell = 0.15;\ x_{ad} = 1.65;\ x_{aq} = 1.12;$$

$$z_2 = 0.04 + j\ 0.25;\ z_o = 0.004 + j\ 0.07.$$

3-17. Consider a synchronous machine running with a slip (that is small compared to unity) with positive-sequence, fundamental-frequency, applied armature currents. Assuming that the field winding is open, determine the form of the voltage e_a. [Note that $p\theta = 1 - s$; the amortisseurs do not disturb flux as the slip is small enough; neglect resistance; the result will be a modulated sine wave as in the case of slip test.]

CHAPTER IV

TRANSIENT PERFORMANCE - II

4.1 Starting Torque

Synchronous motors and condensers are usually started as induction motors with no field excitation, brought up to as near synchronous speed as possible, and then pulled into synchronism by energizing the field. We shall consider computing the starting torque of a salient-pole synchronous machine. It may reasonably be expected that the rotor saliency would cause torque and speed pulsations. However, we shall neglect speed pulsation and also assume the acceleration to be small, both of which may be justified if the rotor inertia is sufficiently large. We shall then consider the machine to be running at various constant rotor speeds, and calculate the torque with no rotor excitation to start with, when balanced armature voltages are applied.

Taking s to be the per-unit slip of the rotor in terms of the rated synchronous speed of the machine, the rotor speed is given by

$$p\theta = 1 - s \tag{4.1.1}$$

We shall analyze the problem as a steady-state one with balanced operation of armature circuit, in terms of d, q, o axis quantities. Thus we have

$$e_d = e \sin \delta = -e \sin st; \quad e_q = e \cos \delta = e \cos st;$$

$$e_o = o \tag{4.1.2}$$

where δ is the angle between the armature open-circuit generated voltage and the bus voltage e, and is continually decreasing according to the relation $\delta = -st$. The armature voltage relations are then given by the following, subject to the constraints of constant rotor speed and no field excitation:

$$e_d = p\,\lambda_d - (1 - s)\,\lambda_q - ri_d = -e \sin st \qquad (4.1.3)$$

$$e_q = p\,\lambda_q + (1 - s)\,\lambda_d - ri_q = e \cos st \qquad (4.1.4)$$

We shall use the usual complex phasor notation and replace p by js since all currents and flux linkages vary as e^{jst}. Thus we have

$$\Lambda_d = -x_d(js)\,I_d; \quad \Lambda_q = -x_q(js)\,I_q \qquad (4.1.5)$$

$$[-js\,x_d(js) - r]\,I_d + (1 - s)\,x_q(js)\,I_q = je \qquad (4.1.6)$$

$$-(1 - s)\,x_d(js)\,I_d - [js\,x_q(js) + r]\,I_q = e \qquad (4.1.7)$$

where from it follows (in the notation of determinants) that

$$I_d = \frac{\begin{vmatrix} je & (1 - s)\,x_q(js) \\ e & -[jsx_q(js) + r] \end{vmatrix}}{\begin{vmatrix} [-jsx_d(js) - r] & (1 - s)\,x_q(js) \\ -(1 - s)x_d(js) & -[jsx_q(js) + r] \end{vmatrix}} \qquad (4.1.8)$$

and

$$I_q = \frac{\begin{vmatrix} [-jsx_d(js) - r] & je \\ -(1 - s)x_d(js) & e \end{vmatrix}}{\begin{vmatrix} [-jsx_d(js) - r] & (1 - s)x_q(js) \\ -(1 - s)x_d(js) & -[jsx_q(js) + r] \end{vmatrix}} \qquad (4.1.9)$$

The average torque may then be calculated from the following relation involving dot products:

$$T_{avg} = \frac{1}{2}\,(I_q \cdot \Lambda_d - I_d \cdot \Lambda_q) \qquad (4.1.10)$$

For the case of zero armature resistance, the above equations get simplified as given below:

$$I_d = -\frac{e}{x_d(js)} \quad ; \quad I_q = \frac{je}{x_q(js)} \qquad (4.1.11)$$

and

$$\Lambda_d = e = j\Lambda_q \quad ; \quad \Lambda_q = -je = -j\Lambda_d \qquad (4.1.12)$$

with which the average torque may be evaluated from the relation

$$T_{avg} = \frac{1}{2}\text{ real part of } [I_q^* \Lambda_d - I_d^* \Lambda_q] \qquad (4.1.13)$$

which may be rewritten as

$$T_{avg} = -\frac{1}{2}\text{ real part of } \left[|I_q|^2 \frac{Z_q}{s} + |I_d|^2 \frac{Z_d}{s}\right] \qquad (4.1.14)$$

or

$$T_{avg} = -\frac{1}{2}\left[|I_q|^2 \operatorname{Re}(Z_q/s) + |I_d|^2 \operatorname{Re}(Z_d/s)\right] \quad (4.1.15)$$

after making use of the relations

$$p\,\Lambda_d = js\,\Lambda_d = -Z_d(js)\,I_d \quad (4.1.16)$$

and

$$p\,\Lambda_q = js\,\Lambda_q = -Z_q(js)\,I_q \quad (4.1.17)$$

For the particular case where Z_d equals Z_q and i_d equals i_q, we have

$$T_{avg} = -|I_d|^2 \operatorname{Re}(Z_d|s) \quad (4.1.18)$$

which gives us the idea of realizing this from the familiar induction motor equivalent circuit and rotor $(i^2\ r/s)$ losses. An approximate way of calculating the average slip torque of a salient-pole machine is by utilizing the induction motor equivalent circuits corresponding to d and q axes separately, calculating the corresponding torques individually, and then averaging the d, q axis quantities. Direct- and quadrature-axis equivalent circuits are shown in Figure 4.1.1, with one amortisseur circuit in each axis.

Let us now investigate the effects of half speed, with negligible armature resistance. With s = 0.5, from Eqs. (4.1.8) and (4.1.9) one obtains

$$I_d = \frac{-je}{j\,\frac{1}{2}\,[x_d(js) + x_q(js)]} \quad (4.1.19)$$

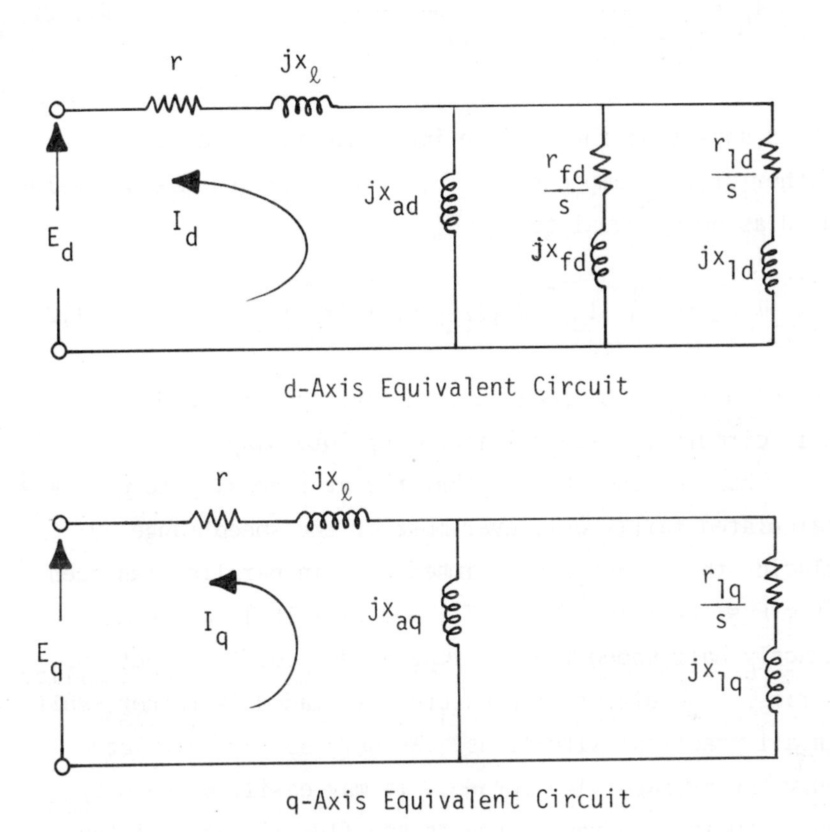

Figure 4.1.1. Equivalent Circuits for Calculating Approximate Average Slip Torque of a Salient-pole Synchronous Machine.

$$I_q = \frac{-e}{j\,\frac{1}{2}\,[x_d(js) + x_q(js)]} = -j\,I_d \qquad (4.1.20)$$

which show that the average impedance is effective in either of the components. The average torque may be evaluated as being equal to

$$T_{avg} = -\frac{1}{2}\,|I_d|^2\,\mathrm{Re}[jx_d(js) + jx_q(js)] \qquad (4.1.21)$$

in which it can be seen that the average of the d- and q-axis circuit resistances is coming into play.

Thus one can observe that the average slip torque is calculated fairly well over most of the speed range by placing the d- anx q-axis impedances in parallel [as seen from Eqs. (4.1.16), (4.1.17), and (4.1.18)], while at exactly half speed the two impedances need to be put in series. The effect of armature resistance is rather small in all practical situations. We have so far calculated only the average slip torque. It may easily be seen by substituting the real currents and flux linkages of the form

$$i_d = a\cos st - b\sin st, \text{ etc.} \qquad (4.1.22)$$

in the expression for the instantaneous torque

$$T = i_q\,\lambda_d - i_d\,\lambda_q \qquad (4.1.23)$$

that a double-slip-frequency component exists besides the average calculated earlier. We have thus completed the approximate analysis of the torque developed by a salient-pole synchronous machine operating at constant asynchronous speed with applied balanced armature voltages and without any rotor excitation.

It is possible for us to develop an exact equivalent circuit for the asynchronous operation of a salient-pole synchronous machine by introducing forward and backward-rotating components, which bear somewhat the same relationship to d, q-axis quantities, as the symmetrical components bear to α, β-axis quantities. Unlike the symmetrical components, these new f and b-quantities are referred to axes fixed in the rotor rather than in the stator.

$$i_f = \frac{1}{\sqrt{2}}(i_d + j\, i_q); \quad i_b = \frac{1}{\sqrt{2}}(i_d - j\, i_q) \qquad (4.1.24)$$

The voltages and flux linkages also satisfy similar relationships as above. The instantaneous torque in terms of the new variables may be seen to be

$$T = j(i_b\, \lambda_f - i_f\, \lambda_b) \qquad (4.1.25)$$

which includes the average torque as well as the pulsating double-slip-frequency components. Leaving the details of analysis as an exercise to the reader, the resultant equivalent circuit is shown in Figure 4.1.2 in which the complex quantities I_f and I_b are related to i_f as

$$i_f = I_f\, e^{jst} + I_b^*\, e^{-jst} \qquad (4.1.26)$$

In case there is no saliency, with $x_d(js) = x_q(js)$, the approximate method becomes exact, regardless of the amount of armature resistance. It may further be observed that for the case of one-half synchronous speed operation, with

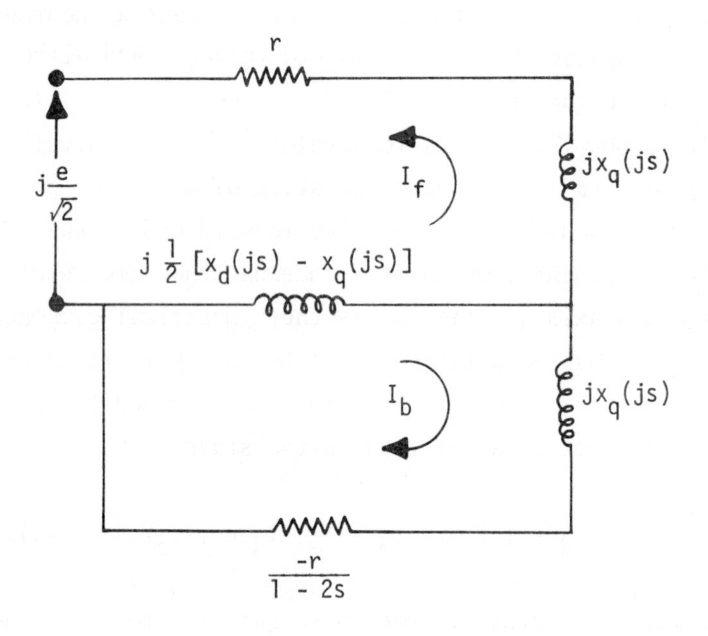

Figure 4.1.2. Exact Equivalent Circuit for Calculating Induction-Motor Torque of a Salient-pole Synchronous Machine.

s = 0.5, the lower mesh of Figure 4.1.2 becomes open-circuited and the direct- and quadrature-axis impedances are connected in series. It can be shown* that the theoretically exact method (on which the equivalent circuit of Figure 4.1.2 is based) would lead to a dip produced by rotor saliency in the torque-speed characteristic in the region of one-half synchronous speed. The approximate method based on independent direct- and quadrature-axis equivalent circuits of Figure 4.1.1 would fail to indicate the dip mentioned above; but otherwise give fairly good

---

*See for details Concordia, C., Synchronous Machines, (Ch. 8), G. E. Series, 1951.

results. In order to show the difference (caused by a combination of both armature resistance and rotor saliency) between the approximate and exact methods, a problem is included as an exercise at the end of this chapter. It suffices here to say that the approximate method yields sufficiently close agreements with test results, even though the dip is not shown.

Returning to the approximate method of analysis, the rotor power loss is given by

$$P_{rotor} = \frac{1}{2}\,|I_d|^2\,\mathrm{Re}(Z_d) + \frac{1}{2}\,|I_q|^2\,\mathrm{Re}(Z_q) = sT \quad (4.1.27)$$

which can be seen from Eq. (4.1.15). This is similar to the case of induction motor, and the mechanical power developed by the motor is given by

$$P_{mech} = (1 - s)\,T \quad (4.1.28)$$

The average rotor power loss may be calculated by using the following relationship:

$$(P_{rotor})_{avg} = \frac{1}{t}\int_0^t sT\,dt \simeq \frac{\Sigma sT\Delta t}{\Sigma\Delta t} \quad (4.1.29)$$

The torque is related to the acceleration through the inertia constant (which is the stored energy in the rotating mass at synchronous speed per kva of the machine rating, expressed as kw-sec. per kva; for additional information, one is referred to Section 4.4):

$$T = 2H\,\frac{d\omega}{dt} \quad (4.1.30)$$

where ω in per-unit is given by $(1 - s)$, t is in seconds, H is the inertia constant in kw-sec. per kva, and T is expressed in per-unit. Equation (4.1.30) may be discretized by choosing Δt sufficiently small and replacing $\frac{d\omega}{dt}$ by $\frac{\Delta\omega}{\Delta t}$, where $\Delta\omega$ represents the change in ω corresponding to decrement of Δs in time Δt.

Field Excitation:

As the machine approaches synchronism, the field is energized and the machine operates asynchronously with rotor excitation for a while before it finally pulls into synchronism. Let us now investigate the effects of rotor excitation on slip torque. From the viewpoint of average torque, the torques due to rotor and stator excitation can be calculated separately and then added. The average torque with rotor excitation acting alone may be seen to correspond with the stator i^2r loss, based on the steady-state operating characteristics:

$$T_{avg} = (i_{do}^2 + i_{qo}^2) \frac{r}{1 - s} \tag{4.1.31}$$

where i_{do} and i_{qo} are the constant steady-state d- and q-axis currents caused by rotor excitation acting alone, with armature circuits short-circuited symmetrically. The torque given by Eq. (4.1.31) is positive and is seen to oppose the induction motor accelerating torque; it is hence referred to as braking torque or speed torque.

Except for the average value of torque, it would not be possible to calculate torques separately due to rotor and stator excitation, and add them up to get the resultant, as there will be pulsating torque components due to interaction of two excitations. However, insofar as the fluxes and currents are concerned, the principle of superposition can be applied since system equations are linear in view of the assumed constant rotor speed.

With rotor excitation acting alone, the armature voltage equations are given by

$$-(1 - s)\,\lambda_{qo} - r i_{do} = 0 \tag{4.1.32}$$

$$(1 - s)\,\lambda_{do} - r i_{qo} = 0 \tag{4.1.33}$$

which are obtained by letting voltages to be zero and the differential operator p to be zero as in the case of steady state. The flux linkage relations are given by

$$\lambda_{do} = E - X_d\, i_{do} \tag{4.1.34}$$

$$\lambda_{qo} = -X_q\, i_{qo} \tag{4.1.35}$$

where the notation adopted is the same as in Chapter II dealing with steady-state theory. The resultant constant (d.c.) currents are given by

$$i_{do} = \frac{(1 - s)^2\, x_q\, E}{r^2 + (1 - s)^2\, x_d x_q} \tag{4.1.36}$$

$$i_{qo} = \frac{(1 - s)\, rE}{r^2 + (1 - s)^2\, x_d x_q} \tag{4.1.37}$$

It may be convenient to transform to the f- and b-axis quantities, in which case the total current i_f produced by both rotor and stator excitations will be of the form:

$$i_f = i_{fo} + I_f e^{jst} + I_b^* e^{-jst} \tag{4.1.38}$$

where i_{fo} is expressed as

$$i_{fo} = \frac{1}{\sqrt{2}} (i_{do} + ji_{qo}) \tag{4.1.39}$$

By completing the analysis and evaluation of torque the details of which are left here as an exercise to the reader, it can be shown that the effect of rotor excitation is to add an average d.c. term and a slip-frequency component to the torque. Thus with both rotor and stator excitation, the torque consists of an average d.c. term, a double-slip-frequency component, and a slip-frequency component. As explained earlier, the average torque terms produced by rotor and stator excitations would oppose each other, thereby reducing the effectiveness of the induction motor accelerating torque.

4.2 Voltage Dip

Let us now consider some aspects of impedance loads suddenly applied to an isolated synchronous machine. We shall be interested in the amount by which the terminal voltage dips and the time needed for the voltage to return to normal. For the purposes of analysis, we shall neglect the amortisseur windings as well as armature transients; we shall also neglect the change of speed that may occur due to the applied load and assume the speed to be the same as its initial value of unity. The basic equations are then given by

$$e_d = -\lambda_q - ri_d \tag{4.2.1}$$

$$e_q = \lambda_d - ri_q \tag{4.2.2}$$

where

$$\lambda_d = G(p)\ E - x_d(p)\ i_d \tag{4.2.3}$$

$$\lambda_q = -x_q\ i_q \tag{4.2.4}$$

where E is the armature open-circuit voltage corresponding to normal speed. The operational impedance $x_d(p)$ is given by

$$x_d(p) = \frac{T'_{do}\ x'_d\ p + x_d}{T'_{do}\ p + 1} \tag{4.2.5}$$

where

$$T'_{do} = \frac{x_{ffd}}{r_{fd}} \tag{4.2.6}$$

The effect of load impedance (R + jX) will be included by increasing the values of $x_d(p)$ and x_q by X; and by adding R to the value of armature resistance r. Then one may analyze the application of load like a short-circuit case and obtain the armature currents. Let us consider the application of load to a machine initially unloaded and normally excited. The armature currents caused by the closing of the armature switch shall be evaluated first and then the currents caused by the change of field excitation. Thus with E = 0, one has the following equations for the changes caused in armature currents:

$$x_q\, i_q - r i_d = 0 \tag{4.2.7}$$

$$-x_d(s)\, i_d - r i_q = -\frac{1}{s} \tag{4.2.8}$$

The solution of the above yields

$$i_d = (i'_d - i_{ds})\, e^{-t/T'_{dz}} + i_{ds} \tag{4.2.9}$$

$$i_q = \frac{r}{x_q}\, i_d \tag{4.2.10}$$

where

$$i'_d = \frac{x_q}{x'_d x_q + r^2} \tag{4.2.11}$$

$$i_{ds} = \frac{x_q}{x_d x_q + r^2} \tag{4.2.12}$$

and

$$T'_{dz} = \frac{x'_d\, x_q + r^2}{x_d\, x_q + r^2}\, T'_{do} \tag{4.2.13}$$

The magnitude of the terminal voltage is given by

$$|e_t| = |i|\, Z \tag{4.2.14}$$

where

$$Z = |R + jX|$$

$$|i| = \sqrt{i_d^2 + i_q^2} = i_d \left(1 + \frac{r^2}{x_q^2}\right)^{1/2} = i_d \frac{z_q}{x_q} \qquad (4.2.15)$$

and $z_q = |r + j\, x_q|$

Immediately following the application of load impedance, the terminal voltage is given by

$$e_{to} = i_d' \frac{z_q\, Z}{x_q} = \frac{z_q\, Z}{x_d'\, x_q + r^2} \qquad (4.2.16)$$

For the case when the resistances may be neglected, one gets

$$e_{to} = \frac{X}{x_d'} \quad \text{and} \quad i_d' = \frac{1}{x_d'} \qquad (4.2.17)$$

which may be rewritten as

$$e_{to} = \frac{X + x_{dm}' - x_{dm}'}{x_{dm}' + X} = 1 - \frac{x_{dm}'}{x_d'} = 1 - x_{dm}'\, i_d' \qquad (4.2.18)$$

where x_{dm}' corresponds to the machine only and X is the applied load reactance. Equation (4.2.18) shows that the machine direct-axis transient reactance determines the initial voltage drop for the assumed zero power-factor load. It may be seen from Eq. (4.2.16) that there may be an initial rise rather than an initial drop in voltage for the case of certain applied loads and machine parameters. The effect of an initial load, which has not been considered here, would only be to lessen in general the severity of the initial voltage dip compared to that of no-load case.

Effects of Field Excitation Changes:

Let us now calculate the armature current components caused by the change in field excitation due to the voltage regulator action. With $e_d = e_q = 0$, and with the change in field excitation of ΔE, one has

$$x_q\, i_q - r i_d = 0 \tag{4.2.19}$$

$$-x_d(s)\, i_d - r i_q = -G(s)\, \Delta E(s) \tag{4.2.20}$$

The solution of the above leads to

$$i_d = \frac{x_q}{x_d\, x_q + r^2} \frac{1}{T'_{dz}\, s + 1} \Delta E(s) \tag{4.2.21}$$

where T'_{dz} is as defined in Eq. (4.2.13), and further solution depends on the functional form of ΔE. Assuming a linear form

$$\Delta E = kt \qquad \text{or} \qquad \Delta E(s) = k/s^2 \tag{4.2.22}$$

one obtains

$$i_d = \frac{k\, x_q}{x_d\, x_q + r^2} \left\{ (t - t_1) + T'_{dz} \left[e^{-(t-t_1)/T'_{dz}} - 1 \right] \right\} \tag{4.2.23}$$

where t_1 is the time delay in the regulator action before the field voltage starts to increase. The terminal voltage is then given by

$$e_t = i_d \frac{z_q Z}{x_q} = \frac{k z_q Z}{x_d x_q + r^2} \cdot$$

$$\cdot \left\{ (t - t_1) + T'_{dz} \left[e^{-(t-t_1)/T'_{dz}} - 1 \right] \right\} \quad (4.2.24)$$

This may be added on to the previously calculated armature terminal voltage due to the armature current components caused by the closing of the armature switch. Thus, one has for $0 < t < t_1$,

$$e_t = K_1 \left[(K_2 - 1) e^{-t/T'_{dz}} + 1 \right] \quad (4.2.25)$$

for $t > t_1$,

$$e_t = K_1 \left[(K_2 - 1) e^{-t/T'_{dz}} + 1 \right] + K_1 t \left\{ (t - t_1) + \right.$$

$$\left. + T'_{dz} \left[e^{-(t-t_1)/T'_{dz}} - 1 \right] \right\} \quad (4.2.26)$$

where

$$K_1 = \frac{z_q Z}{x_d x_q + r^2} \quad (4.2.27)$$

and

$$K_2 = \frac{x_d x_q + r^2}{x'_d x_q + r^2} \quad (4.2.28)$$

The time corresponding to minimum voltage following the sudden application of an impedance load, which can be seen to occur after $t = t_1$, may be obtained by setting de_t/dt equal to zero. On differentiating and solving, one gets

$$t_{\text{for minimum voltage}} = T'_{dz} \log_e \left\{ \frac{(K_2 - 1)}{k\, T'_{dz}} + e^{t_1/T'_{dz}} \right\} \tag{4.2.29}$$

The corresponding minimum voltage is then given by

$$e_{t_{min}} = K_1 + K_1 k \left\{ T'_{dz} \log_e \left(\frac{(K_2 - 1)}{k\, T'_{dz}} + e^{t_1/T'_{dz}} \right) - t_1 \right\} \tag{4.2.30}$$

It may be noted here that there is no guarantee that the minimum voltage will be attained before the exciter voltage reaches its ceiling.

Assuming the exponential variation of excitation, including the effects of both rate of rise and ceiling voltage, given by

$$\Delta E = (E_c - E_o)\left(1 - e^{-t/T_e}\right) \tag{4.2.31}$$

where

E_c is the excitation ceiling,

E_o is the initial excitation

and

T_e is the excitation-system time constant,

the solution of Eq. (4.2.21) in combination with Eq. (4.2.31) leads to

$$i_d = \frac{(E_c - E_o)\, x_q}{x_d\, x_q + r^2} \left[1 + \frac{T_e\, e^{-(t-t_1)/T_e} - T'_{dz}\, e^{-(t-t_1)/T'_{dz}}}{T'_{dz} - T_e} \right] \tag{4.2.32}$$

The terminal voltage is then given by

$$e_t = i_d \frac{z_q\, Z}{x_q} = K_1\, (E_c - E_o) \cdot \left[1 - \frac{T'_{dz}\, e^{-(t-t_1)/T'_{dz}} - T_e\, e^{-(t-t_1)/T_e}}{T'_{dz} - T_e} \right] \tag{4.2.33}$$

This may be added on, as before, to the previously calculated armature terminal voltage due to the armature current components caused by closing of the armature switch. It may be pointed out here that the voltage would continually decrease if the exciter ceiling is not high enough, and a minimum voltage may not even occur since the applied field voltage is limited to the ceiling value E_c. In cases where the ceiling is the dominant regulating-system characteristic, one may simply set T_e equal to zero in the above analysis and obtain the relevant results, in which case the minimum voltage occurs either at $t = t_1$ or not at all.

For the voltage to recover to its normal value without change of load impedance, the condition to be satisfied is

$$[K_1 (E_c - E_o) + K_1\, E_o] > 1$$

or

$$K_1 E_c > 1$$

or

$$E_c > \frac{x_d x_q + r^2}{z_q Z} \tag{4.2.34}$$

or, neglecting the resistances,

$$E_c > \frac{x_{dm} + X}{X} \tag{4.2.35}$$

A complete voltage-time characteristic may be plotted and the voltage recovery time read. Only simple cases are presented here in order to given an understanding of the phenomena and of the factors involved. Anderson (quoted in Bibliography) has analyzed the dynamic response on a differential analyzer, including time lags, limits, regulator stabilizer, and saturation characteristics of the alternator and exciter. Saturation, which does affect machine reactances and whose effect may appear in the determination of the final steady-state terminal voltage with excitation ceiling, is not considered here as it is beyond the scope of this chapter. In the case of applied motor loads, if the voltage does not recover unless the motor starts, one may have to resort to a step-by-step or differential-analyzer solution.

Field Current:

Let us next consider the variation of field current with time. The armature flux linkage equations are

$$\lambda_d = x_{ad}\, i_{fd} - x_d\, i_d = E - x_d\, i_d \tag{4.2.36}$$

and

$$\lambda_q = -x_q\, i_q \tag{4.2.37}$$

Then the armature voltage equations are given by

$$E - x_d\, i_d - r i_q = 0 \tag{4.2.38}$$

$$x_q\, i_q - r i_d = 0 \tag{4.2.39}$$

in which the terminal voltages are set equal to zero. One can then obtain

$$E = \frac{x_d\, x_q + r^2}{x_q}\, i_d \tag{4.2.40}$$

From Eqs. (4.2.14) and (4.2.15), one has

$$i_d = \frac{x_q}{z_q\, Z}\, e_t \tag{4.2.41}$$

So it follows that

$$E = \frac{x_d\, x_q + r^2}{z_q\, Z}\, e_t \tag{4.2.42}$$

which shows that the field current is simply proportional to the terminal voltage.

4.3 Synchronizing Out-of-Phase

Let us now consider synchronizing-out-of-phase of a synchronous machine (normally excited and brought to the correct synchronous speed) to an infinite bus. We shall neglect the resistances, damping, and assume no amortisseur circuits in attempting to plot the first-half cycle of torque as a function of δ, the synchronizing-out-of-phase closing angle. Figure 4.3.1 shows the conditions and notation corresponding to the problem under consideration.

The equations of performance of a synchronous machine are given by

$$e_d = p\lambda_d - \lambda_q\, p\theta - ri_d \tag{4.3.1}$$

$$e_q = p\lambda_q + \lambda_d\, p\theta - ri_q \tag{4.3.2}$$

$$e_o = p\lambda_o - ri_o \tag{4.3.3}$$

$$e_{fd} = p\lambda_{fd} + r_{fd}\, i_{fd} \tag{4.3.4}$$

where, with no amortisseur windings,

$$\lambda_d = -L_d\, i_d + L_{ad}\, i_{fd} \tag{4.3.5}$$

$$\lambda_q = -L_q\, i_q \tag{4.3.6}$$

$$\lambda_o = -L_o\, i_o \tag{4.3.7}$$

$$\lambda_{fd} = -L_{ad}\, i_d + L_{fd}\, i_{fd} \tag{4.3.8}$$

Noting that $p\theta$ is equal to unity at synchronous speed,

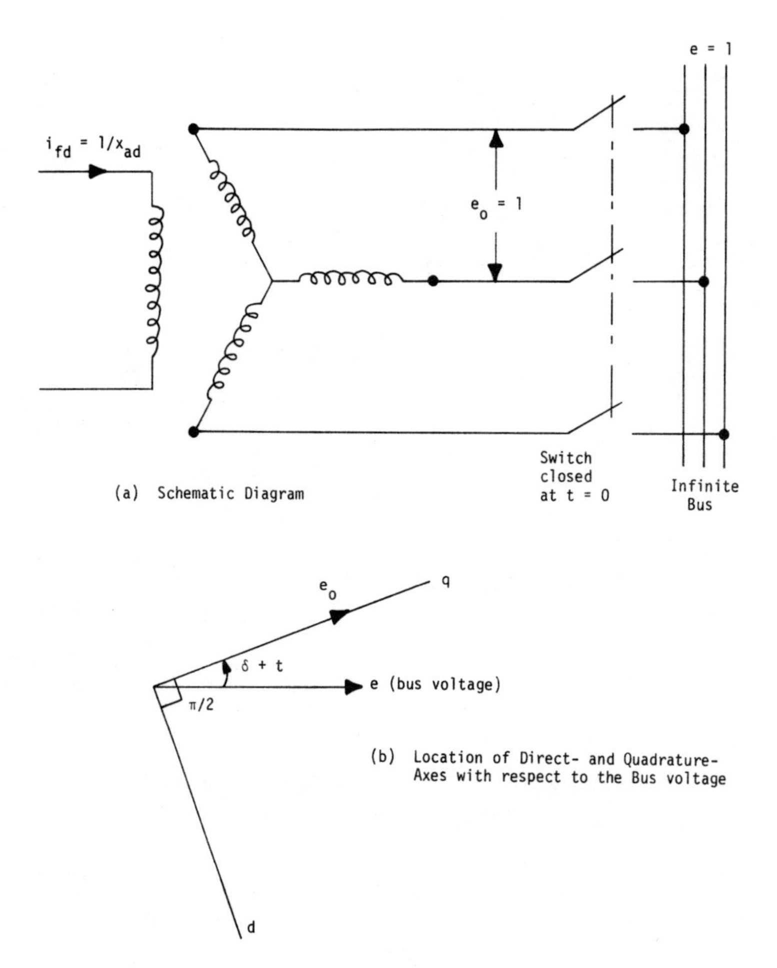

Figure 4.3.1. Synchronizing-out-of-phase of a synchronous machine.

resistance effects are neglected here, and the zero-sequence components do not exist for balanced cases, one has then in per-unit notation:

$$e_d = -p\, x_d\, i_d + x_q\, i_q + p\, x_{ad}\, i_{fd} \qquad (4.3.9)$$

$$e_q = -x_d\, i_d - p\, x_q\, i_q + x_{ad}\, i_{fd} \qquad (4.3.10)$$

$$e_{fd} = -p\, x_{ad}\, i_d + p\, x_{fd}\, i_{fd} \qquad (4.3.11)$$

$$\text{at } t = o_-,\ e_d = o;\quad e_q = 1.0;\ e_{fd} = \frac{r_{fd}}{x_{ad}} \quad \text{just before closing the switch.} \qquad (4.3.12)$$

$$\text{at } t = o_+,\ e_d = \sin\delta;\ e_q = \cos\delta;\ e_{fd} = \frac{r_{fd}}{x_{ad}} \quad \text{just after closing the switch.} \qquad (4.3.13)$$

The field circuit has reached its steady state and is normally excited before the synchronizing switch is closed. The differences are given by

$$\Delta e_d(s) = \frac{\sin\delta}{s};\quad \Delta e_q(s) = \frac{\cos\delta - 1}{s};\quad \Delta e_{fd} = o \qquad (4.3.14)$$

Applying the performance equations to the difference voltages obtained above, one can solve for the currents:

$$\Delta i_d(s) = \frac{\frac{1}{s}\begin{vmatrix} \sin\delta & x_q & sx_{ad} \\ (\cos\delta - 1) & -sx_q & x_{ad} \\ 0 & 0 & sx_{fd} \end{vmatrix}}{\begin{vmatrix} -sx_d & x_q & sx_{ad} \\ -x_d & -sx_q & x_{ad} \\ -sx_{ad} & 0 & sx_{fd} \end{vmatrix}} \qquad (4.3.15)$$

$$= \frac{-s\sin\delta + (1 - \cos\delta)}{x_d' \, s(s^2 + 1)}$$

$$= \frac{1 - \cos\delta}{x_d' \, s} - \frac{s(1 - \cos\delta) + \sin\delta}{x_d' \, (s^2 + 1)} \qquad (4.3.16)$$

where

$$x_d' = x_d - \frac{x_{ad}^2}{x_{fd}} \qquad (4.3.17)$$

or

$$\Delta i_d(t) = \frac{1}{x_d'} [(1 - \cos\delta)(1 - \cos t) - \sin\delta \sin t] \qquad (4.3.18)$$

Since the initial $i_d(t)$ is zero before the switch is closed, the total current is given by

$$i_d(t) = \frac{1}{x_d'} [(1 - \cos\delta)(1 - \cos t) - \sin\delta \sin t] \tag{4.3.19}$$

Similarly, the quadrature-axis and field currents may be obtained as

$$i_q(t) = \frac{1}{x_q} [(1 - \cos t) \sin\delta + (1 - \cos\delta) \sin t] \tag{4.3.20}$$

$$\Delta i_{fd}(t) = \frac{x_{ad}}{x_{fd}\, x_d'} [(1 - \cos\delta)(1 - \cos t) - \sin\delta \sin t] \tag{4.3.21}$$

It may be observed that the above solution of field current is

$$\left[\frac{x_{ad}}{x_{fd}} i_d(t)\right] .$$

Recognizing that an initial field current equal to $1/x_{ad}$ (such that $e_{qo} = x_{ad}\, i_{fd} = 1.0$) exists before the switch is closed, one gets the following for the total field current:

$$i_f(t) = \frac{1}{x_{ad}} + \frac{x_{ad}}{x_{fd}\, x_d'} [(1 - \cos\delta)(1 - \cos t) + - \sin\delta \sin t] \tag{4.3.22}$$

Utilizing the results of Eqs. (4.3.19), (4.3.20) and (4.3.22), the flux linkages are obtained from Eqs. (4.3.5) and (4.3.6):

$$\lambda_d(t) = 1 - [(1 - \cos\delta)(1 - \cos t) - \sin\delta \sin t] \tag{4.3.23}$$

$$\lambda_q(t) = -[(1 - \cos t)\sin\delta + (1 - \cos\delta)\sin t] \tag{4.3.24}$$

Then the instantaneous torque may be expressed as

$$T = \lambda_d\, i_q - \lambda_q\, i_d$$

or

$$T = f(\delta,t) = \left\{\frac{1}{x_q}[(1 - \cos t)\sin\delta + (1 - \cos\delta)\sin t]\right\} \cdot \left\{1 - [(1 - \cos\delta)(1 - \cos t) - \sin\delta\sin t] \cdot \left(1 - \frac{x_q}{x_d'}\right)\right\} \tag{4.3.25}$$

The torque, expressed here as a function of δ and t, can be seen to be symmetrical in δ and t; replacing δ by t and t by δ, the expression may be seen unaltered. It may then be concluded that the time at which maximum torque occurs is equal to the closing angle δ. So, substituting $t = \delta$ in Eq. (4.3.25), one gets

$$T = f(\delta) = \frac{2}{x_q}(1 - \cos\delta)\sin\delta + \frac{2}{x_q}\left(1 - \frac{x_q}{x_d'}\right)(1 - \cos\delta)^2 \cdot \sin 2\delta \tag{4.3.26}$$

The condition for maximum torque may simply be evaluated by setting $dT/d\delta$ equal to zero; the result is given by

$$\frac{x_q}{x'_d} = 1 + \frac{1 + 2\cos\delta}{2(\cos 2\delta - \cos 3\delta)} \tag{4.3.27}$$

δ may have quite a few values satisfying the above relation, for a given saliency ratio of (x_q/x'_d); hence, appropriate δ is to be chosen for finding the maximum torque. A plot of the saliency ratio as a function of δ is considered to be helpful; it is shown in Figure 4.3.2 for a necessary and sufficient range of δ = o to π, as Eq. (4.3.27) is unaltered by replacing δ by either $-\delta$, $(\pi - \delta)$, or $(\pi + \delta)$. The positive side of the graph is drawn with due clarity, as the negative region is not of any interest in our problem. The curve may be seen to be very steep in the region of δ = 140° to 144°; for all normal machines with saliency ratios of 1.5 or above, the δ or t corresponding to the peak T_{max} is near about 140° (between $\frac{\pi}{2}$ and π).

Considering a typical machine with x_q = 0.8 and x'_d = 0.2909, one has 2.75 for the saliency ratio, corresponding to which it may read from Figure 4.3.2 as δ = 43°, 56°, 142°; out of which the peak T_{max} occurs at δ = t = 142°, and the corresponding T_{max} works out as 16.35. The expression of $T(\delta,t)$ is plotted as a function of t in Figure 4.3.3, for value of δ = 142°; it is seen that the peak value occurs at t = 142° or $\frac{142\pi}{180}$ radians.

Having analyzed the effect of synchronizing-out-of-phase, it is obvious why the synchronizing switch is not to be closed at random and why a random closing in is not recommended.

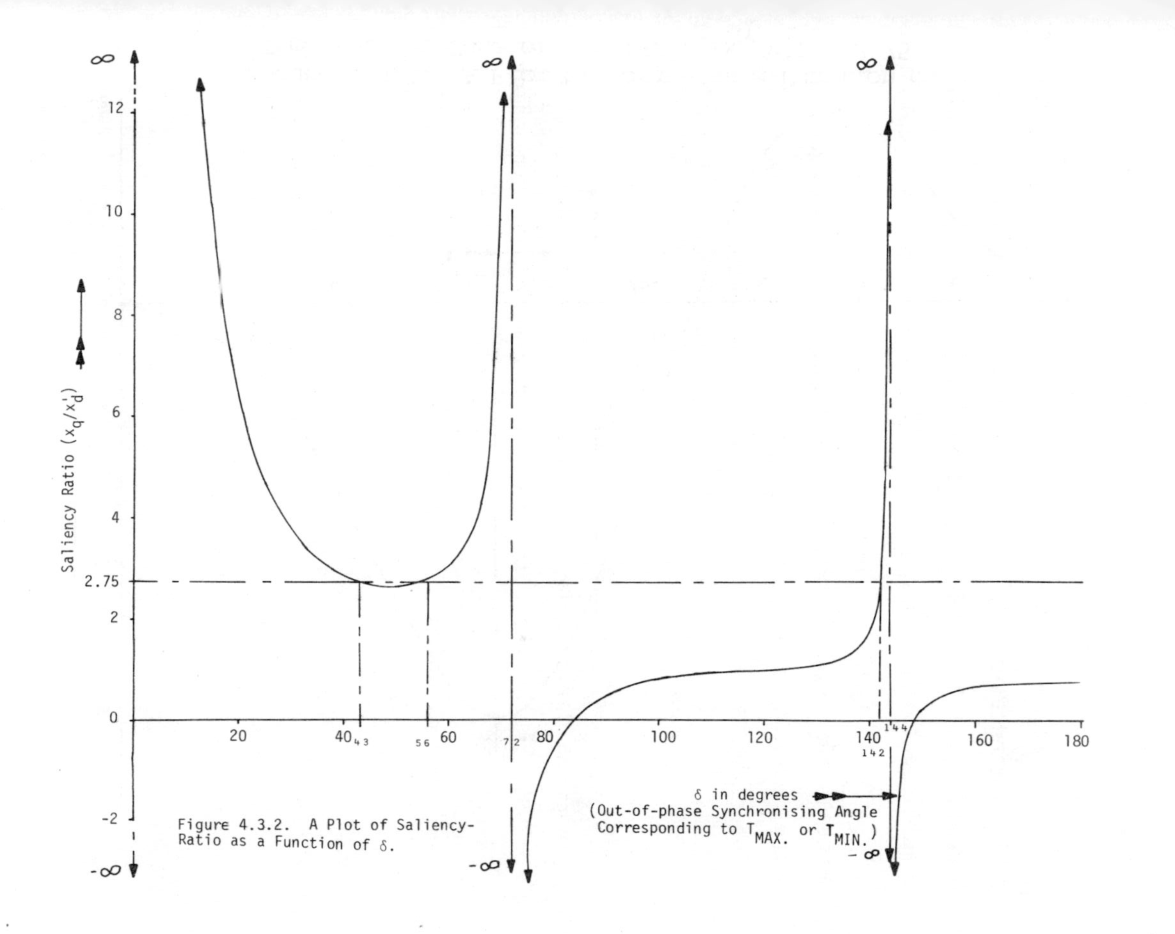

Figure 4.3.2. A Plot of Saliency-Ratio as a Function of δ.

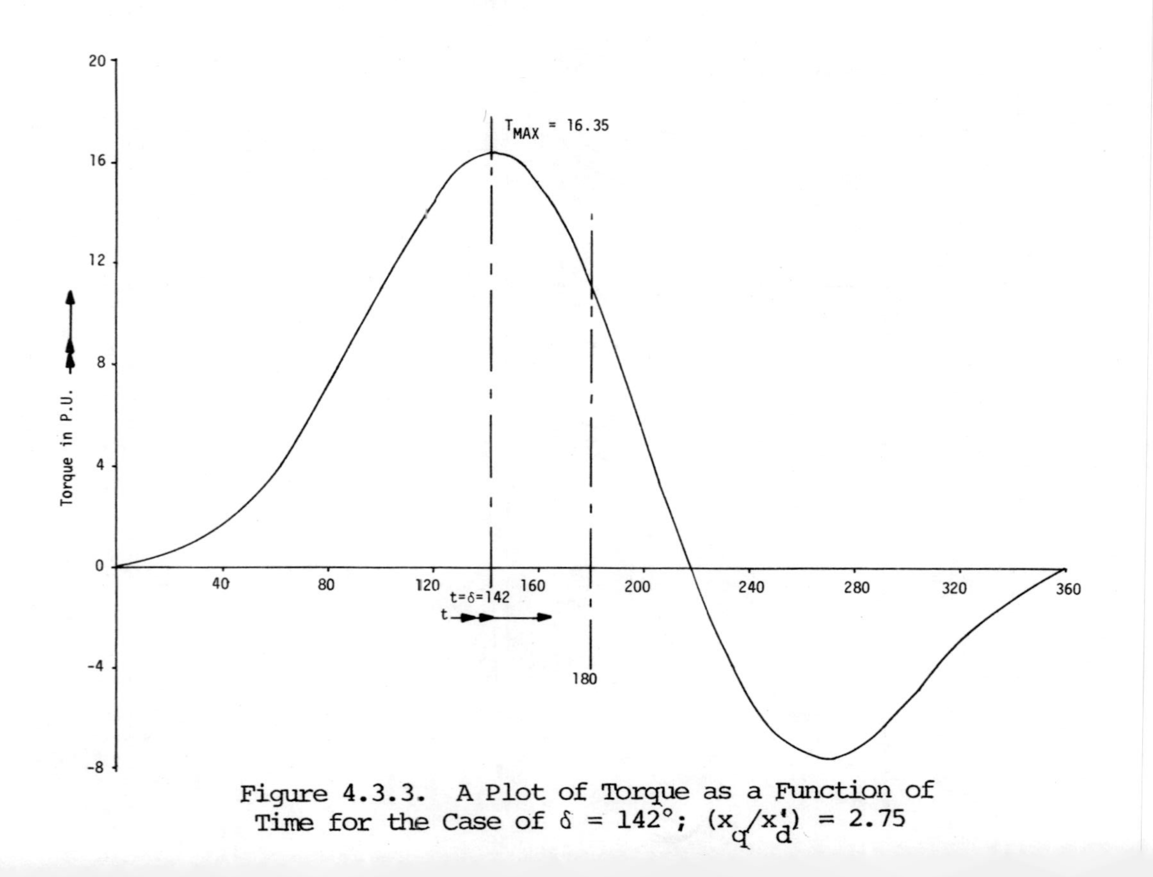

Figure 4.3.3. A Plot of Torque as a Function of Time for the Case of $\delta = 142°$; $(x_q/x'_d) = 2.75$

4.4 Synchronizing and Damping Torques

We shall consider a synchronous machine connected to an infinite bus, operating at a torque angle of δ. Let Δ be the amplitude of swing above and below normal, such that

$$\delta = \delta_o + \Delta \sin \omega_h t = \delta_o + \delta_1 \tag{4.4.1}$$

where ω_h is the angular velocity of swinging or hunting in radians per second; ω_h is much less than the radian frequency of the bus. The motion can be analyzed as simple harmonic as in the case of a pendulum swinging, only if the angle of swing is very small, in which case the following approximations are justified for typical values of Δ of 0.1 radian:

$$\sin (\Delta \sin \omega_h t) \simeq \Delta \sin \omega_h t; \cos (\Delta \sin \omega_h t) \simeq 1.0 \tag{4.4.2}$$

Otherwise, elliptical integrals will show up in the solution. It may be remarked here that Δ is in general a function of time and it gets damped eventually, if the system is stable. We shall develop an expression for the torque of the form given below, while neglecting damping and considering the swinging to be simple harmonic motion:

$$T = T_o + T_S(\Delta \sin \omega_h t) + T_D \left[\frac{d}{dt} (\Delta \sin \omega_h t) \right] \tag{4.4.3}$$

where

T_o is the average steady-state non-hunting torque;

$T_S(\Delta \sin \omega_h t)$ is the synchronizing or restoring torque, with its maximum occurring at the extremes of the swing;

and

$T_D [\frac{d}{dt} (\Delta \sin \omega_h t)]$ is the damping torque developed by induction machine action, with its maximum occurring at the middle of the swing.

Since the torque is given by

$$T = \lambda_d i_q - \lambda_q i_d \tag{4.4.4}$$

we shall first attempt to obtain expressions for λ_d, λ_q, i_d and i_q. Referring to Figure 4.4.1, we have

$$\lambda_d = e \cos\delta = e \cos\delta_o - e(\Delta \sin \omega_h t) \sin\delta_o \tag{4.4.5}$$

$$\lambda_q = -e \sin\delta = -e \sin\delta_o - e(\Delta \sin \omega_h t) \cos\delta_o \tag{4.4.6}$$

Since

$$\lambda_q = -x_q i_q \tag{4.4.7}$$

one gets for i_q the following:

$$i_q = \frac{e}{x_q} \sin\delta_o + \frac{e \cos\delta_o}{x_q} \Delta \sin \omega_h t \tag{4.4.8}$$

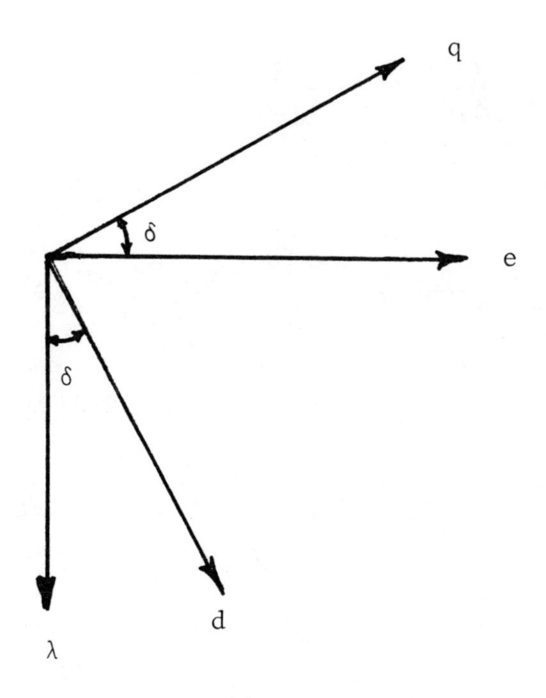

Figure 4.4.1. Relative location of various axes.

Since

$$\lambda_d = G(p)\, e_{fd} - x_d(p)\, i_d \tag{4.4.9}$$

it follows that

$$i_d = \frac{G(p)}{x_d(p)} e_{fd} - \frac{\lambda_d}{x_d(p)} \tag{4.4.10}$$

or substituting the result of Eq. (4.4.5), one has

$$i_d = \frac{G(p)}{x_d(p)} e_{fd} - \frac{e \cos\delta_o}{x_d(p)} + \frac{e \sin\delta_o}{x_d(p)} \Delta \sin \omega_h t \tag{4.4.11}$$

where G(p) and $x_d(p)$ are in general given by

$$G(p) = \frac{x_{ad}}{px_{ffd} + r_{fd}} \tag{4.4.12}$$

and

$$x_d(p) = x_d - \frac{px_{ad}^2}{px_{ffd} + r_{fd}} = \frac{x_d + x_d' T_{do}' p}{p T_{do}' + 1} \tag{4.4.13}$$

in which the following relations have been used:

$$x_d - x_d' = \frac{x_{ad}^2}{x_{ffd}} \qquad \text{and} \qquad T_{do}' = \frac{x_{ffd}}{r_{fd}} \tag{4.4.14}$$

Observing that the first two terms of the right-hand-side of Eq. (4.4.11) are the d.c. quantities and the third one is an a.c. component, we shall set p equal to zero in the first two terms and p equal to '$j\omega_h$' in the third term, in order to obtain steady-state solution while neglecting damping.

Thus we shall have

$$i_d = \frac{x_{ad}\, e_{fd}}{x_d\, r_{fd}} - \frac{e \cos\delta_o}{x_d} + e \sin\delta_o \frac{1 + j\omega_h T'_{do}}{x_d + j\, x'_d\, T'_{do}\, \omega_h} \cdot$$

$$\cdot\ \Delta \sin \omega_h t \qquad (4.4.15)$$

which may be rewritten as follows after some manipulations and simplifications:

$$i_d = \frac{e_o - e \cos\delta_o}{x_d} + \frac{e \sin\delta_o}{x_d} \Delta \sin \omega_h t +$$

$$+ \frac{e \sin\delta_o (x_d - x'_d)}{x_d\, x'_d} \frac{(\omega_h T'_d)^2}{1 + (\omega_h T'_d)^2} \Delta \sin \omega_h t +$$

$$+ \frac{e \sin\delta_o (x_d - x'_d)}{x_d\, x'_d} \frac{T'_d}{1 + (\omega_h T'_d)^2} \omega_h \Delta \cos \omega_h t \qquad (4.4.16)$$

where e_o stands for

$$\frac{x_{ad}\, e_{fd}}{r_{fd}} ,$$

and $T'_{do} = T'_d \frac{x_d}{x'_d}$ has been used.

From the results of Eqs. (4.4.5), (4.4.6), (4.4.8) and (4.4.16), one can now obtain the expression for torque from Eq. (4.4.4) in the form of Eq. (4.4.3) with the following results, after neglecting the Δ^2 terms:

$$T_o = \frac{e\,e_o}{x_d}\sin\delta_o + e^2\frac{x_d - x_q}{2x_d\,x_q}\sin 2\delta_o \tag{4.4.17}$$

$$T_S = \frac{e_o\,e}{x_d}\cos\delta_o + \frac{x_d - x_q}{x_d\,x_q}e^2\cos 2\delta_o +$$

$$+ e^2\sin^2\delta_o\frac{x_d - x_d'}{x_d\,x_d'}\,\frac{(\omega_h\,T_d')^2}{1 + (\omega_h\,T_d')^2} \tag{4.4.18}$$

$$T_D = e^2\sin^2\delta_o\frac{x_d - x_d'}{x_d\,x_d'}\,\frac{T_d'}{1 + (\omega_h\,T_d')^2} \tag{4.4.19}$$

The bus voltage e has been retained in the above analysis to show the explicit dependence of torque on the bus voltage. Although the damping is neglected in the analysis presented here, it may be pointed out that the closed field will provide some damping and additional damping would be provided if the amortisseur circuits are also present. In practice, the one having the faster swing has to be damped; since it is easier to handle the governor which is mechanically damped with dashpot or other mechanism, the governor is damped and hence it ought to be built to swing faster.

Angular Velocity of Hunting:

The restoring torque in synchronous kilowatts is given by

$$T_S\,\delta_1 = T_S\,\alpha_1\,P/2 \tag{4.4.20}$$

where δ_1 is in electrical radians, α_1 is in mechanical radians, and P is the number of poles of the machine. It is related to the moment of inertia through the following equation:

$$T_S \alpha_1 \frac{P}{2} \frac{33000}{0.746 \ (RPM) \ 2\pi} = \frac{WR^2}{g} \left(- \frac{d^2 \alpha_1}{dt^2} \right) \tag{4.4.21}$$

where W is the weight in pounds of the rotating parts of the machine, R is the radius of gyration in feet, and g is the acceleration due to gravity given by 32.2 ft/sec^2. Equation (4.4.21) may be rearranged as

$$\frac{d^2 \alpha_1}{dt^2} + T_S \frac{P}{2} \frac{33000 \times 32.2}{0.746 \ (RPM) \ 2\pi \ WR^2} \alpha_1 = 0 \tag{4.4.22}$$

Assuming α_1 to be of the form ($A \sin \omega_h t$), one gets the expression for the angular velocity of hunting in radians per second as

$$\omega_h = \sqrt{\frac{T_S}{WR^2}} \sqrt{\frac{P}{2} \frac{33000 \times 32.2}{0.746 \ (RPM) \ 2\pi}} \tag{4.4.23}$$

which may be rearranged as

$$\omega_h = \sqrt{\frac{T_S \ f}{WR^2}} \sqrt{\frac{33000 \times 32.2 \times 60}{0.746 \times 2\pi} \frac{1}{RPM}}$$

$$= \frac{3688}{RPM} \sqrt{\frac{T_S \ f}{WR^2}} \quad \text{in radians/sec.} \tag{4.4.24}$$

or

$$\omega_h = \frac{9.78}{RPM} \sqrt{\frac{T_S \, f}{WR^2}} \qquad \text{in per-unit} \tag{4.4.25}$$

where T_S is often approximated as

$$T_S = T/\delta \tag{4.4.26}$$

in which T is expressed in synchronous kilowatts and δ is in electrical radians.

Per-Unit Moment of Inertia and Inertia Constant:

The per-unit moment of inertia may be expressed as

$$I_{pu} = \frac{WR^2}{g} \frac{(RPM)^2 (0.746)(4\pi^2)(2\pi f)}{(KVA)(33000)(60)} \tag{4.4.27}$$

which is justified from the dimensional analysis given below:

$$\text{ft-lbs sec}^2 \frac{\text{rev}^2}{\text{min}^2} \frac{1}{\text{KVA}} \frac{\text{KW}}{\text{HP}} \frac{\text{min HP}}{\text{ft-lbs}} \frac{\text{min}}{\text{sec}} \frac{\text{rad}^2}{\text{rev}^2} \frac{\text{rad}}{\text{sec}} = \frac{\text{KW}}{\text{KVA}} \tag{4.4.28}$$

which is dimensionless in nature. Equation (4.4.27) may be rearranged to read as

$$I_{pu} = 4\pi f \left[0.231 \frac{WR^2 \, (RPM)^2}{KVA} 10^{-6}\right] = 4\pi f H = (2\pi f)(2H) \tag{4.4.29}$$

where the inertia constant H is given by

$$H = 0.231 \frac{WR^2 \ (RPM)^2}{KVA} 10^{-6} \text{ KW-sec/KVA} \qquad (4.4.30)$$

Typical inertia constants of synchronous machines are given in Table 4.4.1.

The torque in per-unit may be seen to be related to the inertia constant through the relation:

$$T = 2H \frac{d\omega}{dt} \qquad (4.4.31)$$

where H is dimensionally in seconds, t is in seconds, and ω is expressed in per-unit. If rated full-load torque is applied, the machine would be accelerated to its rated speed from standstill in (2H) seconds.

A Numerical Example:

Let us now consider a 15-KVA, 6-pole, 60-HZ synchronous machine with the following constants operating at full-load, rated voltage and 0.8 pf lagging:

$x_d = 0.85 \qquad x_q = 0.50 \qquad x_d' = 0.24$

$T_d' = 44$ (all in per unit)

$WR^2 = 20.8 \text{ lb-ft}^2$

It is also given that

$\delta_o = 17° = 0.298$ elec. radians; $e_o = 1.65$ p.u.

Neglecting damping, the first trial for the per-unit angular velocity of hunting given by Eq. (4.4.25) yields:

Table 4.4.1

Typical Inertia Constants of Synchronous Machines[*]

Type of Machine	Inertia constant H[†] KW-Sec/KVA
Turbine generator:	
Condensing, 1,800 rpm	9-6
3,600 rpm	7-4
Noncondensing, 3,600 rpm	4-3
Waterwheel generator:	
Slow-speed, <200 rpm	2-3
High-speed, >200 rpm	2-4
Synchronous condensor:[§]	
Large	1.25
Small	1.00
Synchronous motor with load varies from 1.0 to 5.0 and higher for heavy flywheels	2.00

[*]Reprinted by permission of the Westinghouse Electric Corporation from Electrical Transmission and Distribution Reference Book.

[†]Where range is given, the first figure applies to machines of smaller kilovoltampere rating.

[§]Hydrogen-cooled, 25% less.

$$\omega_h = \frac{9.78}{1200}\sqrt{\frac{12}{0.298}\,\frac{60}{20.8}} = 0.088 \text{ p.u.}$$

Now T_S is calculated from Eq. (4.4.18) as

$$T_S = \left(\frac{1.65}{0.85} \times 0.96\right) + \left(\frac{0.35}{0.425} \times 0.83\right) +$$

$$+ \left(0.085 \times \frac{0.61}{0.205} \times \frac{15}{16}\right) = 2.78 \text{ p.u.}$$

or

$$T_S = 2.78 \times 15 = 41.7 \text{ KW/rad.}$$

which is seen to be slightly larger than $\frac{0.8}{0.298} = 2.68$ p.u.

The second trial for the per-unit angular velocity of hunting yields

$$\omega_h = \frac{9.78}{1200}\sqrt{\frac{41.7 \times 60}{20.8}} = 0.089 \text{ p.u.}$$

The process may be repeated until satifactory agreement is reached.

4.5 Governor Action

While the general problem of stability would be presented formally in more detail under Chapter VI, it is considered appropriate to include the analysis of governor action at the present stage. For the governor-turbine-generator system shown in Figure 4.5.1, the differential equations of motion for the alternator and the governor are given by

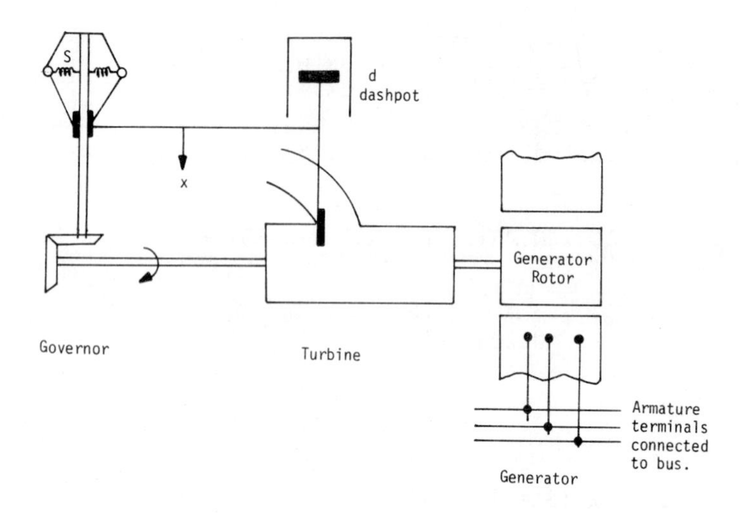

Figure 4.5.1. A sketch of the governor-turbine-generator system.

$$I\,p^2\,\delta + T_D\,p\,\delta + T_S\,\delta = K\,x \tag{4.5.1}$$

and

$$m\,p^2\,x + d\,p\,x + S\,x - c\,p\,\delta \tag{4.5.2}$$

where (Kx) represents the prime-mover torque with K as a constant; p is the operator $\frac{d}{dt}$; m, d, S and c correspond to the governor mechanism. From Eq. (4.5.1) one has

$$x = \frac{I}{K}\,p^2\,\delta + \frac{T_D}{K}\,p\,\delta + \frac{T_S}{K}\,\delta \tag{4.5.3}$$

Then Eq. (4.5.2) may be rewritten as

$$p^4 \delta \frac{mI}{K} + p^3 \delta \left(\frac{m T_D}{K} + \frac{dI}{K} \right) + p^2 \delta \left(\frac{m T_S}{K} + \frac{d T_D}{K} + \frac{SI}{K} \right) +$$

$$+ p \delta \left(\frac{d T_S}{K} + \frac{S T_D}{K} + c \right) + \frac{S T_S}{K} \delta = 0 \qquad (4.5.4)$$

from which the characteristic equation is given by

$$p^4 + \left(\frac{T_D}{I} + \frac{d}{m} \right) p^3 + \left(\frac{T_S}{I} + \frac{d T_D}{mI} + \frac{S}{m} \right) p^2 +$$

$$+ \left(\frac{d T_S}{mI} + \frac{S T_D}{mI} + \frac{cK}{mI} \right) p + \frac{S T_S}{mI} = 0 \qquad (4.5.5)$$

Let the maximum travel of the governor sleeve be x_{max} corresponding to a speed change of $\Delta\omega$, so that

$$c = \frac{S x_{max}}{\Delta\omega} = \frac{S x_{max}}{\omega\rho} \qquad (4.5.6)$$

where $\rho (= \frac{\Delta\omega}{\omega})$ is the governor regulation which is usually of the order of 3 percent. Letting x_{max} also correspond to full action of the inlet valve, let T_o be the normal full-load torque, while not taking overload into consideration. Then

$$K = \frac{T_o}{x_{max}} \qquad (4.5.7)$$

so that one can write

$$\frac{cK}{mI} = \frac{S x_{max}}{\omega\rho} \frac{T_o}{x_{max}} \frac{1}{mI} = \frac{S T_o}{\omega\rho \, mI} = \frac{\omega_G^2}{\tau\rho} \qquad (4.5.8)$$

where

$$\omega_G = \sqrt{\frac{S}{m}} \quad \text{and} \quad \tau = \frac{\omega I}{T_o} = 2H \tag{4.5.9}$$

From the characteristic equation of the alternator alone, one can obtain approximately:

$$P_A \simeq -\frac{T_D}{2I} \pm j\sqrt{\frac{T_S}{I}} = -a \pm j\,\omega_A \tag{4.5.10}$$

where ω_A is the natural angular velocity of hunting of the alternator given by $\sqrt{T_S/I}$. The negative real root of Eq. (4.5.10) suggests that the swinging will be damped out if the alternator is hunting by itself. Similarly the characteristic equation of the governor alone yields approximately

$$p_G = -\frac{d}{2m} \pm j\sqrt{\frac{S}{m}} = -b \pm j\,\omega_G \tag{4.5.11}$$

where ω_G is the natural angular velocity of hunting of the governor given by $\sqrt{S/m}$. The characteristic equation (4.5.5) may now be written as

$$p^4 + 2(a + b)\,p^3 + (\omega_A^2 + 4ab + \omega_G^2)\,p^2 + \left(2b\,\omega_A^2 + 2a\,\omega_G^2 + \frac{\omega_G^2}{\rho\tau}\right)p + \omega_A^2\,\omega_G^2 = 0 \tag{4.5.12}$$

The combined system would be stable if the real roots are all negative and the real parts of all complex roots are also negative. An equation of the sixth degree may sometimes result if the alternator is driven by reciprocating engine, in which case the torque is of a pulsating nature.

Illustrative Example:

Let ω_A = 20 rad/sec; ρ = 0.025; τ = 2H = 10 sec. Then Eq. (4.5.12) can be written as

$$p^4 + 2(a + b)\, p^3 + 400(1 + R^2 + \frac{ab}{100})\, p^2$$
$$+ 400(2a\, R^2 + 2b + 4R^2)\, p + 16 \times 10^4\, R^2 = 0 \qquad (4.5.13)$$

where R is the ratio of (ω_G/ω_A). This is of the form

$$A\, p^4 + B\, p^3 + C\, p^2 + D\, p + E = 0 \qquad (4.5.14)$$

with A = 1; B = 2(a + b); C = $400(1 + R^2 + \frac{ab}{100})$; D = $400(2a\, R^2 + 2b + 4R^2)$; and E = $16 \times 10^4\, R^2$.

The Routh criterion* for stability is given by the condition

$$B\, C\, D - A\, D^2 - B^2\, E \geq 0 \qquad (4.5.15)$$

which is equivalent to the following for the system under consideration

$$(2a - 2b - ab)\, R^2(1 - R^2) - 4R^4 +$$
$$+ [-ab + (a + b)(a + 2)(\frac{ab}{100})]\, R^2 +$$
$$+ [ab + (a + b)(\frac{ab^2}{100})] \geq 0 \qquad (4.5.16)$$

The stability regions for different ratios R of governor to

*See Appendix E for details on Stability Criteria (E-7).

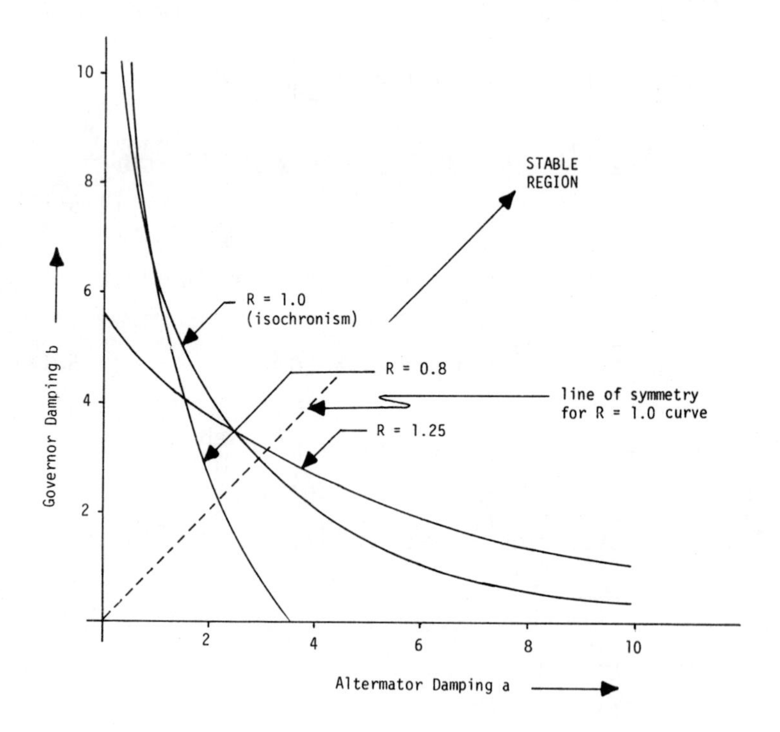

Figure 4.5.2. Stability Regions for Different Ratios R of Governor-Turbine-Generator System

Note: 1. $R = \omega_G/\omega_A = \dfrac{\text{Natural angular frequency of the governor}}{\text{Natural angular frequency of the alternator}}$

2. The stable region includes the limit curve and beyond, in the direction of arrow indicated.

alternator natural frequencies are shown in Figure 4.5.2 based on the above criterion. It may be observed that an increase in the damping of one of the two components reduces the damping requirement on the other component for stable operation. The one that swings faster must be damped. It is possible to achieve stable operation with zero damping on the component having lower natural frequency. Thus for R = 1.25 one can operate with no alternator damping, and for R = 0.8 with no governor damping. However, from a practical point of view, providing required damping by means of a governor dashpot is more economical and is amenable to better control than attempting to control alternator damping. In such a case one will have to make sure that the governor does swing faster than the alternator and that adequate damping is provided in the governor mechanism. Since ω_A (the natural angular frequency of the alternator) is faster at no-load and is reduced as the load is increased, it is better to test at no-load to check whether the governor is still faster.

Problems

4 1. Starting from the armature voltage equations (4.1.3) and (4.1.4), utilizing the definitions given by Eq. (4.1.24), develop the equivalent circuit shown in Figure 4.1.2, and show in terms of f- and b-axis quantities that the resultant torque of Eq. (4.1.25) [with no rotor excitation and with applied balanced armature voltages] comprises of an average term as well as a double-slip-frequency component.

4-2. Consider the equivalent circuits shown in Figures 4.1.1 and 4.1.2. Calculate and plot the average torque vs. slip for a typical motor (having the data given below) by both the approximate and exact methods in order to compare the results of the two methods.

$x_d = x_{ad} + x_\ell = 1.17$; $x_{11d} = x_{ad} + x_{1d} = 1.122$;

$x_{ffd} = x_{ad} + x_{fd} = 1.297$;

$x_{afd} = x_{a1d} = x_{f1d} = x_{ad} = 1.03$;

$x_q = x_{aq} + x_\ell = 0.75$; $x_{11q} = x_{aq} + x_{1q} = 0.725$;

$x_{a1q} = x_{aq} = 0.61$; $r = 0.0121$; $r_{1d} = 0.0302$;

$r_{fd} = 0.00145$ (or 0.03 including discharge resistance);

$r_{1q} = 0.039$.

Repeat the calculations and plotting with r = 0.0363, triple the value of armature resistance given above.

4-3. A 2-pole, 26-KV, 908-MVA turbine generator, while on turning gear, is inadvertently connected to the lines by the accidental closing of the main generator line breaker; the field is shorted through a discharge resistance. The average slip torques are given below for various values of slip.

(a) Calculate the time required for the machine to accelerate to rated speed, given the inertia constant to be 3.14 seconds.

(b) Compute the average power absorbed in the

generator rotor during the period of acceleration.

Slip	$T_d = \|I_d\|^2 \, Re(Z_d/S)$	$T = \|I_q\|^2 \, Re(Z_q/S)$
1.0	-0.076	-0.167
0.8	-0.086	-0.184
0.6	-0.100	-0.208
0.5	-0.110	-0.227
0.4	-0.122	-0.250
0.3	-0.133	-0.284
0.2	-0.141	-0.336
0.1	-0.135	-0.435
0.03	-0.172	-0.651
0.01	-0.374	-0.739
0.003	-0.828	-0.557

4-4. Starting from Eqs. (4.1.36) and (4.1.37), transform to the f- and b-axis quantities and calculate the total current i_f (produced by both stator and rotor excitations) as well as the flux density. Show then that the effect of rotor excitation is to add an average d.c. term and a slip-frequency component to the torque.

4-5. Compute the average torque of a synchronous machine (whose data is given below) during its asynchronous operation corresponding to starting, 45 percent speed, and 50 percent speed, when the machine is started with applied rated terminal voltage and no field excitation.

$x_d = 1.2$; $x_q = 0.8$; $x_{ad} = 1.0$; $x_{ffd} = 1.1$;

$r_{fd} = 0.0011$; $r = 0.005$; no amortisseur.

4-6. Equations (4.2.7) and (4.2.8) are written for a machine initially unloaded and normally excited. Suppose the initial excitation corresponds to e_{qo} so that the right-hand-side of Eq. (4.2.8) becomes e_{qo}/s. Follow through the solution procedure suggested in the text and show the effect of e_{qo} on the rest of the development.

4-7. Assuming the exponential variation of excitation given by Eq. (4.2.31), obtain Eqs. (4.2.32) and (4.2.33); then write down the expression for the total terminal voltage including the effect due to the armature current components caused by closing of the armature switch. By differentiating the complete expression for terminal voltage, solve for the time t_d corresponding to a mimimum voltage and show that

$$t_d = \frac{T'_{dz} T_e}{T'_{dz} - T_e}\left\{\frac{t_1}{T_e} - \log_e\left[e^{t_1/T'_{dz}} - (K_2 - 1)\cdot\right.\right.$$

$$\left.\left.\cdot \frac{1}{E_C - E_o}\left(\frac{T'_{dz} - T_e}{T'_{dz}}\right)\right]\right\}$$

For the case of $T_e = t_1 = r = o$ and $E_o = 1.0$, for a minimum voltage to occur (i.e., $t_d < \infty$), show that the approximate condition to be satisfied by the excitation ceiling is

$$E_c > \frac{x_d}{x'_d}$$

4-8. A 3-phase, 60-Hz synchronous generator (with the data given below) is operating on open circuit with rated voltage, and with a field voltage of 50 volts. A reactance load of X = 1.0 per unit is suddenly applied.

$x_d = 1.2;\quad x'_d = 0.15;\quad x_{afd} = 1.1;\quad x_{ffd} = 1.15;$

$r_{fd} = 0.001;\ x_q = 1.2.$

(a) Compute and plot the terminal voltage as a function of time for a linearly rising excitation voltage of the form $\Delta E = kt$, where k is to be taken such that the rate of rise is 100 volts per second.

(b) Determine the exciter ceiling voltage required to permit the generator terminal voltage to recover to normal.

(c) Find the value of k required to limit the minimum voltage to 85 percent.

4-9. For a saliency ratio of $x_q/x'_d = 2.75$, Figure 4.3.2 gives δ as 43°, 56°, and 142° at which maximum torque due to synchronizing out-of-phase occurs. Plot the torque as a function of time for the cases of $\delta = 43°$ and 56°, and compare the results with those of Figure 4.3.3, which is drawn for the case of $\delta = 142°$.

4-10. Following the analysis presented in the text, check the results given by Eqs. (4.4.17) to (4.4.19).

4-11. Starting from the constraint Eq. (4.5.16), obtain the necessary data to plot the limit curves shown in

Figure 4.5.2 for different values of R equal to 0.8, 1.0, and 1.25 respectively.

CHAPTER V

FLUX DISTRIBUTION AND SATURATION IN SYNCHRONOUS MACHINES

5.1 Field Equations

The fundamental laws governing all electromagnetic fields can be expressed by the well-known Maxwell's equations. Assuming that the dielectric properties of any of the materials are of no significance in comparison with the conduction properties, as is true in low frequency problems, the dielectric effects or the displacement currents will be neglected, and the material regions will be considered as void of volume charge density. All the materials will be taken to be isotropic and homogeneous. Then the relevant partial differential equations may be written as follows:

$$\nabla \cdot \bar{B} = 0; \; \nabla \times \bar{H} = \bar{J}; \; \nabla \times \bar{E} = -\frac{\partial \bar{B}}{\partial t}; \; \nabla \cdot \bar{J} = 0 \qquad (5.1.1,2,3,4)$$

where $\bar{B}$ is the magnetic induction, $\bar{H}$ the magnetic field intensity, $\bar{J}$ the current density, $\bar{E}$ the electric field intensity, and t is the time. Rationalized MKSA system of units* has been assumed. The constituent relations are given by

$$\bar{B} = \mu\bar{H}; \; \bar{J} = \sigma\bar{E} \qquad (5.1.5,6)$$

where μ is the permeability of the ferromagnetic material,

*See Appendix A for details on Units, Constants, and Conversion Factors.

and σ is the conductivity. It may be observed here that the current continuity relation given by Eq. (5.1.4) is a consequence of Eq. (5.1.2).

The general boundary conditions to be satisfied at the interfaces of stationary dissimilar media may be derived from the limiting integral forms of Eqs. (5.1.1) to (5.1.4), and they are given in terms of the normal and tangential components as

$$B_{n1} - B_{n2} = 0; \; H_{t1} - H_{t2} = J_s; \; E_{t1} - E_{t2} = 0; \qquad (5.1.7,8,9)$$

$$J_{n1} - J_{n2} = 0 \qquad (5.1.10)$$

where J_s is the surface current density, which will exist at the interface separating region 1 from region 2 only if one of the regions has infinite conductivity. Otherwise, if both regions have finite (including zero) conductivity, then Eq. (5.1.8) takes the form

$$H_{t1} - H_{t2} = 0 \qquad (5.1.8a)$$

It must be borne in mind that the boundary conditions need to be modified at the interfaces between dissimilar media, if they are in relative motion.

To allow for the magnetic nonlinearity, the permeability μ of the ferromagnetic material will be taken as a single-valued function of the magnetic induction $\overline{B}$, while the hysteresis effects are neglected. Constant conductivity σ of the medium will be assumed. One can express $\overline{B}$ as the curl of $\overline{A}$, the magnetic vector potential:

$$\bar{B} = \nabla \times \bar{A} \tag{5.1.11}$$

thereby satisfying Eq. (5.1.1). From Eqs. (5.1.3) and (5.1.11), one obtains

$$\bar{E} = -\frac{\partial \bar{A}}{\partial t} - \nabla\Phi \tag{5.1.12}$$

where Φ is the electric scalar potential. Defining the reluctivity ν as the reciprocal of the magnetic permeability μ, one can obtain the following from Eqs. (5.1.2), (5.1.5), (5.1.6), (5.1.11), and (5.1.12):

$$\nabla \times [\nu(\nabla \times \bar{A})] = \bar{J} \tag{5.1.13}$$

where $\bar{J}$ represents either the impressed current density sources or the conduction current densities existing due to induction phenomena given by

$$\bar{J} = -\sigma\left(\frac{\partial \bar{A}}{\partial t} + \nabla\Phi\right) \tag{5.1.14}$$

It follows then, as a consequence of Eqs. (5.1.13) and (5.1.14), that Φ has to satisfy

$$\nabla^2\Phi = -\frac{\partial}{\partial t}(\nabla \cdot \bar{A}) \tag{5.1.15}$$

unless σ is zero. So far, the divergence of $\bar{A}$ and the gradient of Φ have not been defined. In fact, they may be defined in any convenient manner as long as the basic equation (5.1.13) is not violated.

Magnetostatic Field Problems with Impressed Current Densities:

For the particular case of magnetostatic field problems with impressed current densities, only Eqs. (5.1.1) and (5.1.2) are needed. Equation (5.1.4) and hence Eq. (5.1.10) will be satisfied through the specification of the current densities.

For the case of linear magnetostatic problems in which the permeability is a constant, it can easily be seen from Eq. (5.1.13) that the Poisson's equation of the vector $\overline{A}$ applies, with the arbitrary definition of the divergence of $\overline{A}$ being equal to zero, which is in fact the familiar Lorentz condition:

$$\nabla \cdot \overline{A} = 0 \tag{5.1.16}$$

$$\nabla^2 \overline{A} = -\mu \overline{J} \tag{5.1.17}$$

Air regions with no current density fields have to satisfy the well-known Laplace's equation of the vector potential:

$$\nabla^2 \overline{A} = 0 \tag{5.1.18}$$

In two-dimensional problems including discrete current carrying regions with the current flow in only one direction (which is perpendicular to the plane under consideration), one uses with advantage the concept of the magnetic vector potential $\overline{A}$ which has only one component (in the same direction as that of the current) and whose divergence is trivially zero. It can be shown that the following partial differential equations apply in such cases:

In right-handed rectangular coordinate system*

$$\frac{\partial}{\partial x}\left(\nu \frac{\partial A_z}{\partial x}\right) + \frac{\partial}{\partial y}\left(\nu \frac{\partial A_z}{\partial y}\right) = -J_z \tag{5.1.19}$$

and in right-handed cylindrical coordinates*

$$\frac{1}{r}\,\nu\,\frac{\partial A_z}{\partial r} + \frac{\partial}{\partial r}\left(\nu \frac{\partial A_z}{\partial r}\right) + \frac{1}{r^2}\frac{\partial}{\partial \theta}\left(\nu \frac{\partial A_z}{\partial \theta}\right) = -J_z \tag{5.1.20}$$

Eddy-Current Problems:

For charge-free nonlinear ferromagnetic material regions of finite conductivity, in which induction phenomena causes conduction current densities, Eqs. (5.1.13) to (5.1.15) apply. For linear material regions of finite permeability and conductivity, the divergence of $\overline{A}$ is conveniently defined as

$$\nabla \cdot \overline{A} = -\mu\sigma\Phi \tag{5.1.21}$$

in which case the following partial differential equation results:

$$\nabla^2\overline{A} = \mu\sigma \frac{\partial \overline{A}}{\partial t} \tag{5.1.22}$$

Two-dimensional problems, with either linear or nonlinear magnetic media, can be solved rather easily with the assumptions that the vector potential $\overline{A}$ has only one component and the scalar potential Φ does not exist.

---

*See Appendix F for Vector Relations in Orthogonal Curvilinear Coordinate Systems.

All the field equations associated with the eddy currents can be seen to contain the partial derivative term with respect to the time variable. Through a suitable transformation of coordinates, it is possible in some cases to replace the time-derivative terms by spatial derivatives so that the time variable need not be considered explicitly in the solution procedure. Sinusoidal dependence of time is usually assumed in analytical solutions and the time derivative is replaced by "$j\omega$".

Boundary Conditions:

The general boundary conditions to be satisfied are given by Eqs. (5.1.7) to (5.1.10), which are in terms of the components of $\bar{B}$, $\bar{H}$, $\bar{E}$ and $\bar{J}$. It is helpful to recognize certain ideal cases such as the boundary conditions at a plane of zero or infinite permeability. The normal z component of the flux density at an xy-plane of zero permeability has to vanish, and the parallel components of the flux density must vanish at a plane of infinite permeability.

If the partial differential equations are set up with the variable $\bar{A}$, the boundary conditions have also to be expressed in terms of $\bar{A}$ using the relations such as (5.1.11). It can be seen that the boundary conditions, when expressed in terms of $\bar{A}$, may not be uniquely specified as the conditions on the components of $\bar{A}$. Thus one is forced to make a wise choice based on the available freedom and intuitive judgment. Three-dimensional field problems become more complex from this point of view.

Boundary conditions are indeed very critical in determining the solution of the field problem and as such have to be taken care of properly. Conditions such as those due to periodicity and rotational symmetry may be taken advantage of in setting up simpler mathematical models for solving the problems. It is common practice to assume the magnetic induction outside the machine surface to be negligible.

In two-dimensional fields, it is easy to show that contours of constant magnetic vector potential are the flux lines. But similar conclusions cannot be drawn in three-dimensional field problems. At no load and purely reactive loads the vector potential is a constant on the direct axis of the north and south poles. The center line of the pole represents a line of flux and therefore is an equipotential. All lines of flux must be orthogonal to the quadrature axis, since it forms an axis of symmetry. Thus it is enough to consider only one-half of a pole pitch for analysis. At general balanced loads the vector potential is a periodic function of the double-pole pitch. Thus, the vector potential A_p at a point P in the general area of a north pole must be equal in magnitude, and opposite in sign, to the vector potential $A_{p'}$, at a point P' at the same location under a south pole as point P under the north pole. It is then necessary to consider one full pole pitch for analysis.

When the value of the vector potential is given to be fixed on a part of a boundary, such a condition is called the first boundary value problem, or Dirichlet's problem. When the normal derivative of the vector potential is given on parts of a boundary, it is known as the second boundary value problem, or the Neumann problem.

The nature of the magnetic field of distributed currents is such that it contains a kernel about which the flux circulates; and at which the "lines of no-work" (which are perpendicular to the flux lines) meet. For a symmetrical conductor in air, the kernel is at the center of the conductor. When the conductor is influenced by a permeable boundary (or current of like sign), the kernel moves towards the boundary. When influenced by an impermeable boundary (or current of opposite sign) the kernel moves away from the boundary. Also, a flux line crossing the boundary of a current-carrying region does not suffer a sudden change in direction unless the conductor material has a permeability different from the surrounding medium; but it experiences a change in curvature.

Mathematical Models:

All physical fields are, of course, three-dimensional; but for most cases of practical interest exact analytical solutions are not possible, and numerical solutions often involve a prohibitive amount of computer time. However, approixmate solutions of quite sufficient accuracy can be obtained by using a two-dimensional treatment, in which the variation of the field in one direction is neglected. As a result, analysis is made possible in many cases and also the computer time required of numerical solutions is greatly reduced. As an example, the distribution of the main magnetic field within the air-gap region of a synchronous machine can be found with negligible error by analyzing the field at a cross-section perpendicular to the axis, while the variation along the length of the machine is neglected. The field in the end region of a synchronous machine can be found by analyzing the field in the axial plane, while neglecting the peripheral variations. However, such an analysis yields less accurate results, as the field distribution is inherently three-dimensional in nature;

the field is excited by a complicated pattern of currents and is modified by boundaries of intricate shape and of varying values of permeability, which carry modifying eddy currents.

Rectangular cartesian coordinates system is often employed for the sake of convenience and simplicity, when such an assumption of neglecting the curvature can be justified for the analysis. Cylindrical coordinates are also used in two-dimensional analysis; however, their use in three-dimensional problems may lead to complicated coupled component partial differential equations. A combination of rectangular and cylindrical coordinates is also used sometimes to follow the complicated contours involved in an electromechanical energy converter.

Discrete currents in the conductors are usually replaced by a uniform current-density field over the cross-sections of the armature and field coils while neglecting the skin effects. Simplifying assumptions are made in order to bring the solution into manageable proportions, consistent with the expected accuracy of the end results. After the mathematical model is set up, it is analyzed by a suitable method of solution as a boundary value problem satisfying relevant partial differential equations in different regions.

A Few Comments:

Before concluding this section, it is considered necessary to make a few relevant remarks at this stage. In view of the scope and limitations of the book, only magnetic field problems with the vector potential formulations have been discussed. However, field problems may also be approached with the concept of a scalar potential.

Electric field problems may also be solved with the aid of vector potential or scalar potential concepts. Such ideas would be incorporated in the problems given at the end of this chapter. The formulation and solution procedures can be made directly applicable with appropriate interpretation to other physical phenomena (such as heat flow and fluid flow) which are mathematically analogous to electric and magnetic field phenomena. Hysteresis, non-isotropic properties of the material (such as grain-oriented sheet steels), nonlinear magnetic as well as electric materials may be taken care of in the formulation of partial differential equations.

5.2 Field-Plotting Methods

An accurate knowledge of the magnetic as well as other field distributions is of great importance in design optimization. The continuing rapid increase in the rating of the generating plant and the consequent increased electric and magnetic loadings of electrical machines are forcing manufacturers to work with greatly reduced safety factors in designing alternators. In order to meet the specifications of an electrical apparatus, the designer must be able to analyze and modify several fields of interest (like an electric field, a magnetic field, a current field, a field of mechanical forces, a field of heat flow, a field of elastic forces in the material stressed by mechanical forces, a field of fluid flow of the cooling medium, etc.) with the ultimate aims of safety, reliability, simplicity, efficiency, and economy. The accurate knowledge of these fields is very useful for the power engineer, who writes the specifications and should make the best use of the characteristics of the apparatus.

The fields of common engineering interest satisfy in general Laplace's or Poisson's equations of either a scalar potential or a vector potential. If the nonlinearities of the materials involved are to be considered, the equations to be solved would be nonlinear partial differential equations, whose component equations may in general be coupled. The problems to be solved require the integration of these equations and the determination of values of the unknown function (scalar or vector) at any point of the region of interest, for assigned values on the boundary.

The endeavor to find the flux distribution in an electrical machine is as old as the art of rotating machine design. Scanning the various methods available in potential theory, they can be grouped readily into four main sections:

(1) Mathematical - Direct mathematical solutions with the concept of images, technique of separation of the variables and Green's functions; Direct mathematical approximations using Fourier Series; Transformation methods involving conformal transformation and conjugate functions, as well as elliptic integrals and functions.

(2) Graphical - Curvilinear square techniques; Free-hand flux plotting; Semigraphical methods.

(3) Experimental - Resistor networks; Stretched membrane of rubber or soap film; Hydraulic analogies; Electrically conducting sheet; Electrolytic Tank.

(4) Digital Computer - Numerical methods with iterative relaxation procedures using finite-difference or finite-element techniques; Monte Carlo statistical method.

A purely mathematical "exact" solution is possible only for a very limited number of cases. A mathematical method never solves the actual problem found; but only an idealization of it. This may appear at first as a disadvantage. But an analogue that simulates the actual problem only gives immediate answers, and by reason of the fact that all the complications present are included in the set-up, the analogue does not give the engineer a satisfactory basis for mental picture of the factors at work. To the designer, the idealized apparatus forms the standard form which the actual one is a deviation, and such an idealization gives a better idea of the relative influences of the various factors in the configuration. The details of various mathematical methods are outside the scope of this book and the interested reader is referred to the bibliography.

An engineer is not fully equipped unless he is familiar with all the flux-plotting methods available and can apply the most appropriate one for the problem at hand. Ever-increasing requirements for faster, more accurate and rigorous solutions and predictions in the design of electric apparatus replace always the empirical and semi-empirical "rules of thumb" used in the past.

The graphical and semigraphical methods are of utmost value in all engineering design problems, and probably constitute the tools most widely used for field mapping in the early part of this century, even up to the 1930's or so. The graphical method for linear conditions is based on Cauchy-Riemann conditions, according to which the solution of Laplace's equation can be established by a function of a complex variable. Cauchy-Riemann's solution, represented by two conjugate functions, yields two systems of orthogonal equipotentials forming the well-known curvilinear squares.

For the air-gap two-dimensional field of electrical machines, these two orthogonal sets are represented by the flux lines (or vector equipotentials) and the scalar potential function equipotentials. The ferromagnetic material is generally assumed to be infinitely permeable so that the contours of the iron regions become scalar equipotentials, and the lines of force are perpendicular to the iron boundaries, if no current densities exist on the iron surface. The linear Laplace equation and Cauchy-Riemann's conditions do not apply to the nonlinear ferromagnetic regions. Flux lines can, however, still be drawn. The highly cherished principle of superposition becomes also invalid for saturated, nonlinear magnetic circuits.

Simple problems, such as distribution of flux in a slot shown in Figure 5.2.1*, may be solved by either mathematical or graphical methods. Regarding Figure 5.2.1, it should be pointed out that the scale to which the field is mapped increases at locations where a flux line (represented by solid line) is drawn only up to a point where it crosses a scalar equipotential line (shown by dashed line). When this is borne in mind, it is easily seen that the flux density in the slot is far lower than in the tooth, as shown in Figure 5.2.2*. The general nature of the flux-density wave with slot effects superimposed is shown in Figure 5.2.3*. Slot effects are exaggerated here because of the use of only a few relatively wide slots per pole. The distribution of flux around a salient pole, as obtained by a graphical solution, is given in Figure 5.2.4*, in which solid lines represent flux lines. The flux-map is drawn

---

* *Adapted by permission of the Publisher:* Fitzgerald, A. E. et al, Electric Machinery, 3rd Ed., Ch. 4, McGraw-Hill Book Company, 1971.

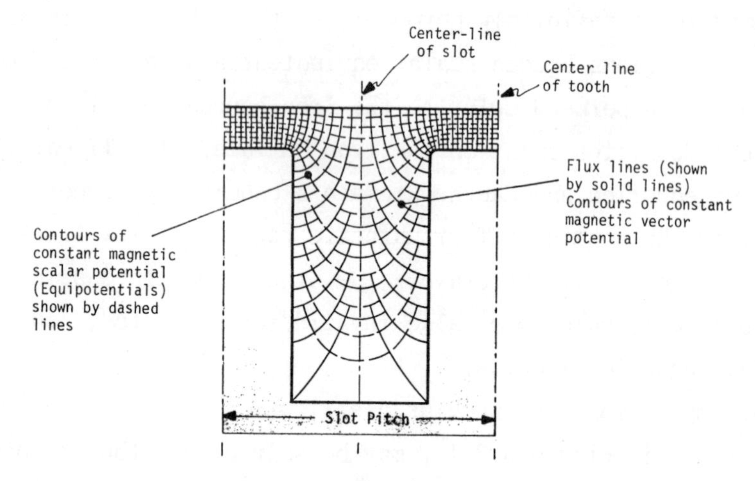

Figure 5.2.1. Flux-Distribution in a Slot

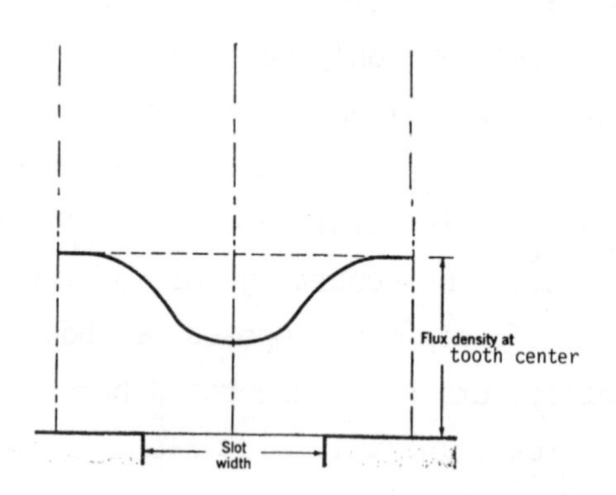

Figure 5.2.2. General Effect of a Slot on the Air-gap Density

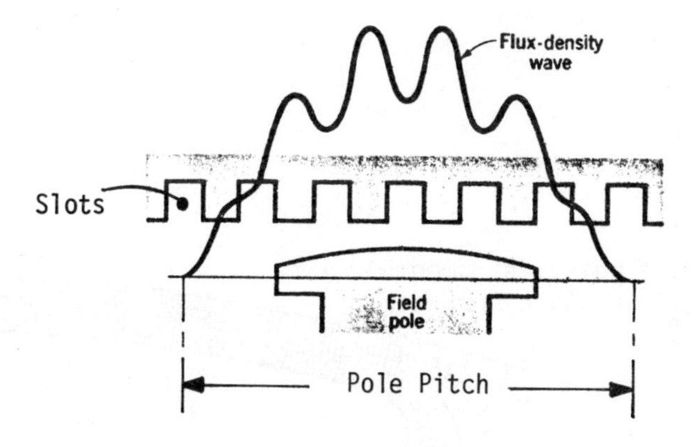

Figure 5.2.3. Effect of Slots on the Flux Distribution of the Main Field

(*Note:* Slot effects are exaggerated; only a few wide slots per pole are chosen.)

with the assumed infinite permeability of the iron and a smooth armature surface with the air-gap width increased to compensate for the effect of armature slots on the flux per pole. The flux-density wave corresponding to Figure 5.2.4 is shown in Figure 5.2.5* with its fundamental and third-harmonic components.

The disadvantage of the graphical methods lies in the amount of labor involved to obtain correct field pictures, necessity for frequent redrawing and the fact that each pictures holds for one individual case only. Since the hand-plotted curves must be drawn somewhat by eye, a certain amount of personal judgment and skill is involved. In regard to accuracy, hand-plotting is rather inferior.

**Adapted by permission of the Publisher:* Fitzgerald, A. et al, Electric Machinery, 3rd Ed., Ch. 4, McGraw-Hill Book Company, 1971.

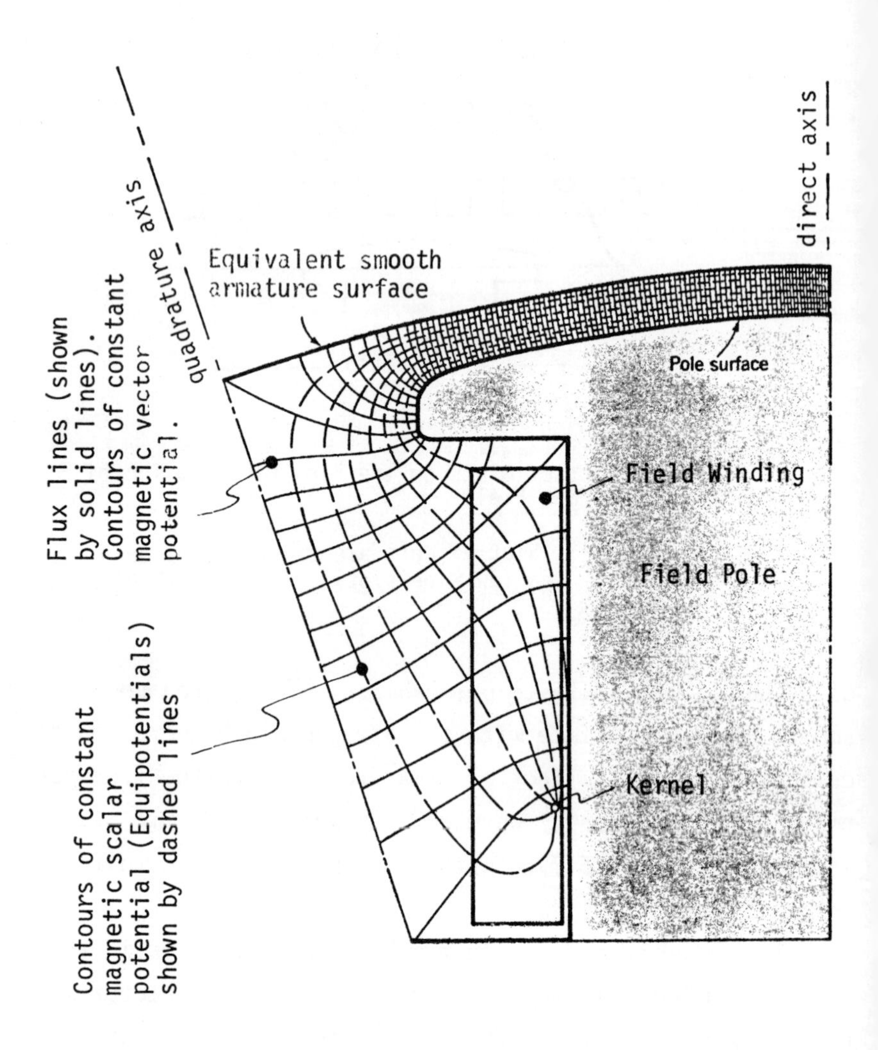

Figure 5.2.4. Nature of flux distribution around a salient pole of a synchronous machine.

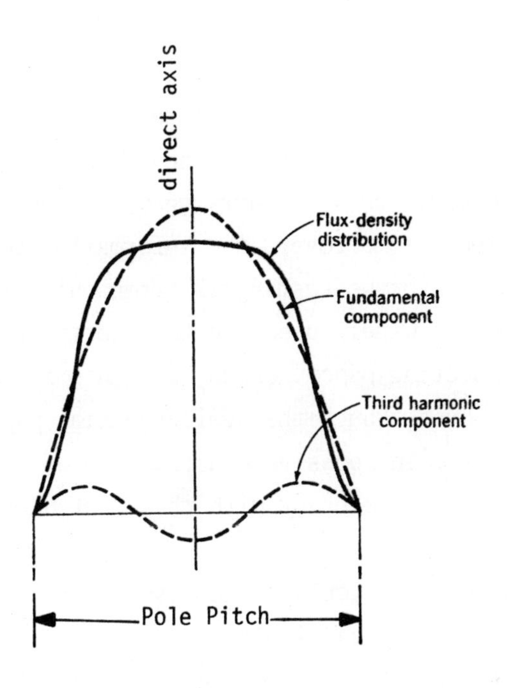

Figure 5.2.5. Typical flux-density wave of a salient-pole synchronous machine (with its fundamental and third-harmonic components).

However, for certain simple problems of translational symmetry, hand-plotting may be better on the basis of time. The ability to match the unknown fields by the familiar technique of curvilinear squares is still extremely useful when coupled with the knowledge of the field configuration obtained by other mapping methods. It is desirable to compare the different methods of approximate solution from the viewpoint of accuracy of results, labor involved, and expenditure. In this comparison, allowance need be made for the time and expense necessary to build the models wherever necessary.

Field plotting by analog models is one of the most practical methods of investigating the complicated field problems involving complex geometries, and has made free-hand plotting more or less obsolete. As an aid in industrial and scientific investigations, the analogue method is quite inexpensive compared with carrying out actual tests on a prototype model. Moreover, the information that can be obtained by prototype testing is often only limited. Two types of models widely used are the electrolytic tank and the semiconducting paper. Each model has its advantages and drawbacks. Even though the semiconducting paper lends itself to fast, inexpensive construction of a model, it cannot compete in accuracy with the electrolytic tank; but may be recommended as an effective tool for preliminary investigations, particularly in cases where the electrolytic tank is not readily available.

The electrolytic tank method is of great value in most of the electrostatic field problems involving complex geometries of the boundary shapes. It is quite suitable to represent different dielectric materials in one model. Three-dimensional models, in particular those with rotational symmetry, can be represented rather easily. Refinements have been recently introduced with the techniques of conformal mapping and double-tank arrangements. However, the initial investment is rather high for the tank, power supply, null detector, etc.; it may be difficult to build models on short notice and to change them in the course of a study, and the models themselves may be relatively expensive. But the electrolytic field analyzer is very flexible and indeed of great value in obtaining solutions of field problems with highly complex geometries. The significant feature is the ease with which the boundaries and/or boundary conditions may be altered, and the results due to these changes observed and analyzed. The application of the electrolytic field analyzer is rather limited for magnetic field problems, as nonlinearities and/or time dependency cannot possibly be adequately represented with the tank method. Magnetic fields, eddy currents and forces have been estimated by some researchers with the use of RC-network analogues based on difference techniques; the consideration of the field-strength dependent permeability is practically impossible with RC-network analogues.

Digital Computer Methods:

While the electrolytic tank method has been in common use for the past thirty years or so, it is only within the last few years that numerical methods have been greatly stimulated by the advent of high-speed digital computational robots which have made routine solutions possible to a high degree of accuracy for many types of problems for which the solution would otherwise be extremely or even prohibitively laborious. Finite-difference or finite-element methods can be used to obtain numerical solutions of any desired accuracy to almost any kind of field problems of common engineering interest. Though some theoretical considerations associated with their development are rather difficult, their application is quite easy. The main disadvantage of numerical methods is that the process of solution must be repeated for each set of parameters in a problem.

Numerical methods can be used to solve any steady-state, two-dimensional field problem (or indeed, any three-dimensional or transient one, linear or nonlinear, static or time-dependent). In many practical problems of engineering interest, the desired accuracy is achieved rather quickly by computation through the iterative process. It is often possible to save time and money by making the final solutions of one problem the starting point of the computation for a similar one. An iterative procedure becomes much more economical when a general computer program, requiring only a few changes of data, is available for the application to a class of problems. As computer speeds and storage capacities increase, and as iterative techniques improve, such general programs will be more developed and be used almost universally. In fact, even highly nonlinear field problems in two dimensions are being solved these days within a very reasonable time and expense with the aid of numerical methods and various techniques of accelerating the convergence of the solution. Formulations for solving three-dimensional field problems (linear or nonlinear) have been recently developed. However, computer storage limitations as well as the length of computational time required for reasonable convergence seem to be the present limiting factors. In fact, these are the kind of problems (such as the three-dimensional field determination in the end zone of an alternator) which cannot probably be solved by any other available method anyway. Thus, the future of the numerical methods for use in solving engineering field problems appears to be very bright.

5.3 Formulation for Computer-aided Analysis by Numerical Methods

Two examples will now be presented in order to illustrate the basic application of finite-difference and finite-element methods for the determination of flux distribution in synchronous machines. The development and details of these numerical methods are so involved that a separate book on the subject would be justified. The case of "no load" is only considered here, while the methods can easily be extended to the conditions of load. The following simplifying assumptions are made in order to permit the problem to be approached in a practical way:

(a) The generator is infinitely long in the direction of the shaft, and the magnetic vector potential has only one component along the axis of the machine. The problem is then reduced to two dimensions.

(b) The magnetic materials are isotropic and the B-H characteristics are single-valued. Hysteresis effects are thus neglected. An approximate representation of the saturation curve is done in a piecewise linear fashion.

(c) The discrete currents are replaced by uniform current density fields of the vector $\overline{J}$ over the cross-section of the armature and field coils.

(d) The magnetic field outside the contours of the machine is negligible.

The nonlinear partial differential equation (in rectangular coordinates) to be satisfied is given by

$$\frac{\partial}{\partial x}\left(\nu \frac{\partial A_z}{\partial x}\right) + \frac{\partial}{\partial y}\left(\nu \frac{\partial A_z}{\partial y}\right) = -J_z \tag{5.3.1}$$

subject to the specified boundary conditions. The flux density can be evaluated from the vector potential values as

$$B_x = \frac{\partial A_z}{\partial y} \quad ; \qquad B_y = -\frac{\partial A_z}{\partial x} \tag{5.3.2}$$

The solution of Eq. (5.3.1) yields the discrete values of the vector potential of the magnetic field. The contours of constant values of A_z are the lines of magnetic induction, and such vector equipotentials plotted for uniform increments bound flux tubes containing the same flux. The z-subscript may be dropped as we have only one component of vector potential.

Finite-Difference Method:

Figure 5.3.1 shows the cross-section of half a pole pitch of a typical salient-pole machine*, in which the cylindrical surfaces of the stator and rotor on both sides of the air gap are represented as planes. The cross-section is developed along the middle line of the air gap so that rectangular cartesian coordinates can be used. The boundary conditions to be satisfied may easily be specified from the discussion presented in Section 5.1, and it is left out as an exercise to the reader.

The two-dimensional continuum of Figure 5.3.1 is subdivided by horizontal and vertical grid lines. The vector potential is to be evaluated at the discrete points given by the intersection of the grid lines. A small area bounded by two adjacent vertical grid lines and two adjacent horizontal gridlines is denoted as a mesh. Uniform square meshes over the whole cross-section would offer great

*See Ahamed, S. V., and Erdelyi, E. A., "Nonlinear Theory of Salient Pole Machines", IEEE Trans. PAS-85, No. 1, pp. 61-70, Jan. 1966. (Figs. 5.3.1 through 5.3.5 are adapted from this publication by permission of IEEE).

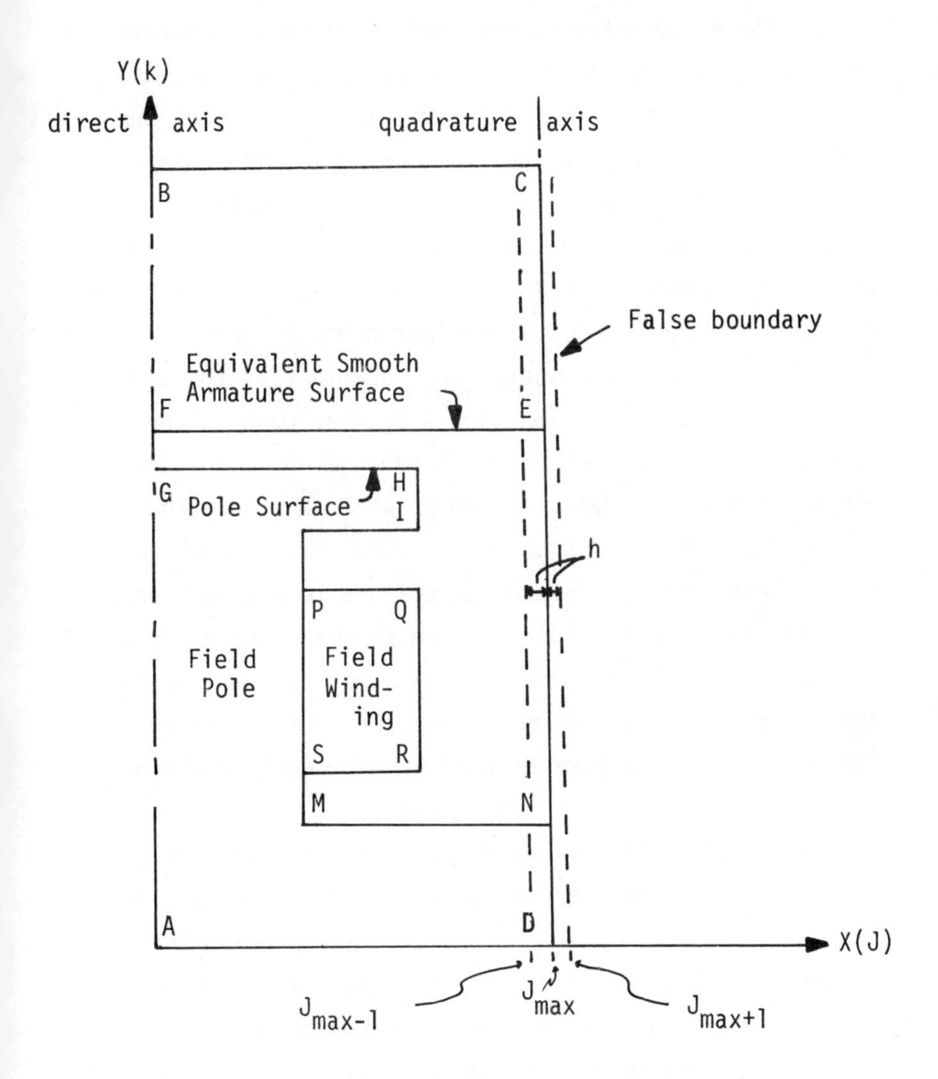

Figure 5.3.1. Typical cross-section of half a pole pitch of a salient-pole synchronous machine (in rectangular coordinates).

computational advantages, but such a scheme is not entirely possible. In order to satisfy boundary conditions, a very fine grating of the grid would have to be chosen at and near the boundaries. If such fine meshes were chosen everywhere for the sake of uniformity, the number of grid points would become prohibitively large. It becomes advisable to settle for sets of nonuniformly spaced gridlines, as shown in Figure 5.3.2. This arrangement makes it possible to choose a finer mesh where great accuracy is required to satisfy boundary conditions and a coarser mesh in the interior of the regions where the change of reluctivity is not too large. The grid system is such that each row is cut by the same columns and each column is, in turn, cut by the same rows. This procedure leads to a great simplification in the programming of the numerical solution on a digital computer. In order to take care of possible nonuniform air gaps, it becomes necessary to replace any flaring of the pole tips by a stair-type contour which closely approximates the pole shape.

The reluctivity may be defined at the center of each mesh, as being common to the area of the mesh. In order to avoid a separate location in the computer program for the reluctivity, the reluctivity of each mesh may be conveniently assigned to one of the corners of the mesh consistently. The current density may also be specified in a similar manner. As an example, referring to Figure 5.3.3, the x and y components of the flux density at the center of the rectangle $\overline{0184}$ are calculated as

$$B_x = \frac{A_1 - A_8 + A_0 - A_4}{2h_4}\,; \quad B_y = -\frac{A_1 - A_0 + A_8 - A_4}{2h_1} \tag{5.3.3}$$

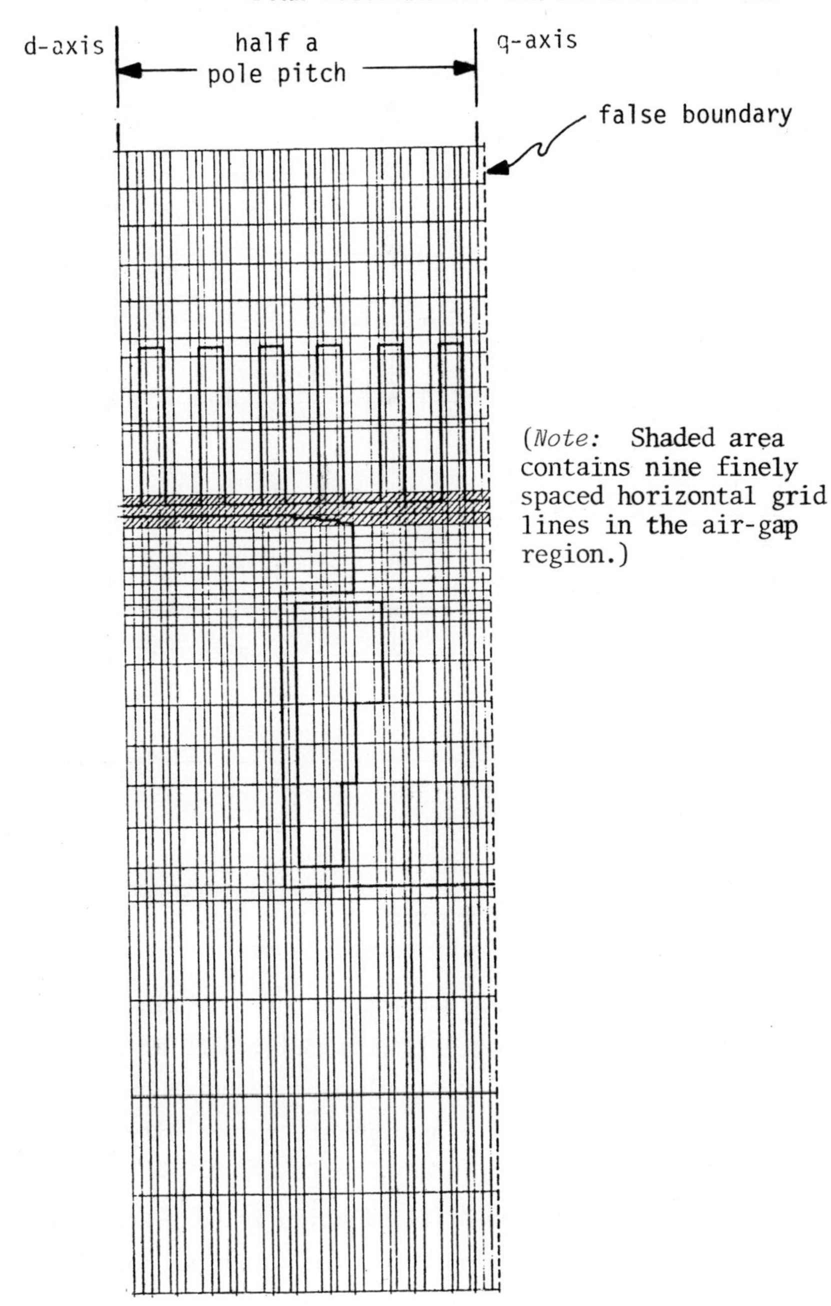

Figure 5.3.2. Grid system for numerical analysis by finite-difference method.

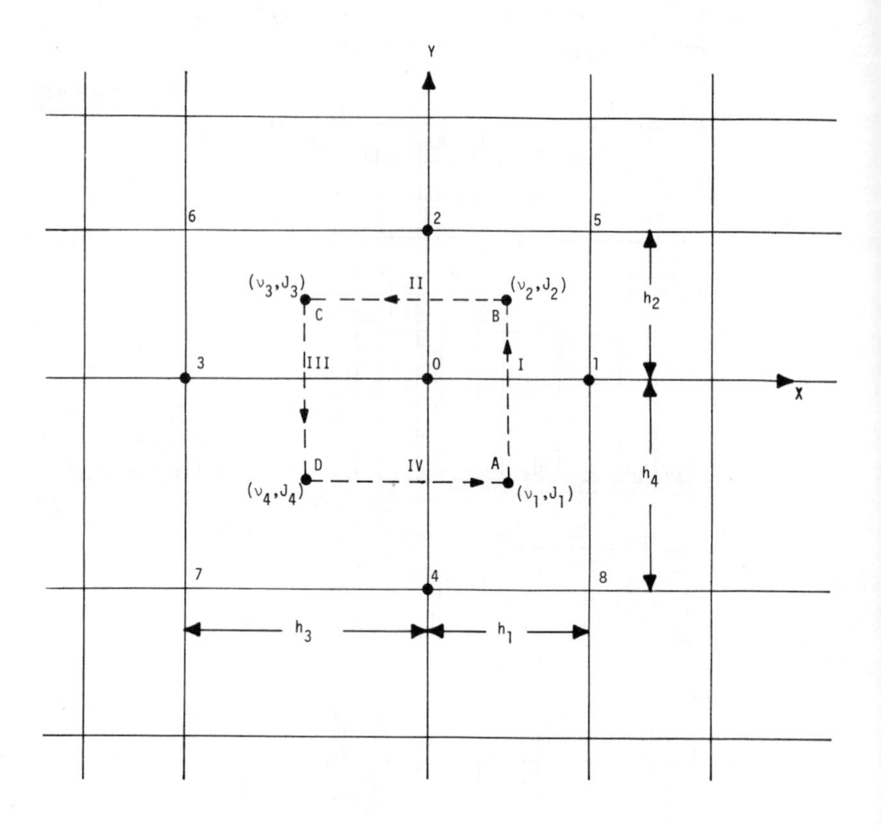

Figure 5.3.3. Typical grid point 0, its surrounding meshes and grid points.

where A_0, A_1, A_8, and A_4 are the vector potentials at the grid points 0, 1, 8, and 4 respectively. The absolute value of the flux density can then be computed and be used for the evaluation of the appropriate reluctivity in the mesh $\overline{0184}$.

The partial differential equation is next transformed into a difference equation to obtain numerical solution. In order to find the finite difference expression for the vector potential A_0 at gridpoint 0 in terms of the vector potentials A_1, A_2, A_3, and A_4 at the gridpoints 1, 2, 3, and 4, the partial derivatives in Eq. (5.3.1) are replaced by partial difference quotients, with the use of the first two terms of the Taylor's series expansion of the vector potential A around point 0. The details of this method are left out as a desirable exercise to the reader. However, the difference expression for the vector potential at a typical gridpoint 0 may also be obtained by the application of Ampere's law around the closed rectangular contour $\overline{ABCDA}$ shown in Figure 5.3.3, passing through the center points of the surrounding meshes:

$$\oint_C \overline{H} \cdot \overline{d\ell} = I \text{ (enclosed by the closed contour C)} \tag{5.3.4}$$

Eq. (5.3.4) may be expressed as

$$\int_{AB} \nu_{AB} B_{y_I} \, dy + \int_{BC} \nu_{BC} B_{x_{II}} \, dx + \int \nu_{CD} B_{y_{III}} \, dy +$$

$$+ \int \nu_{DA} B_{x_{IV}} \, dx = I_0 \tag{5.3.5}$$

where I_0 is the current enclosed by the area $\overline{ABCDA}$. Using Eq. (5.3.2), one can write

$$B_{y_I} = \frac{A_0 - A_1}{h_1} \; ; \quad B_{x_{II}} = \frac{A_2 - A_0}{h_2}$$

$$B_{y_{III}} = \frac{A_3 - A_0}{h_3} \; ; \quad B_{x_{IV}} = \frac{A_0 - A_4}{h_4} \tag{5.3.6}$$

$$
\left.
\begin{array}{l}
\int_{AB} \nu_{AB}\, dy \text{ may be replaced by } \dfrac{\nu_1 h_4 + \nu_2 h_2}{2} \\[2ex]
\int_{BC} \nu_{BC}\, dx \text{ may be replaced by } \left\{- \dfrac{\nu_2 h_1 + \nu_3 h_3}{2}\right\} \\[2ex]
\int_{CD} \nu_{CD}\, dy \text{ may be replaced by } \left\{- \dfrac{\nu_3 h_2 + \nu_4 h_4}{2}\right\} \\[2ex]
\int_{DA} \nu_{DA}\, dx \text{ may be replaced by } \dfrac{\nu_4 h_3 + \nu_1 h_1}{2}
\end{array}
\right] \quad (5.3.7)
$$

Substituting Eqs. (5.3.6) and (5.3.7) in Eq. (5.3.5), one obtains the following after some simplification:

$$
A_0 = \frac{I_0 + \sum_{i=1}^{4} \alpha_i A_i}{\sum_{i=1}^{4} \alpha_i} \qquad (5.3.8)
$$

where

$$
I_0 = \frac{1}{4} [J_1 h_4 h_1 + J_2 h_1 h_2 + J_3 h_2 h_3 + J_4 h_3 h_4] \qquad (5.3.8a)
$$

$$
\alpha_1 = \frac{\nu_1 h_4 + \nu_2 h_2}{2h_1} \quad ; \quad \alpha_2 = \frac{\nu_2 h_1 + \nu_3 h_3}{2h_2} \qquad (5.3.8b)
$$

$$
\alpha_3 = \frac{\nu_3 h_2 + \nu_4 h_4}{2h_3} \quad ; \quad \alpha_4 = \frac{\nu_4 h_3 + \nu_1 h_1}{2h_4} \qquad (5.3.8c)
$$

The gradient boundary condition along the quadrature axis, that all lines of flux must be orthogonal, is satisfied by adding a so-called false boundary denoted as gridline $J_{max + 1}$ such that equal small distances are maintained on either side of the q-axis in the x-direction, and by imposing the condition in finite difference form that

$$A_{J_{max} + 1, K} = A_{J_{max} - 1, K} \tag{5.3.9}$$

Equation (5.3.8) is applicable at all gridpoints except those on the outer boundaries, where the Dirichlet boundary condition that A is equal to zero is chosen. Thus, N simultaneous equations would be obtained for the vector potentials at N internal gridpoints. An initial guess of the vector potentials at all gridpoints needs to be made at the beginning of the numerical solution. The solution of the set of N simultaneous equations yields the numerical value of the vector potentials at the N gridpoints. Such a solution could be carried out in a number of possible iterative methods, the details and relative merits of which are outside the scope of this book for discussion. Experience has shown that it is advantageous to carry out the iterations in two steps. In the first step, the reluctivities are assumed to be constants, and the vector potentials are relaxed successively using Eq. (5.3.8) by the method known as successive point over-relaxation, in which an appropriate over-relaxation factor of about 1.3 to 1.6 has been used. In the second step, the reluctivities are recalculated using the numerical values of the vector potentials found in the first step of the relaxation. Some under-relaxation of reluctivities (with an under-relaxation factor of about 0.1) has also been found advantageous for rapid convergence of the solution. In order to accelerate the convergence of the numerical solution, other accelerating techniques such as block-relaxation and alternating-direction methods may be used. The iterative procedure is continued until the vector potentials are converged to values that are, for practical engineering purposes, near enough to final values. Special-purpose computer programs are developed in order to accomplish the optimum solution.

A typical flux-distribution plot in the cross-section is shown in Figure 5.3.4. Considerable increase in the radial magnetic induction at the toothtips may be observed, as it should be in accordance with theoretical considerations. Figure 5.3.5 shows the tangential component of the magnetic induction in the air on the quadrature axis representing the leakage between poles. The results could be utilized to predict all the normal shop-floor test results, such as the no-load saturation curve, radial air-gap induction, and the iron losses.

Finite-Element Method:

An alternate approach is to formulate the field problem in variational terms so that it is reduced to one of extremizing an energy functional rather than that of solving differential equations. This method has found wide application in elasticity problems that occur in civil engineering, but its application to the analysis of nonlinear field problems has recently been explored*. It can be

*See Chari, M. V. K., and Silvester, P., "Analysis of Turbo-alternator Magnetic Fields by Finite Elements", IEEE Trans. PAS-90, No. 2, pp. 454-64, March/April 1971. (Figs. 5.3.6 and 5.3.7 are reproduced by permission of IEEE).

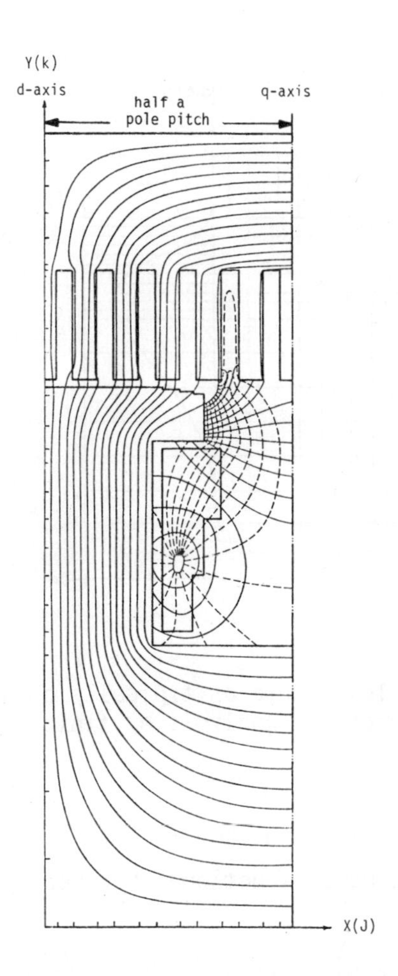

Figure 5.3.4. Flux distribution in the cross-section of Figure 5.3.1 of a salient-pole synchronous machine (contours of constant magnetic vector potential are drawn in solid lines).

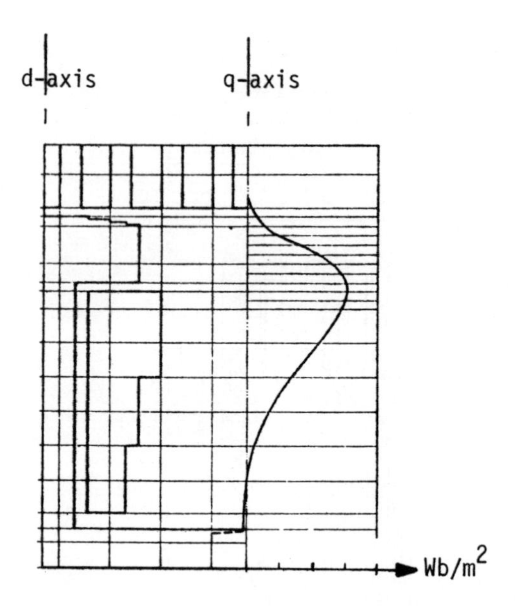

Figure 5.3.5. Pole leakage as typically obtained by numerical analysis by finite-difference method.

shown* that the solution of Eq. (5.3.1) is always such that it minimizes the energy functional F given by

$$F = \iint_R \left(\int_0^B \nu b\,db \right) dx\,dy - \iint_R JA\,dx\,dy \qquad (5.3.10)$$

where R is the area of definition of the problem. Equation (5.3.1) is the Euler equation of the functional defined by Eq. (5.3.10), satisfying homogeneous Dirichlet- and Neumann-

---

*See Silvester, P. and Chari, M. V. K., "Finite Element Solution of Saturable Magnetic Field Problems", IEEE Trans. PAS-89, No. 7, pp. 1642-51, September/October 1970.

type boundary conditions, if no others are specified. The essence of any variational method is to search directly for a function A that minimizes the functional. In order to search efficiently, however, it is first necessary to discretize the problem. Figure 5.3.6 shows the cross-section of one-quarter of a 2-pole turbo-generator with rotor and stator slots, while neglecting the wedge grooves. The boundary conditions may easily be specified from the previous discussions. The entire problem region is divided up into triangles (first-order finite elements) in any convenient fashion (as shown in Figure 5.3.6), ensuring that all iron-air interfaces coincide with triangular sides. An approximation to the vector potential solution A is assumed in each triangle, such that its value is a linear interpolate of its values at the triangular vertices. This is concisely expressed as

$$A(x,y) = \frac{1}{2\Delta} \sum_i (p_i + q_i x + r_i y) A_i \qquad (5.3.11)$$

where the index i ranges over the triangle vertices k, m, n, and the following relations hold:

$$p_k = x_m y_n - y_m x_n \; ; \quad q_k = y_m - y_n \; ; \quad r_k = x_m - x_n \qquad (5.3.12)$$

Δ represents the triangular area, and A_k, A_m, A_n are the vertex values of A. On differentiating Eq. (5.3.11), one obtains the values of flux density within the triangle as

$$\overline{B} = \frac{1}{2\Delta} \sum_k (\overline{i}_x q_k A_k + \overline{i}_y r_k A_k) \qquad (5.3.13)$$

where $\overline{i}_x$ and $\overline{i}_y$ are the unit vectors in the x- and y-

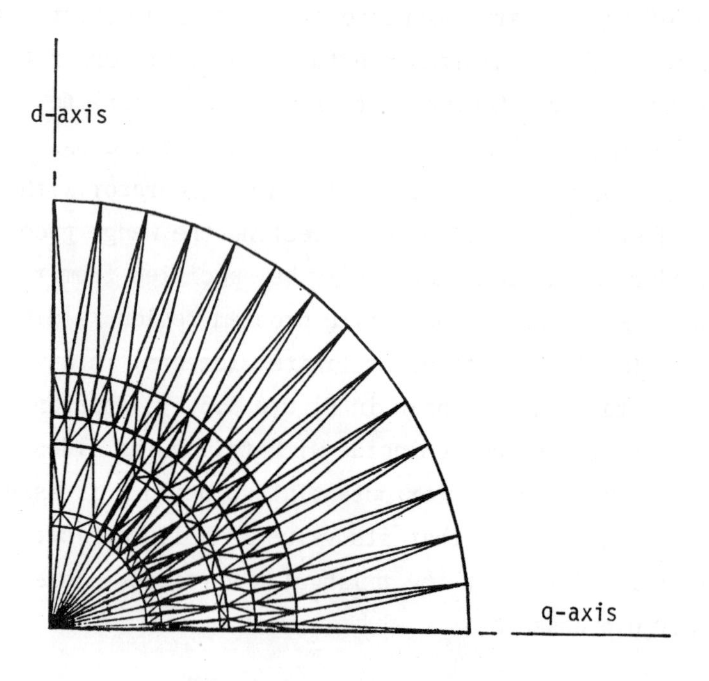

Figure 5.3.6. Typical cross-section of half a pole pitch of a 2-pole turbo-generator, subdivided into finite-element triangles.

directions respectively. The flux density is assumed to be a constant within each triangle, and hence also the reluctivity. The approximate expressions given by Eqs. (5.3.11) to (5.3.13) are valid for one triangle, namely Δkmn, only. For a complete region as shown in Figure 5.3.6, it is necessary to write equations of this form for each triangle in turn.

Minimization of the energy functional is achieved by setting its first derivative with respect to each of the vertex values of the potential to zero, so that, for all k,

$$\frac{\partial F}{\partial A_k} = 0 \tag{5.3.14}$$

It is convenient to rewrite the surface integral in Eq. (5.3.10) in such a way as to permit its evaluation over one triangle at a time:

$$F = \sum \iint_{\Delta} \left(\int_0^B \nu b \, db \, dx \, dy \right) - \iint_{\Delta} JA \, dx \, dy \tag{5.3.15}$$

the summation being taken over all the triangles in the region. A typical term in the above summation is given by

$$F_t = \iint_t \left(\int_0^B \nu b \, db - JA \right) dx \, dy \tag{5.3.16}$$

Differentiating Eq. (5.3.16) with respect to each A_k and summing over all triangles, one gets

$$\frac{\partial F}{\partial A_k} = \iint_t \left(\nu \, B \frac{\partial B}{\partial A_k} - \frac{\partial A}{\partial A_k} \right) dx \, dy \tag{5.3.17}$$

Unless k is one of the vertices of the triangle t considered, the differentiation with respect to A_k clearly produces a zero. Equation (5.3.17) is, therefore, an expression in three variables only. Substituting for B from Eq. (5.3.13) and assuming J to be constant throughout the triangle, one obtains for Eq. (5.3.17) as

$$\frac{\nu}{4\Delta} \left[\Sigma (q_k q_m + r_k r_m) \, A_k \right] = \left[\frac{\Delta}{3} J \right] \tag{5.3.18}$$

which represents a set of three equations for each triangle.

When equations similar to Eq. (5.3.18) are written for each triangle in the region of integration and the corresponding terms added, one can obtain a single matrix equation for the entire region of the form

$$[S][A] = [\,\frac{\Delta}{3} J] \tag{5.3.19}$$

where [S] is the combined coefficient matrix, [A] is the column vector of vector potentials, and [J] is the vector of current densities. The order of the matrix S is the same as the number of vertex potentials in the region. The set of nonlinear algebraic equations given by Eq. (5.3.19) are now to be solved. The sparse nature of the coefficient matrix, along with its symmetry and band properties, can be taken advantage of, and it is sufficient to store only the lower half of the matrix and solve the resulting set of equations by Gaussian elimination. Experience has shown that an iterative scheme consisting of some initial simple chord iterations along with a Newton-Raphson iterative method is very helpful for a faster convergent solution. In each iteration the resulting set of algebraic equations for a given value of ν may be solved by any method such as Gaussian elimination. The details of the iterative schemes are not presented here as they are considered to be outside the scope of this book. The interested reader is referred to the relevant literature quoted in the Bibliography. Special-purpose computer programs have been developed in order to accomplish the optimum solution.

Figure 5.3.7 shows the flux distribution in a turbo-alternator at no-load, as obtained typically from the application of the finite-element method. The results could be utilized to predict the normal shop-floor test data.

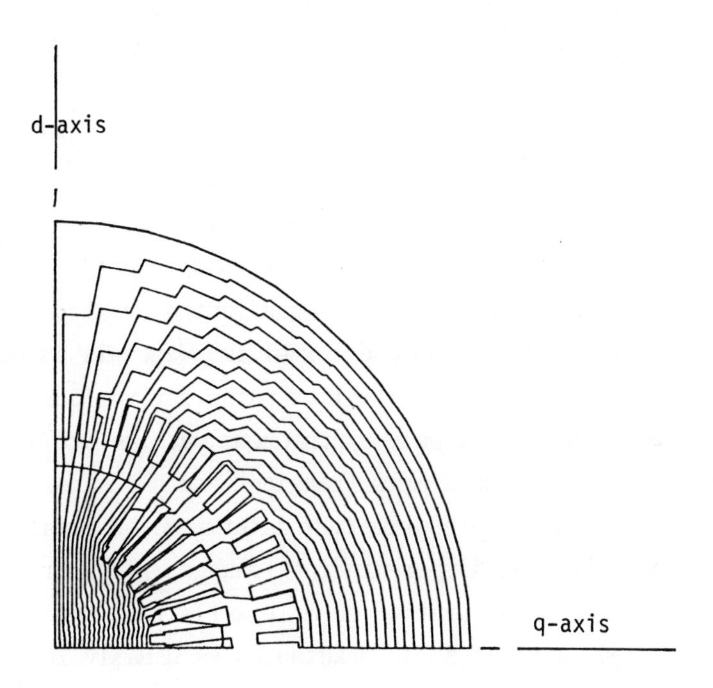

Figure 5.3.7. No-load flux distribution in the cross-section of Figure 5.3.6 of a 2-pole turbo-alternator (as typically obtained by numerical analysis by finite-element method).

Considerable development is taking place these days regarding the numerical methods of solution of nonlinear electromagnetic field problems, both in two and three dimensions. The relative merits of finite-difference and finite-element methods are yet rather unclear, as of this book's publication time, but would no doubt be resolved in the near future.

5.4 Effect of Saturation on Reactances and Regulation

When a synchronous machine is loaded, magnetic conditions are determined by the combined influence of field and armature mmfs. The open-circuit characteristic of a machine is essentially a d-axis magnetization curve. The dependence of the direct synchronous reactance x_d on saturation is greater, while the quadrature synchronous reactance x_q is less dependent on saturation. This is because the interpolar air-gap linearizes the flux distribution and remains almost independent of the load regime. The reactance x_q may be assumed to be unaffected by saturation, especially where appreciable external impedance is associated with the machine. Approximate flux paths for various reactances of a salient-pole synchronous machine are shown in Figure 5.4.1*. The d-axis transient and subtransient reactances x_d' and x_d'' are primarily determined by armature leakage and field or damper-winding leakage. They are therefore influenced by saturation to a lesser extent than is x_d. The saturated value of these reactances is usually 10 to 15 percent lower than the unsaturated value. The q-axis transient and subtransient reactances x_q' and x_q'' also vary with saturation. It may be noted here that x_q and x_q' are equal for salient-pole machines, and x_q' and x_d' are approximately equal for solid-rotor turboalternators.

The assumptions usually made are that the armature leakage reactance is a constant independent of saturation, and the saturation is a function of the resultant air-gap flux only. Thus the effects of armature leakage flux on the saturation of the armature iron and of changes in field-leakage flux on the saturation of the field iron are neglected. Also neglected are the effect of the armature mmf on the wave form of the synchronously rotating air-gap

---

*See Prentice, B. R., "Fundamental Concepts of Synchronous Machine Reactances", AIEE Trans., Vol. 56, Supplement, pp. 1-21, 1937.

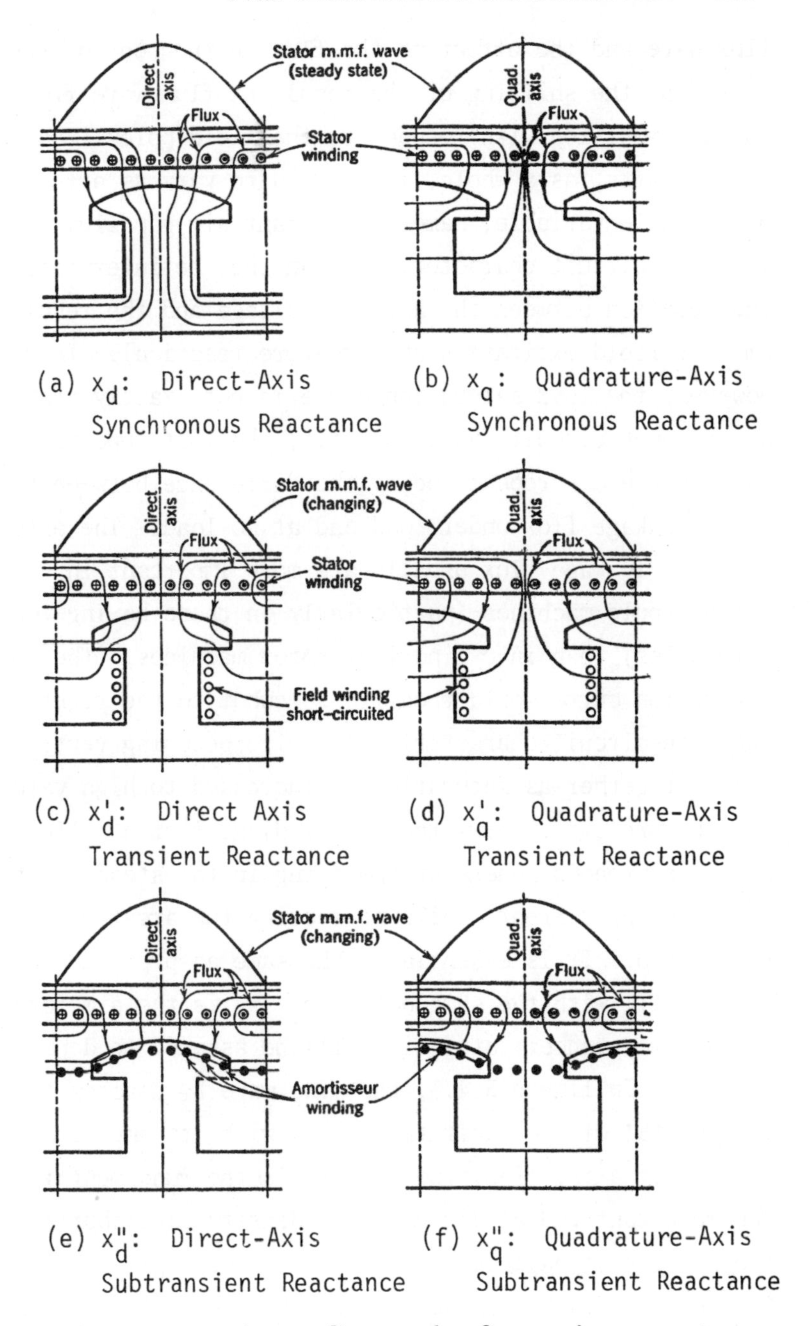

Figure 5.4.1. Approximate flux paths for various reactances of salient-pole synchronous machine.

flux wave and the effect on the flux in the body of the rotor caused by the shifting of the resultant flux wave from its no-load position with respect to the field poles. Saturation of all parts, as a whole, is assumed to vary as a function of a single variable, namely resultant air-gap flux, only. The open-circuit characteristic can then be interpreted as the relation between the air-gap voltage and the resultant mmf (of field excitation and armature reaction). In reality, however, the load saturation curve is not exactly the same as the open-circuit characteristic, the most important factor causing the discrepancy being the difference between the field-leakage flux under load and at no-load. The effects of field-leakage flux usually are more important in salient-pole machines (particularly in those having long, slim poles) than in cylindrical-rotor machines. The load saturation curve would then lie somewhat to the right of the open-circuit characteristic, while becoming vertically nearer together as saturation is increased to high values.

Figure 5.4.2 shows the vector diagram of a salient-pole synchronous generator operating in the steady state with lagging current, while neglecting the armature resistance drop. It is essentially the same as presented in Chapter II, with the same notation. x_ℓ is the armature leakage reactance; saturation will be assumed to depend upon E_p. In figure 5.4.3, the sine wave B_f denotes the fundamental of the excitation field with its amplitude in the direct axis. The sine wave B_r is the fundamental of the resultant radial air-gap flux density distribution in the air-gap at balanced loads. The space angle ψ shown in Figure 5.4.3 is the angle between the center line of the phase belt carrying the instantaneous maximum current and the direct axis, and is also known as the internal power-

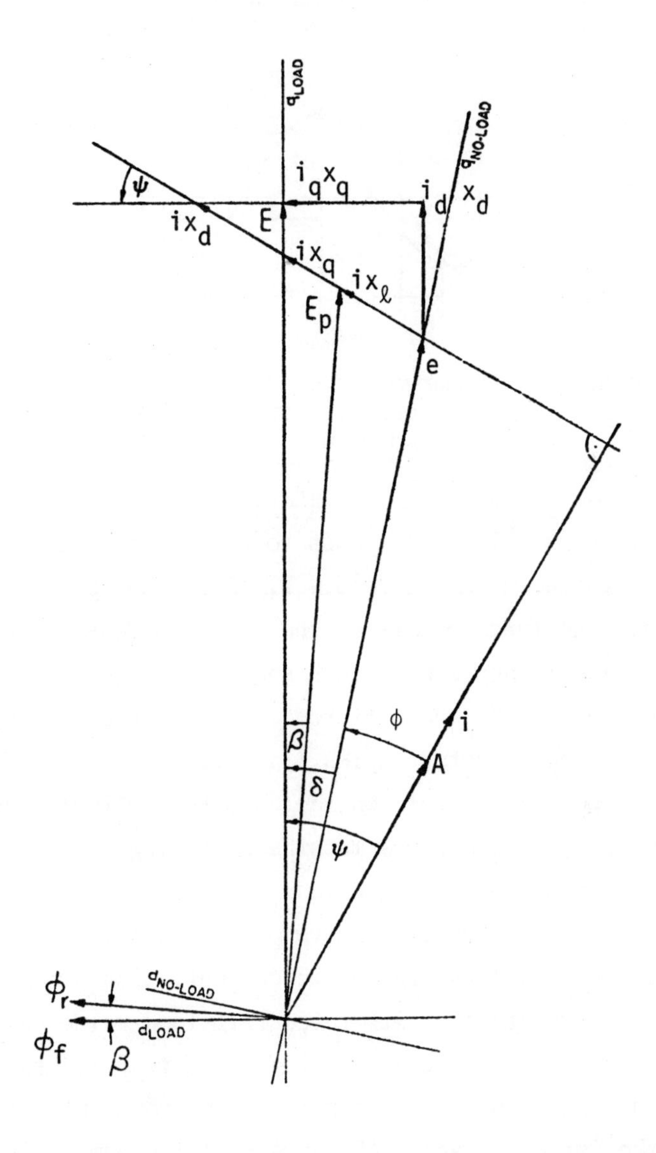

Figure 5.4.2. Vector diagram of a salient-pole synchronous generator operating in the steady state with lagging current (armature-resistance drop neglected).

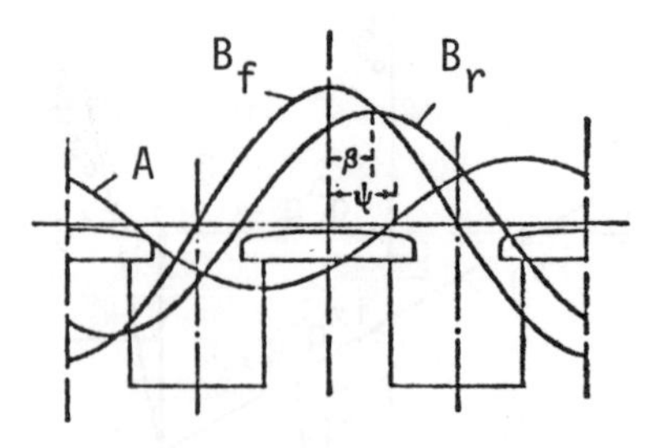

Figure 5.4.3. Fundamentals of spatial armature-mmf (A) and flux-density waves corresponding to no-load (B_f) and resultant on load (B_r).

factor angle. It corresponds to the phase angle between E and i, as shown in Figure 5.4.2. β is the space angle between the fundamentals of the B_f and B_r waves, and is also the same angle between E and E_p of Figure 5.4.2. δ can be seen to be the difference between the internal and external power-factor angles ψ and ϕ.

A special case of considerable theoretical importance is that of a synchronous machine operating at zero or near-zero power factor, overexcited. The zero-power-factor rated-current saturation curve* is a graph of armature terminal voltage vs. the field current, with armature current held constant at its rated full-load value and power-factor angle constant at 90° lag. Figure 5.4.4 shows the no-load and zero-power-factor rated-current saturation curves, and the Potier triangle ABC or A'B'C' (so named after its inventor). With the assumptions that the leakage reactance is a constant, the armature resistance is negligibly small,

---

*See Kimbark, E. W., Power System Stability; Synchronous Machines, Ch. XII, Dover Publications, Inc., New York, 1956/1968. (Figures 5.4.4 and 5.4.5 are adapted from this publication).

and the open-circuit characteristic is also the load saturation curve, the zero-power-factor characteristic would be a curve of exactly the same shape as the no-load saturation curve shifted vertically downward by an amount equal to the leakage-reactance voltage drop and horizontally to the right by an amount equal to the armature-reactance mmf. O'A' in Figure 5.4.4 is parallel to OA, which is a part of the air-gap line. The Potier reactance x_p, as determined from the vertical side (AC or A'C' = $x_p i$) of the Potier triangle replaces the leakage reactance x_ℓ and the voltage behind Potier reactance E_p (obtained by the vector combination of e and $jx_p i$ as shown in Figure 5.4.5) is the air-gap voltage, of which saturation is assumed now to be a function. E of Figure 5.4.2 is proportional to the field current of an unsaturated machine. E_s, the correction for saturation, is a function of E_p as shown in Figure 5.4.4 and would be parallel to E_p in the vector diagram. E_f is a vector summation of the voltages E and E_s. However, for convenience, E_s is sometimes taken parallel to E itself rather than to E_p as shown in Figure 5.4.5, the theoretical justification of which is possibly greater with the salient-pole machine than with the non-salient-pole machine. E_f, a fictitious armature voltage, is now proportional to the field current of a saturated machine. Figure 5.4.5 shows the vector voltage diagram of a saturated salient-pole synchronous generator in the steady state, while that of a non-salient-pole synchronous generator is left to the reader's imagination. The per-unit Potier reactance x_p is of the order of 0.15 for the case of turboalternators and of 0.32 for water-wheel generators as well as synchronous condensers.

The regulation of a synchronous generator at any given power factor is the percentage rise in voltage, under the

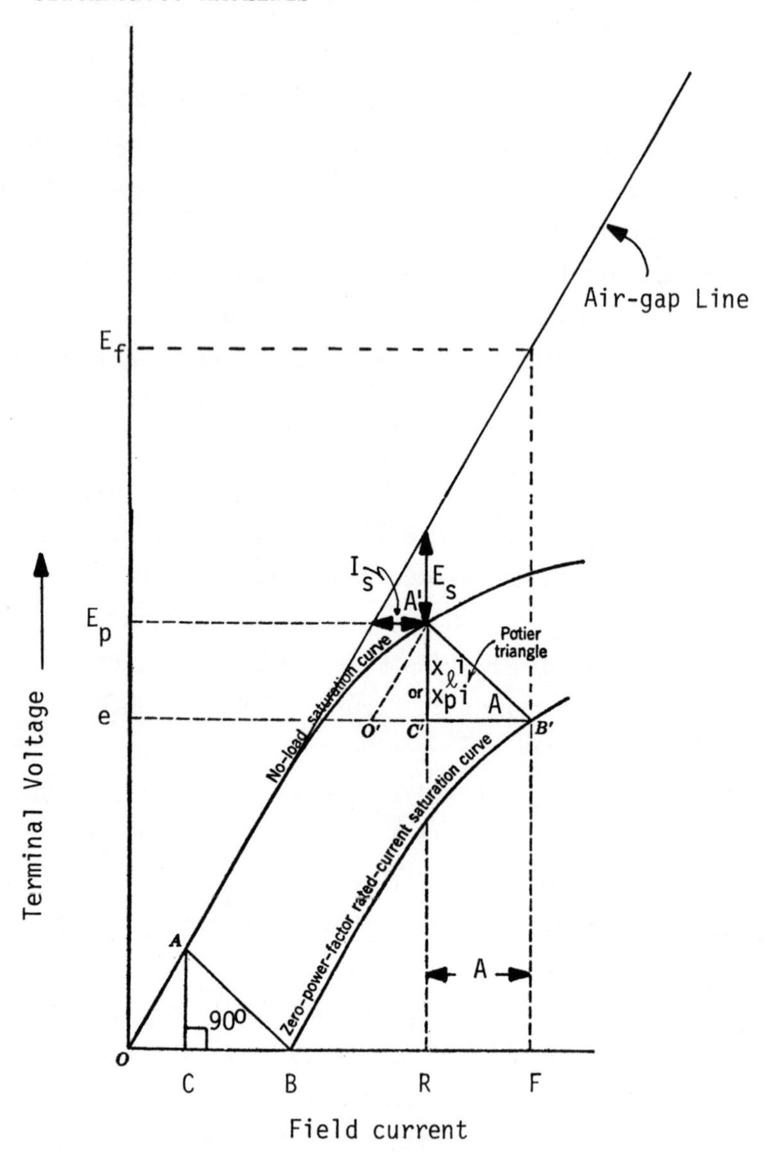

(*Note:* E_p in general is a vector combination of e and $jx_p i$ as shown in Figure 5.4.5)

Figure 5.4.4. No-load and zero-power-factor rated-current saturation curves of a synchronous machine and the Potier Triangle.

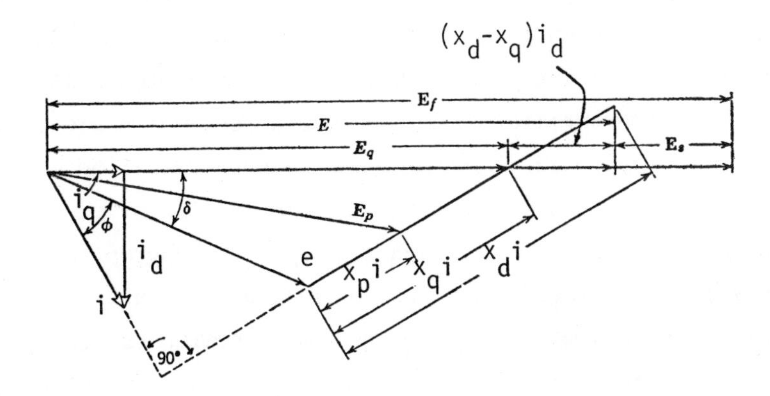

Figure 5.4.5. Steady-state vector-voltage diagram of a salient-pole synchronous generator, while taking effect of saturation into account approximately.

conditions of constant excitation and frequency, when the rated kva load at the given power factor is removed. The regulation is positive for both a noninductive and an inductive load, while it may be negative for capacitive loads, if the angle of lead is sufficiently great. Since it is now an almost universal practice to operate synchronous generators with automatic voltage regulators to hold the voltage constant under varying loads, a knowledge of the voltage regulation of an alternator is of less importance than the knowledge of the change in the field excitation which is required to maintain constant voltage under varying loads. However, both involve the same factors as can be seen from the discussion presented above in this section.

Saturation is an important factor in steady-state stability studies and will be treated in further chapters. However, it is usually neglected in transient stability problems. The effect of saturation can approximately be taken care of by modifying suitably the vector voltage diagrams developed earlier for an unsaturated machine. An alternate and apparently more straightforward approach is to evaluate the appropriate saturated reactances, which are in fact functions of load. However, such predetermination of the saturated reactances is unfortunately very involved and rather complicated, as it depends upon the appropriate flux distribution inside the nonlinear saturated machine. Attempts have been made recently by some researchers* to do just the same by obtaining various flux plots and computing the relevant flux linkages. The saturated direct synchronous reactance has been found from a simulated static zero power factor condition ($\cos \psi = 0$) and the saturated quadrature synchronous reactance from a static unity power factor condition ($\cos \psi = 1.0$). Field methods have also been extended to find the steady-state short-circuit current in synchronous generators, taking saturation into account, and saturated transient as well as subtransient reactances through simulation of dynamic conditions. The procedure is relatively complicated and will not be treated here.

5.5 Relation Between Field Theory and Circuit Theory

In engineering practice it is generally required to find solutions to problems in a convenient form that are interpretable and easy to visualize. This accounts for the

*See Fuchs, E. F., "Numerical Determination of Synchronous, Transient and Subtransient Reactances of a Synchronous Machine", Ph.D. Thesis, Univ. of Colorado, Boulder, Colorado, 1970; Siegl, M., "Transient and Subtransient Behavior of Homopolar Inductor Alternators", Ph.D. Thesis, Univ. of Colorado, Boulder, Colorado, 1972.

almost exclusive use of the circuit concept as distinct from the field concept. Although all phenomena occur in space and time, and these are field phenomena, it is possible to approximate them by time relations only, summarizing the actual physical constants by means of parameters, preferably of constant value. However, it is becoming increasingly evident that a re-examination of the basis of the circuit concept and its replacement by the more refined methods of field theory should be considered. Thus, field theory as opposed to circuit theory is becoming increasingly important in many engineering problems. Fortunately, the fundamental laws of circuit theory can be obtained from Maxwell's equations by applying an approximation called the quasi-static field approximation.

The conventional linear circuit analysis of synchronous machines starts with output quantities such as terminal voltage, load current, and power factor; the nonlinear field analysis, unfortunately, has to start with the input quantities such as field excitation, armature currents and the internal power-factor angle. Recent advances made in the computation of the flux distribution inside the synchronous machinery reveal a fundamental change in outlook by necessity that it is the $\overline{B}$ field which is determined to satisfy a given $\overline{J}$ field. An accurate calculation of the flux distribution, including the effects of iron saturation, make it possible to understand and appreciate better the physical phenomena taking place inside the electromagnetic device, and paves the way for improving design optimization processes and decreasing the need for use of design safety factors.

The two-reaction theory will certainly retain its proper place as one of the available linear mathematical techniques for the analysis of synchronous machines. Better understanding of the fundamental phenomena will be gained substantially by moving in the right direction towards the determination of the flux distribution at various conditions of operation. However, field theory methods of analysis seem to be very involved (as judged from the present state of the art) so that average power-system engineers cannot consider such complicated methods, and even the designer of low rating machines may not be able to apply such methods because of economic considerations. Thus there is need to make the field-theory methods of analysis accessible for everyday design use. Considering the many different materials used in the design of alternators, different operating conditions and saturation of the ferromagnetic materials, computations and compilation of a book of generally useful curves (such as plots of the flux distribution coefficients, including the effects of saturation as a function of major machine parameters) would be a long and costly task, which will undoubtedly be attempted in the near future. Such curves would have the advantage of presenting the results on a generalized basis in terms clearly understood by machine designers, so that they may serve as a bridge to greater familiarity with the calculation procedure and lead to their more ready acceptance. While two-dimensional fields are quite easy to be plotted, much work needs to be done in the meaningful representation of three-dimensional flux distribution.

Another important area that needs immediate attention of the researchers appears to be the establishment of clear and correct interpretation of the results of the field-theory analysis in terms of the circuit concepts and conventially-used terminal parameters, and vice versa. It is only then the full impact and advantage of the recent field-theory analyses would be felt.

Calculation of Inductance:

A familiar fundamental definition of the inductance of a single turn coil is given by the flux linkages per ampere of the current in the turn:

$$L = \lambda/I \tag{5.5.1}$$

λ is the flux which links, or is enclosed by, the current loop. It may be expressed as a surface integral involving the flux-density vector:

$$\lambda = \iint_S \overline{B} \cdot d\overline{s} \tag{5.5.2}$$

For closed loops, by replacing $\overline{B}$ as curl of $\overline{A}$ and using Stoke's theorem* it can be shown that

$$L = \frac{1}{I} \oint_C \overline{A} \cdot d\overline{\ell} \tag{5.5.3}$$

where C is the closed contour enclosing the open surface S. The inductance for a loop with a small gap may be approximated as

$$L = \frac{1}{I} \int_C \overline{A} \cdot d\overline{\ell} \tag{5.5.4}$$

where the line integral is evaluated along the entire length

*See Appendix F for Vector Identities.

of the loop. If a coil is increased to N turns of fine wire very closely spaced, a good approximation is

$$L = N\lambda/I \tag{5.5.5}$$

The inductance for a single loop is called self-inductance because the flux which links the conductor is produced by the current in the conductor.

When more than one loop are present, the flux produced by current in one loop may link another loop, thus inducing a current in that loop. This is called mutual inductance. Let us consider two loops with currents I_1 and I_2; let B_{12} be the field produced at loop 1 by I_2 and B_{11} be the field produced at loop 1 by I_1. Then one has

$$\lambda_{12} = \iint_{S_1} \bar{B}_{12} \cdot d\bar{s} = \oint_{C_1} \bar{A}_{12} \cdot d\bar{\ell} \tag{5.5.6}$$

$$\lambda_{11} = \iint_{S_1} \bar{B}_{11} \cdot d\bar{s} = \oint_{C_1} \bar{A}_{11} \cdot d\bar{\ell} \tag{5.5.7}$$

and the mutual inductance is given by

$$L_{12} = \lambda_{12}/I_2 \tag{5.5.8}$$

where λ_{12} is the magnetic flux produced by the current I_2 which links the current loop I_1. Positive sign is assigned to L_{12} if the flux λ_{12} links loop 1 in the same direction as the flux λ_{11}, or equivalently, if $(\bar{B}_{12} \cdot d\bar{s})$ has the same sign as $(\bar{B}_{11} \cdot d\bar{s})$. By changing the direction of either I_1

or I_2, the sign of the mutual inductance can be changed.

Since inductance is directly related to the magnetic field, an expression for the inductance may be obtained from the considerations of stored energy in the magnetic field. For the calculation of the field inductance of a linear synchronous machine, such a definition of inductance given below by energy relations can be seen to be particularly useful.

$$L = \frac{1}{I^2} \iiint_V (\overline{B} \cdot \overline{H}) \, dv \tag{5.5.9}$$

where dv is the differential volume. Replacing $\overline{B}$ by the curl of $\overline{A}$ and employing vector identities*, one obtains

$$L = \frac{1}{I^2} \left[\iiint_V \nabla \cdot (\overline{A} \times \overline{H}) \, dv + \iiint_V \overline{A} \cdot (\nabla \times \overline{H}) \, dv \right] \tag{5.5.10}$$

After applying the divergence theorem* to the first integral and using one of the Maxwell's equations giving the current density $\overline{J}$ by the curl of $\overline{H}$, the inductance may be written as

$$L = \frac{1}{I^2} \left[\oint\!\!\oint_S (\overline{A} \times \overline{H}) \cdot d\overline{s} + \iiint_V (\overline{A} \cdot \overline{J}) \, dv \right] \tag{5.5.11}$$

It can be shown that the contribution due to the surface integral goes to zero in cases where the region of the field extends to infinity and where the current sources are small in volume compared with the whole region. Even if the

---

*See Appendix F for Vector Identities.

region of integration is finite as in the case of electromechanical energy converters, the contribution of the surface integral becomes usually negligible compared with that of the volume integral due to the nature and size of the terms involved. In particular, for the models in which $\bar{A}$ and $\bar{H}$ are made zero on the bounding surface S, it is obvious that the surface integral vanishes trivially. The inductance of the arrangement may then be obtained simply as

$$L = \frac{1}{I^2} \iiint_V (\bar{A} \cdot \bar{J}) \, dv \tag{5.5.12}$$

It is thus only necessary to evaluate the volume integral of the scalar product of the vector potential and current density in order to calculate the inductance. For cases where the current density is uniform and the model is only two-dimensional, further simplifications result.

If ferromagnetic materials are present in a magnetic circuit, no single definition of saturation-independent reactance can be given. However, as an engineering approximation, the nonlinear relation between the excitations and the resultant fluxes may be assumed to be linearized in a small neighborhood of the operating point. For example, linearization may be accomplished by assuming that the reluctivity in each finite-difference mesh or a finite-element triangle calculated for an operating condition remains constant in a small region around the operating point. The total flux linkages may then be obtained by adding the flux linkages of each turn, the turns being considered as filaments because the cross-section of a turn is small in comparison to the area of the coil. The inductance may then be evaluated from the concept of the appropriate flux linkages per ampere.

Problems

5-1. Consider a magnetostatic field problem in rectangular-coordinate two dimensions. Assuming that no current-density fields are present and the permeability of the ferromagnetic material is a function of the magnetic induction, develop the nonlinear field equation (of the elliptic form) in terms of a scalar potential V. Show that it reduces to the familiar Laplace's equation for the case of constant permeability. In case current-density fields do exist, discrete currents need to be replaced by equivalent fictitious current sheets in order to solve the problem with the use of scalar potential. Discuss the appropriate boundary condition to be satisfied in such a case.

5-2. Discuss the possible types of boundary conditions to be satisfied in terms of the scalar potential V as applied to the current-free regions of electrical machines; you may consider only two-dimensional models in rectangular coordinates for convenience and simplicity.

5-3. Consider Eq. (5.1.13) in conjunction with Eq. (5.1.14). Write explicitly the corresponding three-component coupled equations in cartesian coordinate system. Define the gradient of ϕ in such a way that the component equations get uncoupled and a typical resultant component equation is of the form given below:

$$\left[\frac{\partial}{\partial x}\left(\nu\,\frac{\partial A_i}{\partial x}\right) + \frac{\partial}{\partial y}\left(\nu\,\frac{\partial A_i}{\partial y}\right) + \frac{\partial}{\partial z}\left(\nu\,\frac{\partial A_i}{\partial z}\right)\right] = \sigma\,\frac{\partial A_i}{\partial t}$$

Explore whether such a procedure is feasible in cylindrical coordinates to uncouple the coupled component equations.

5-4. Formulation in terms of the magnetic vector potential $\overline{A}$ (and electric scalar potential ϕ) has been presented in the text (Section 5.1) for problems including nonlinear magnetic media. Proceeding on similar lines, obtain the formulation in terms of electric vector potential $\overline{F}$ (and magnetic scalar potential Ω) for problems including nonlinear electric materials; the conductivity σ may be taken as a single-valued function, being dependent on $\overline{J}$ and $\overline{E}$, to allow for the electric nonlinearity, and constant permeability of the medium may be assumed.

Consider then the particular case of a linear medium for which the conductivity is independent of $\overline{E}$ and $\overline{J}$, and specialize the field equation to be satisfied. Check whether it is possible to alleviate the calculation of the potential Ω and to formulate in terms of $\overline{F}$ alone.

5-5. Discuss the possible types of boundary conditions to be satisfied in terms of the vector potential $\overline{A}$ as applied to Figure 5.3.1 showing the cross-section of half a pole-pitch of a typical salient-pole machine (in rectangular coordinates), for which the no-load field distribution is sought.

5-6. In two-dimensional fields, show that the contours of constant magnetic vector potential A are the flux lines.

5-7. Obtain Eq. (5.3.8) in difference form from the partial differential equation (5.3.1) by replacing the partial derivatives with partial difference quotients, while using the first two terms of the Taylor's series expansion of the vector potential A around point 0 in Figure 5.3.3.

5-8. The true functional formulation for the nonlinear electromagnetic field problem is that for which the Euler equation yields the nonlinear partial differential equation (5.3.1). Show from the principles of variational calculus that the correct solution of Eq. (5.3.1) is always such that it minimizes the energy functional F given by Eq. (5.3.10).

5-9. Compute the field current required for a power factor of 0.80 leading current when a 45-KVA, 3-phase, Y-connected, 220-volt (line-to-line), 6-pole, 60-Hz synchronous machine is running as a synchronous motor at a terminal voltage of 230-volts and with a power input to its armature of 45-KW. Given that the unsaturated value of the synchronous reactance (based on cylindrical-rotor theory) is 0.92 per-unit, Potier reactance of 0.227 per-unit, effective armature resistance of 0.040 per-unit, and E_s, the correction for saturation, corresponding to E_p of 1.2 per-unit (see Figure 5.4.4) is 0.45 per-unit. Further, it is given that 1.00 per-unit excitation is 2.40 field amperes, when unit excitation is defined as the value corresponding to unit voltage on the air-gap line of the open-circuit characteristic of the machine.

5-10. A 4-pole, 60-Hz, 3-phase, star-connected, 13,200-volt (line-to-line), 93,750-KVA turbogenerator is delivering rated load at 0.8 power-factor lagging. The Potier reactance of the generator is given to be 0.248 ohm per phase; effective armature resistance is 0.00402 ohm per phase; unsaturated synchronous reactance is 2.13 ohms per phase; E_s, the correction for saturation, corresponding to E_p of 8,280-volts per phase (see Figure 5.4.4) is 1,155-volts per phase; a field current of 425 amperes corresponds to the rated voltage on the air-gap line of the open-circuit saturation curve of the machine.

Determine the field current required for the given operating condition of the turbo-alternator.

5-11. Consider a long straight wire and square loop shown in Figure 5.5.1 on the next page.
Obtain an expression for the mutual inductance L_{12}.

5-12. Starting from Eq. (5.5.9), justify the expression given for the inductance in Eq. (5.5.11) by giving through the procedure indicated in the text.

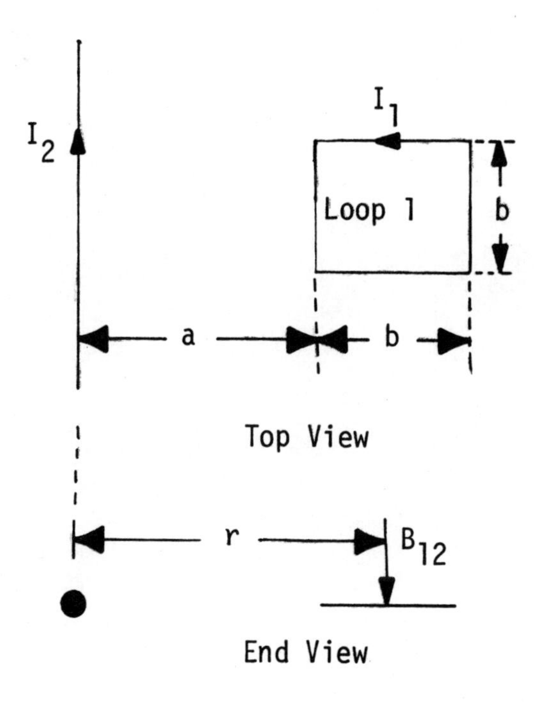

Figure 5.5.1

CHAPTER VI

STABILITY STUDIES

6.1 Parallel Operation of Interconnected Synchronous Generators

In order to ensure the uninterrupted supply of electrical energy within set limits of frequency and voltage to various loads scattered over a wide geographical area, it becomes necessary to operate several synchronous generators in parallel, interconnected by many transmission lines. With the current rise in the demand of electrical energy, present-day power systems which are large and complex will continue to grow both in size and complexity.

Modern power systems represent the largest man-made system in existence from the viewpoint of invested capital and offer a great challenge to a systems engineer. Power-system engineers should be able to perform studies such as load-flow, economic dispatch, fault analyses and system stability. The availability of the computer is drastically changing the role of the human operator and the computer is taking over most of the decision-making job. Computers play an ever-increasing role in power-systems engineering in both on-line and off-line modes of operation. As more and more of the various sub-controls have been made automatic, man's role has changed accordingly.

A typical power system may require thousands of state variables for adequate description. The changing character of the system on a daily and yearly scale, the vulnerability to environmental influences of statistical nature in view of the geographical spread, difficulties in data transmission, impossibility to conduct meaningful full-scale system tests due to the 24-hour operation, the mixed nature (continuous as well as discrete) of control inputs, power-system reliability not attaining the 100% level, and the range of system dynamics over a very broad bandwidth (line transients of the order of milli- and micro-seconds; short-circuit phenomena of the order of milliseconds to seconds; generator-rotor swings and frequency transients of the order of seconds to minutes) are but some of the unique features of a modern power system. The analysis and control of such power systems are many-faceted complex problems, including several operational actions which have as their final objective to maintain the system in its normal operating state and deliver uninterrupted supply of electrical power within set limits of frequency and voltage levels.

Since parallel alternators must run at exactly the synchronous steady-state speed corresponding to the system frequency and the number of poles of each generator, speed-power characteristics of their prime movers (usually of drooping nature) decide almost wholly the way in which the active power gets divided amidst them. The system frequency and the division of active power amidst the generators can be controlled by means of the prime-mover throttles. If the prime-mover throttle is not shifted, the real power output as well as the in-phase component of the armature current of the corresponding alternator will not be changed. Changes in excitation, controlled by means of the field rheostats, affect the terminal voltage and reactive-power distribution among the generators.

It is a common practice to control the prime-mover throttles by governors and automatic frequency regulators so that the system frequency is maintained within certain specified close limits of the rated frequency and optimum division of real power amid the generators is achieved for the economic operation of the power system. Voltage regulators controlling the generator field circuits and automatic tap changing devices on transformers usually regulate automatically the terminal voltage and the reactive-power distribution.

In terms of interconnected synchronous generators, the machines must maintain synchronism during normal as well as abnormal operating conditions. The property of a power system which ensures that it will remain in operating equilibrium under both normal and abnormal conditions is known as power system stability. Steady-state stability is concerned with slow and gradual changes, while transient stability is concerned with severe disturbances such as sudden changes in load and fault conditions. The maximum flow of power possible through a particular point without the loss of stability is known as steady-state stability limit when the power is increased very gradually, while it is referred to as transient stability limit when a sudden disturbance occurs.

Steady-state stability analysis is commonly based on the steady-state equations of the machines, while neglecting the effect of excitation system and voltage regulator action for small disturbances. Such an approach is justified as the majority of the older excitation systems had long enough time delays and sufficient deadband. However, modern excitation systems have been designed without deadband and as such their equations must also be included for small as well as severe disturbances along with the generator equations. The analysis of system behavior following even a small disturbance, including the proper representation of the excitation system equations, is sometimes referred to as dynamic stability analysis. Transient stability analysis involves some of both electrical and mechanical properties of the machines of the system, because the machines must adjust the relative angles of their rotors after every disturbance to meet the conditions of power transfer imposed.

Transient stability studies are usually limited to relatively short time intervals, typically of the order of 1 second or less. They are most often used to determine the stability of a single unit or plant during the initial period of high stress immediately following a nearby fault. Such studies may require the representation of a large system, but primarily to secure accuracy rather than to obtain total system response. Generators, excitation systems, and speed-governing systems are represented in detail in the modern computer-aided transient-stability studies. Dynamic stability studies cover longer real-time intervals, typically of the order of 5 to 10 seconds and even up to 30 seconds occasionally. While attention may be focused on a single unit or plant, these studies are frequently made to obtain large system response. For such analyses where interest centers on the damping of oscillations, dynamic models for generator flux, excitation systems, speed-governing systems, and other apparatus become essential.

To study the performance of one machine connected to a large system, the system may be considered to have constant voltage and frequency thereby assuming an infinite bus at the point of connection of the machine. Thus a multi-machine system can sometimes be reduced to the equivalent of a two-machine problem. The factors affecting the stability of a two-machine system, or the stability of one machine connected to an infinite bus, are the same as those which influence a multi-machine system. The detailed study of multi-machine stability analysis is considered to be outside the scope of this book.

6.2 Steady-State Stability

The power system is subjected to disturbances which are small enough and gradual enough so that the system can be regarded electrically as being in the steady state, and as an approximation the steady-state equations of the machines can be utilized for the steady-state stability analysis, while neglecting the effect of the excitation system. Such an approach is justified in the case of generators equipped with non-continuously acting automatic voltage regulators which cannot act in time to keep the machine stable because of the inherent dead-bands and sufficiently long time delays. The ability of interconnected synchronous machines to remain in synchronism will be analyzed under small and gradual disturbances occurring on the system, without representing the excitation system behavior.

The steady-state power angle characteristics discussed in Section 2.3 may be recalled at this stage. The effect of resistance is usually neglected. When a machine is connected through appreciable impedance to other machines in the system, as is most often the case, the saliency effects can be neglected for all practical calculations. In terms of steady-state stability, the effect of saturation in machines is usually more significant than saliency effects.

For the generator connected to a very large system compared to its own size, the system in Figure 6.2.1 may be used and the corresponding positive-sequence impedance diagram

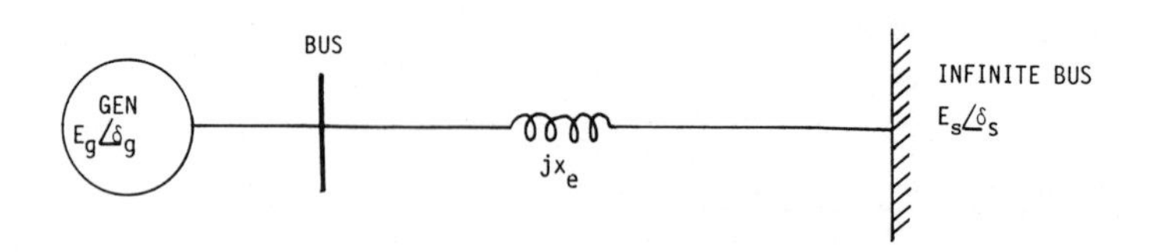

Figure 6.2.1. Generator connected to an infinite bus.

is shown in Figure 6.2.2, assuming x_d to be equal to x_q and neglecting the resistance. The load angle δ is the angle between the terminal voltage and the q-axis, for the case of a single machine as shown in Figure 2.1.1. However, for a system where several machines are interconnected, it is more convenient to measure an angle between the quadrature axis of each machine and a common reference axis which rotates at synchronous speed at all times. Let δ_g and δ_s associated respectively with voltages E_g and E_s of Figure 6.2.1 be such angles measured from a common reference axis. The power-angle equation for the system under consideration becomes

$$P = \frac{E_g\, E_s}{x} \sin \delta_{gs} \tag{6.2.1}$$

where δ_{gs} is the angular difference between δ_g and δ_s. The power-angle characteristic given by a sine wave of Eq. (6.2.1) is plotted in Figure 6.2.3. Let us now examine the operating conditions at three points *a*, *b* and *c* shown in Figure 6.2.3.

At the operating point *a*, the slope of the curve $dP/d\delta_{gs}$ is positive. If there is a small increase of the angle δ_{gs},

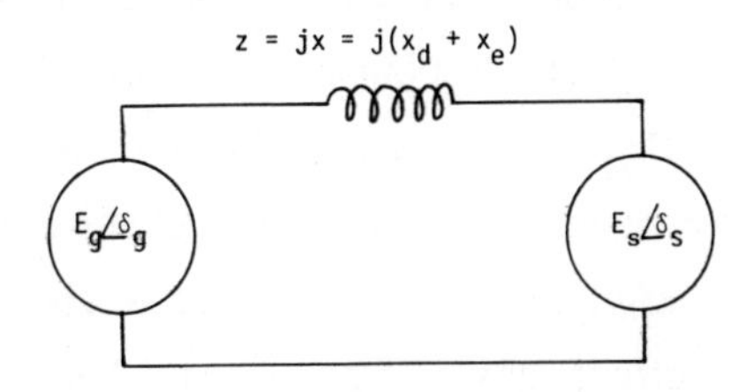

Figure 6.2.2. Positive-sequence impedance diagram corresponding to the system shown in Figure 6.2.1.

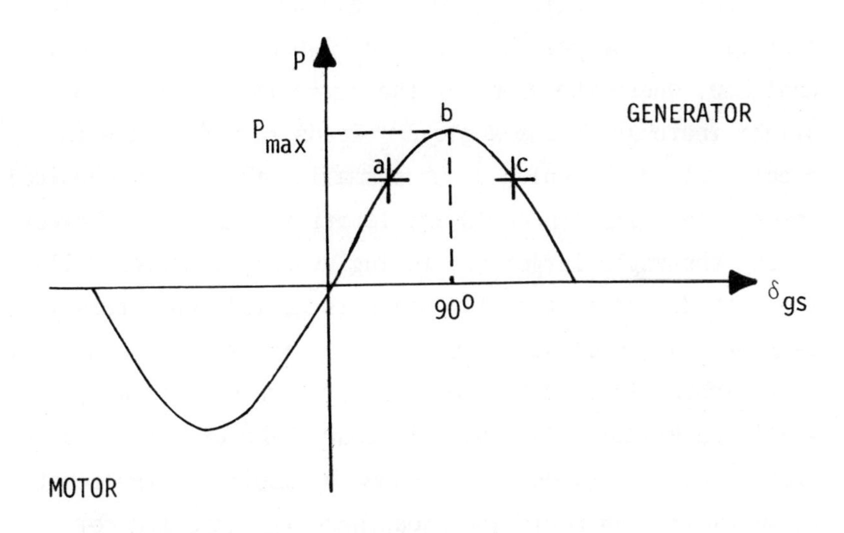

Figure 6.2.3. Power-angle curve of Eq. (6.2.1)

the electrical power can be seen to increase and will exceed the mechanical power input, if the mechanical power input corresponding to the initial power input at the operating point *a* remains constant. This results in deceleration, which will make the angle approach the initial operating point *a*. Then the machine is said to be stable. The peak of the power-angle curve given by P_{max} is known as the steady-state power limit, representing the maximum power that can be theoretically transmitted in a stable manner. A zero-slope may be observed corresponding to such an operating condition shown at point *b* in Figure 6.2.3. A machine is usually operated at less than the power limit, thereby leaving a steady-state margin, as otherwise a slight increase in the angle δ_{gs} would lead to instability. The operation at the point *c* shown in Figure 6.2.3 is clearly unstable, where the slope of the curve is negative. A slight increase in the angle δ_{gs} leads to a decrease in electrical power, which becomes smaller than the mechanical power. This in turn would accelerate the machine, thereby making the angle larger and losing synchronism eventually.

Steady-state stability analysis neglecting saturation leads to unduly pessimistic estimates regarding the ability of a machine in a system to remain in synchronism when small and gradual disturbances occur. The effect of saturation is usually taken into account by modifying the reactance x_d suitably. An equivalent reactance x_{eq} adjusted for saturation is estimated and the representation by a voltage behind x_{eq} amounts to replacing the actual saturated machine by an equivalent unsaturated machine which will have the same behavior at its terminals for small and gradual changes in the system. The saturated synchronous reactance may be obtained from

$$x_{eq} = x_1 + \frac{x_{s(ag)} - x_1}{k} \tag{6.2.2}$$

where $x_{s(ag)}$ is the unsaturated value of the synchronous reactance; k is a saturation factor as a function of air-gap voltage, given by R/R_{ag} or $E_{r(ag)}/E_r$ shown in Figure 6.2.4; and x_1 is the leakage reactance which may be replaced by the Potier reactance x_p when the open-circuit characteristic is assumed to be the saturation curve under load. This procedure amounts to the same as taking x_{eq} as the reciprocal of the short-circuit ratio, which is often used as an approximate value of the equivalent reactance of turbogenerators, especially for underexcited operation. Techniques for estimating the proper value of x_{eq}, discussed in texts on stability, involve not only the saturation characteristics of the generator but also the characteristics of the system of which the generator is a part, and the generator terminal conditions. The value of x_{eq} usually lies within the range 0.4 to 0.8 of the unsaturated synchronous reactance, smaller values being attributed to the newer and larger turbogenerators. It is also a common practice for overexcited operation to choose x_{eq} as $(0.8x_d)$ for light loads and $(0.6x_d)$ for full load.

The slope $(dP/d\delta_{gs})$ of the power-angle curve gives a measure of the steady-state stability of a system, and this is most frequently used to check the steady-state stability of synchronous machines in a system. The test is made between pairs of machines which are electrically most remote from each other, since this is the most likely situation to give rise to instability. If the slopes of both machines are positive, the system is stable; if either slope is negative, the system is unstable. This criterion leads to

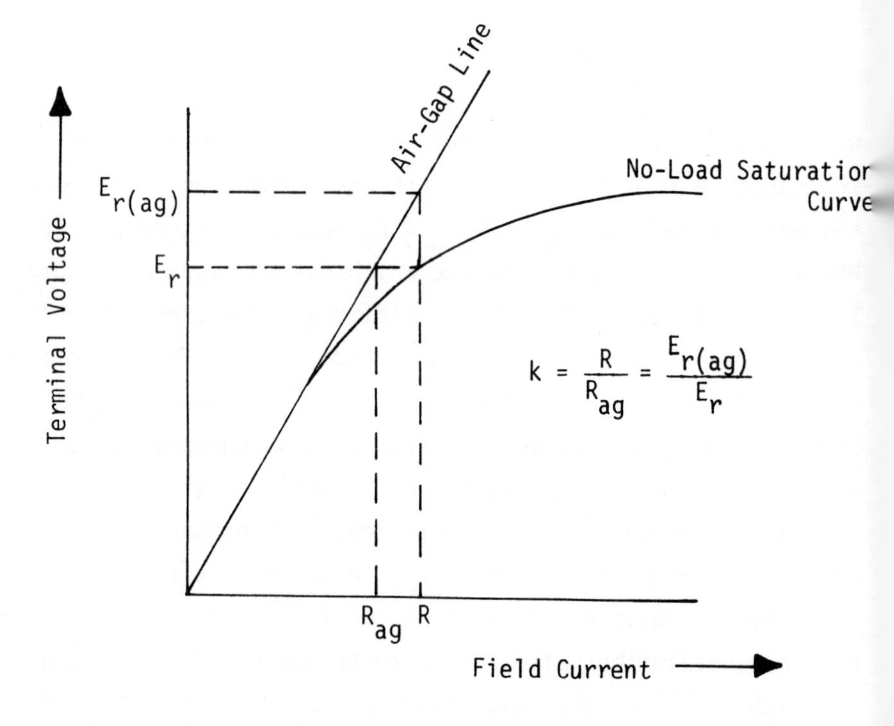

Figure 6.2.4. Saturation factor as a function of air-gap voltage.

conservative estimates, particularly when one of the tested machines is very much larger than the other machine. Refined methods, presented in texts on stability, may then be used.

The steady-state stability limits of a system may be increased either by an increase in the excitation of the generator or by a reduction in the reactance of the network, as can be seen from Eq. (6.2.1). Installation of parallel transmission lines would contribute towards an increase in the stability limit and also increased dependability of the system. Series capacitors are sometimes used in lines either to improve voltage regulation or to raise the stability limit by effectively decreasing the line reactance. Continuously-acting type automatic voltage regulators can raise the steady-state stability limits considerably. The value of x_{eq} may be modified to approximate the effect of the voltage regulator. For excitation systems with rotating exciters, the modified x_{eq} lies between the x_{eq} without the effect of the voltage regulator and the transient reactance x'_d. Steady-state stability with automatic regulators is promoted by low transient reactance of the machines. The continuously-acting voltage regulators may be represented mathematically by differential equations and the overall performance including the regulator, excitation system, generator and system transient characteristics may be evaluated by solving the equations for small changes.

6.3 Transient Stability

Transient stability is concerned with the operation of an interconnected power system subjected to severe disturbances such as faults, switching of circuits, and sudden changes in load. Both electrical and mechanical systems will obviously be in a transient state. The duration of the fault depends on the relay and circuit-breaker characteristics, and it is usually of the order of 10 cycles or less for systems with high speed fault clearing.

The usual assumptions regarding the representation of a synchronous machine that are made for the transient stability analysis are given below: (a) The effects of speed variation will be neglected and the speed will be treated as a constant by setting $(p\theta)$ equal to unity in per-unit notation; (b) The electromagnetic torque T_e is taken to be equal to the electric power developed or the air-gap power P_e for a generator, and the mechanical torque input is assumed to be a constant; (c) Only the fundamental-frequency voltages and currents are considered, and as such symmetrical component theory may be utilized for analyzing unbalanced fault conditions.

The effects of saturation are either neglected or approximately taken care of. The transient direct-axis reactance is sometimes taken to be $(1.1\ X'_{dv})$, where X'_{dv} is the rated-voltage transient reactance as determined when the initial terminal voltage is made equal to the rated terminal voltage. Rated-voltage reactances may appropriately be used for short-circuit studies. When the initial terminal voltage of the machine is adjusted so that the transient current is equal to the rated current upon application of the fault, the reactances determined are known as rated-current values; and these are generally utilized in the analysis of voltage-rise and dip problems.

Amortisseur effects may either be neglected or approximately considered in the transient stability studies. The damping torque due to the slip-frequency currents induced in the rotor circuits is usually beneficial to help the machine regain synchronism by damping out speed variations of the machine. It would be conservative to neglect its effect. Since the principal period of interst in transient stability analysis is after about 3 cycles and most often less than 60 cycles or 1 second, amortisseur effects are not of real significance during the initial period after a fault.

The transient saliency is usually neglected. The effect of transient saliency neglecting damper circuits is shown in Figure 2.1.1 and is briefly discussed in Section 2.1 under "*Field-flux linkage*". The relationships at any given instant of time may be represented by Figure 2.1.1, and constant field-flux linkages may be assumed by treating the magnitude of E'_q (ON of Figure 2.1.1) as a constant. Neglecting transient saliency, one may utilize E' (OM of Figure 2.1.1) as the constant voltage behind transient reactance. While this is usually satisfactory for over-excited operation, it may lead to optimistic stability analysis for under-excited operation. Assuming the field-flux linkage to remain constant, the transient power-angle (or torque-angle) characteristic shown in Figure 2.3.2 is given by Eq. (2.3.12), repeated here for convenience:

$$P = \frac{E'_q\ e}{x'_d} \sin \delta + \frac{x'_d - x_q}{2x'_d\ x_q} e^2 \sin 2\delta \qquad (6.3.1)$$

If E' is used in place of E'_q, neglecting transient saliency, one has

$$T_e = \frac{E'\ e}{x'_d} \sin \delta \qquad (6.3.2)$$

where e is the terminal voltage, and δ is now the angle between E' and e. The assumption of constant field-flux linkages usually leads to satisfactory results for severe disturbances such as three-phase faults near a machine, which are cleared in about six cycles. However, the assumption tends to be conservative for less severe faults.

Variation of field-flux linkage, excitation response, speed changes, and saturation effects can be adequately considered in the step-by-step solution procedure, whenever needed. In particular, let us take a further look regarding the variations of field-flux linkage, in what follows.

Starting from the basic field-circuit equation given by

$$e_{fd} = p\lambda_{fd} + r_{fd}\, i_{fd} \tag{6.3.3}$$

one can obtain the following after multiplying both sides by (x_{ad}/r_{fd}):

$$\frac{x_{ad}}{r_{fd}} e_{fd} = E_{fd} = E + T'_{do}(pE'_q) \tag{6.3.4}$$

where E_{fd} is the exciter voltage referred to the armature circuit; E is $(x_{ad}\, i_{fd})$ as in Eq. (2.1.7); E'_q is given by either Eq. (2.1.31) or (2.1.34); and T'_{do} is the field open-circuit time constant expressed as in Eq. (3.1.16). E_{fd} may either be a constant or be a known function of time, varying due to the voltage-regulator action independent of the other variables in Eq. (6.3.4). Multiplying both sides of Eq. (6.3.4) by (E'_q/E), one gets

$$pE'_q = \frac{\left(\frac{E'_q}{E}\right) E_{fd} - E'_q}{\left(\frac{E'_q}{E_q}\right) T'_{do}} \tag{6.3.5}$$

which may be expressed in the following way for numerical integration with improved accuracy, while replacing the derivative by the average slope:

$$\frac{\Delta E'_q}{\Delta t} = \frac{\frac{E'_q}{E} E_{fd} - E'_q}{\left(\frac{E'_q}{E}\right) T'_{do} + \frac{\Delta t}{2}} \tag{6.3.6}$$

where E_{fd} is evaluated at time $(t + \frac{\Delta t}{2})$. Equation (6.3.6) may then be rearranged by dividing with (E'_q/E) resulting in

$$\frac{\Delta E'_q}{\Delta t} = \frac{E_{fd} - E}{T'_{do} + \left(\frac{E}{E'_q}\right)\frac{\Delta t}{2}} \tag{6.3.7}$$

which shows that the time constant is increased by

$$\left\{\left(\frac{E}{E'_q}\right)\frac{\Delta t}{2}\right\}$$

and in which E_{fd} is computed at time $(t + \frac{\Delta t}{2})$. More elaborate integration schemes are available, if they are found to be necessary.

The assumptions that can be made regarding a salient-pole machine in the order of increasing accuracy are listed below:

(a) The voltage behind direct-axis transient reactance (OM of Figure 2.1.1) may be assumed constant.
(b) The field-flux linkages (proportional to ON of Figure 2.1.1) may be assumed constant.
(c) Variations of field-flux linkages and voltage-regulator action, if any, may be taken into account.

For the case of a solid cylindrical round-rotor machine, the assumptions that can be made are given below:

(a) The voltage behind direct-axis transient reactance may be assumed constant.
(b) The flux linkages of the rotor circuits on both direct and quadrature axes may be assumed constant.
(c) Assuming the quadrature-axis rotor circuit to be open, the variations of field-flux linkages in the direct-axis rotor circuit as well as the voltage-regulator action, if any, may be considered.

(d) Variations of field-flux linkages in both axes and the effect of voltage-regulator action, if any, on the direct-axis linkage may be accounted for.

The Swing Equation

The per-unit mechanical acceleration equation is given by

$$p^2\theta = \frac{1}{H'} (T_m - T_e) \qquad (6.3.8)$$

where p is the operator $\frac{d}{dt}$; θ is the angle in electrical radians between the d-axis and the centerline of the phase-a axis; H' is the per-unit inertia constant in $\frac{\text{KW - radians}}{\text{KVA}}$; T_m represents the per-unit mechanical torque input; and T_e stands for the per-unit electromagnetic torque developed. Since θ is continuously changing with time, it is more convenient to measure the angular position with respect to a synchronously rotating reference axis. θ may then be expressed as

$$\theta = t + \delta \qquad (6.3.9)$$

where δ is the angle between the reference axis and the quadrature axis of the machine. Equation (6.3.8) may then be rewritten as

$$p^2\delta = \frac{1}{H'} (T_m - T_e) \qquad (6.3.10)$$

As it is more common to express the angle in degrees and time in seconds, the acceleration equation may also be given by

$$p^2\delta = \frac{180f_b}{H}(T_m - T_e) \qquad (6.3.11)$$

where p is the operator $\frac{d}{dt}$, with t in seconds; δ is in electrical degrees; f_b is the base rated frequency; H is the per-unit inertia constant in $\frac{KW - sec}{KVA}$; and T_m as well as T_e are the per-unit values. The torque caused by friction, windage, and core loss in a machine is usually disregarded. Since the per-unit electrical torque is equal to the per-unit air-gap power developed, one may rewrite Eq. (6.3.11) as

$$p^2\delta = \frac{180f_b}{H}(P_m - P_e) \qquad (6.3.12)$$

where P_m is the per-unit shaft power input (minus rotational losses, if they are to be considered); and P_e represents the per-unit electrical power developed (electrical power output of the machine plus the armature copper losses). While the equations are developed here with a generator in mind, one may easily modify them appropriately for the case of a motor. Quite often in the literature, the quantities involved are expressed in their natural units without the use of per-unit notation, in which case Eq. (6.3.12) becomes:

$$p^2\delta = \frac{1}{M}(P_m - P_e) \qquad (6.3.13)$$

where p is the operator $\frac{d}{dt}$, with t in seconds; δ is in electrical degrees; M is the angular momentum given by $(\frac{GH}{180f_b})$ kilojoule-seconds per electrical degree; G is the rating of the machine in kilovoltamperes; H is in $\frac{KW - sec}{KVA}$; f_b is the base rated frequency in hertz; P_m and P_e are expressed in kilowatts. Equation (6.3.12) or (6.3.13) is known as the swing equation. It may be remarked here that the inertia constant defined as angular momentum at synchronous speed is truly a constant, while the angular momentum of a machine will not be a constant if the angular velocity is changing. However, M may be treated as a constant since the speed of the machine does not differ much from the synchronous speed unless the stability limit is exceeded. The shaft power input P_m is usually taken as a constant for simplifying the calculations, although it is possible in digital-computer programs to take account of the governor action and hence the change in input from the prime mover. The electrical power developed P_e is of the form given by Eq. (6.3.2). The solution of Eq. (6.3.13) yields the swing curve, which is a graph of δ as a function of time t. The swing equation in the form of Eq. (6.3.13) will be used for further discussion in this section.

Formal analytical solution of the swing equations is almost impossible for a multimachine system. Elliptic integrals result even for the simple case of one machine connected to an infinite bus, when the resistance is neglected and P_m is taken to be zero. Transient stability studies are usually done with the aid of digital computers utilizing numerical integration schemes, and even then a point-by-point solution is attempted. If the swing curve indicates that the angle δ starts to decrease after reaching a maximum value, it is usually assumed that the system will not lose stability and that the oscillations of δ around the equilibrium point will become successively smaller and eventually be damped out. It is possible in certain cases to make use of the equal-area criterion of stability in order to gain an understanding of the transient stability conditions without formally solving the swing equation. The equal-area criterion cannot be used directly in systems where three or more machines exist and the assumption of an infinite bus is not valid.

Equal-Area Criterion for Transient Stability:

For each of two interconnected machines, the acceleration equations are given by:

$$p^2\delta_1 = \frac{P_{a1}}{M_1} ; \qquad p^2\delta_2 = \frac{P_{a2}}{M_2} \tag{6.3.14}$$

in which subscripts 1 and 2 refer to the interconnected machines, and P_a is the accelerating power given by $(P_m - P_e)$. The above may be rearranged as

$$p^2\delta_{12} = p^2(\delta_1 - \delta_2) = \frac{P_{a1}}{M_1} - \frac{P_{a2}}{M_2} \tag{6.3.15}$$

Multiplying the above by $(2p\delta_{12})$ and integrating both sides, one can obtain

$$(p\delta_{12})^2 = 2\int \left(\frac{P_{a1}}{M_1} - \frac{P_{a2}}{M_2} \right) d\delta_{12} \tag{6.3.16}$$

The relative speed bewteen the two machines becomes zero when $(p\delta_{12})$ equals zero, which then forms the basis of the equal-area criterion. The machines will not remain at rest with respect to each other the first time $(p\delta_{12})$ becomes zero; but the fact that δ_{12} has momentarily stopped changing may be taken to indicate stability. This is equivalent to the assumption that the swing curve indicates stability when the angle δ_{12} reaches a maximum and starts to decrease.

For the case of one machine connected to an infinite bus, the equal-area criterion is given by

$$\int_{\delta_o}^{\delta_s} \frac{2P_{a1}}{M_1} d\delta_{12} = 0 \quad \text{or} \quad \int_{\delta_o}^{\delta_s} P_{a1}\, d\delta_{12} = 0 \qquad (6.3.17)$$

in which δ_o is the angle prior to the disturbance when the machine is operating at synchronous speed, and δ_s is the angle after the disturbance when the machine is again operating at synchronous speed and the angle ceases to change. M_2 corresponding to the infinite bus is infinite in view of the infinite inertia of that system. Figure 6.3.1 shows the generator connected to an infinite bus, as well as the equivalent circuit of the system for transient stability study. The accelerating power P_{a1} is given by

$$P_{a1} = P_m - P_e \qquad (6.3.18)$$

where P_m is usually assumed to be a constant equal to the value P_{eo} prior to the disturbance, and P_e is given by the corresponding power angle equation as

$$P_e = \frac{E'_q\ E_s}{(x'_d + x_e)} \sin\delta_{12} = P_{max} \sin\delta_{12} \qquad (6.3.19)$$

which is shown in Figure 6.3.2.

Let us now consider a solid three-phase fault to occur at the generator terminals at $t = 0$, and to have been removed at a time $t = t_s$, without any alteration in the reactance x_e. The generator is isolated from the system during the fault period. The shaded area A_1 of Figure 6.3.2 represents the energy which tends to increase the rotor speed

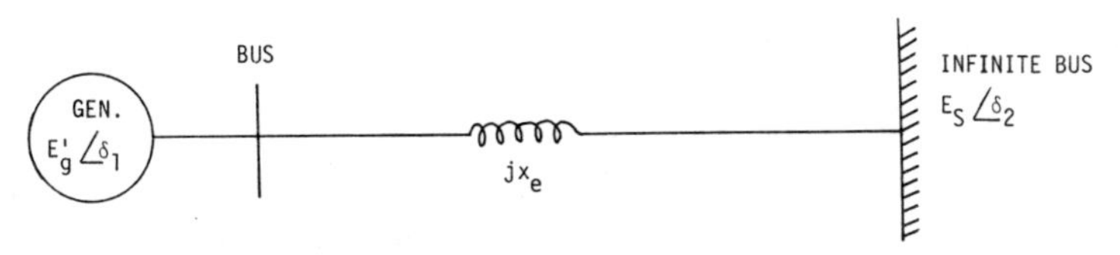

(a) Generator connected to an Infinite Bus

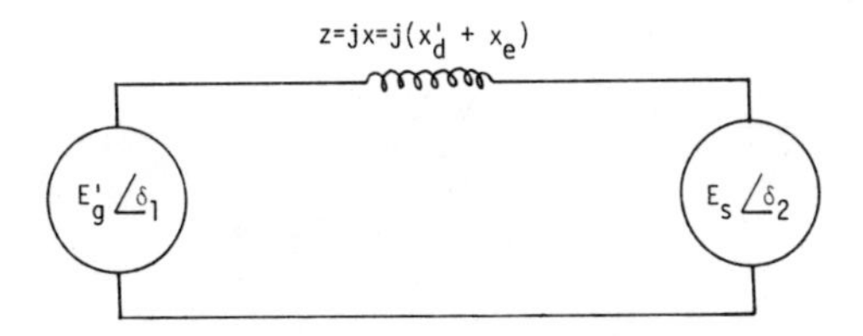

(b) Equivalent Circuit of the System for Transient Stability Study

Note: 1. The angles δ_1 and δ_2 are measured from a common reference axis.

2. E'_g is the voltage behind transient reactance of the generator.

3. Resistance has been neglected here.

Figure 6.3.1.

and cause loss of synchronism between the generator and the infinite bus, while the shaded area A_2 represents the energy which tends to stabilize and restore synchronism. The algebraic sum of the areas producing stability is given by

$$A_2 - A_1 = \int P_{a1}\, d\delta_{12} \tag{6.3.20}$$

When the area A_2 is greater than the area A_1, the machine is transiently stable; when A_2 is less than A_1, the machine is transiently unstable. The limiting case between being

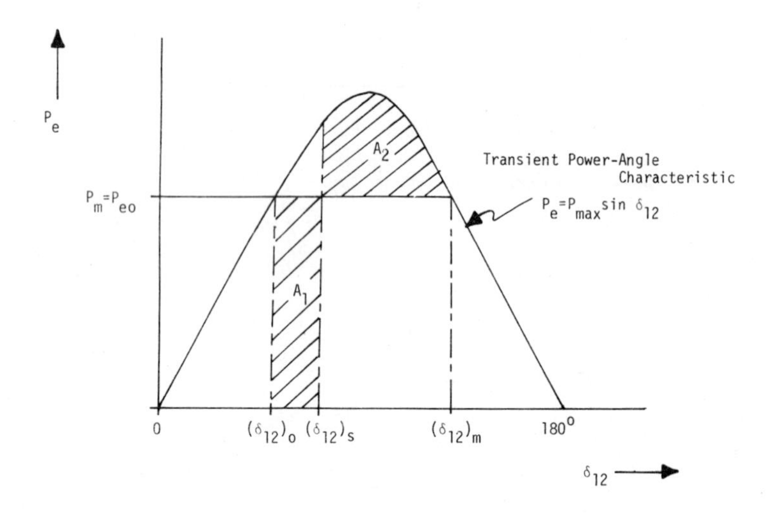

Figure 6.3.2. Equal-area criterion for transient stability of a generator connected to an infinite bus, for critical clearing time.

stable or unstable arises when A_2 is just equal to A_1.

The equal-area criterion may also be easily applied to find the value to which the input power could be suddenly increased without loss of synchronism, for the case of a synchronous generator (connected to an infinite bus) initially operating under steady state conditions while delivering power P_o shown in Figure 6.3.3. When areas A_1 and A_2 are equal in Figure 6.3.3, P_s represents such a value.

It is possible to transmit some power even during the fault, the extent of which depending on the nature of the fault. Such a case is easily analyzed with the equal-area criterion as shown in Figure 6.3.4. Let P_m be the mechanical input from the prime mover; (P_{max} sinδ_{12}) be the electric

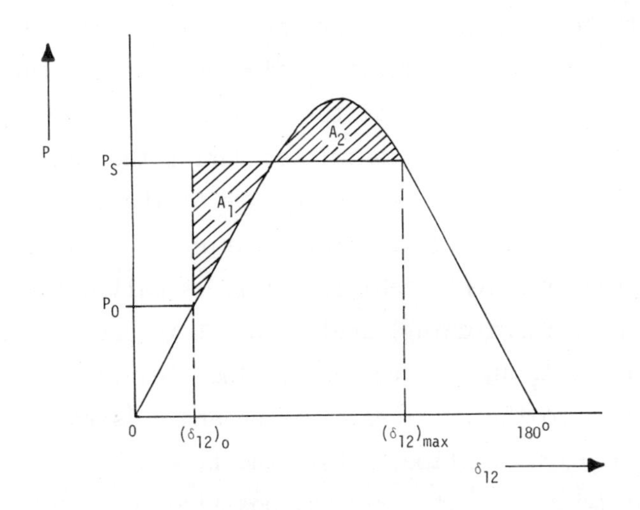

Figure 6.3.3. Equal-area criterion for determination of power limit.

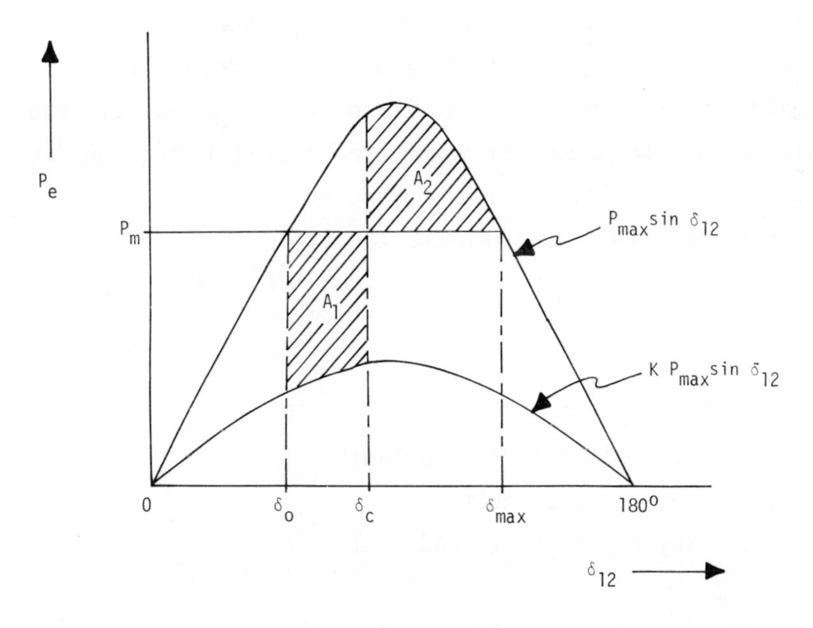

Figure 6.3.4. Equal-area criterion applied to fault-clearing, when power is transmitted during the fault.

[*Note:* Areas A_1 and A_2 are equal.]

power that can be transmitted by the generator before the fault and also after the fault is cleared by switching at the instant when δ_{12} is equal to δ_s; and $(K\ P_{max}\ \sin\delta_{12})$ be the power that can be transmitted during the fault, where K is a constant depending on the nature of the fault. The largest possible value of δ_{12} for clearing to occur without exceeding the transient stability limit is called the critical clearing (or switching) angle, and it is given by δ_c when the areas A_1 and A_2 are equal. The clearing time corresponding to the critical clearing angle is known as the critical clearing time. It is the time taken for the machine to swing from its original position to its critical clearing angle. If the fault clearing should occur at a value of δ_{12} greater than δ_c, the area A_2 will be less than the area A_1 and δ_{12} will continue to increase beyond δ_{max}; the speed will increase further and the generator cannot regain synchronism. The severity of the fault affects the value of K, the amount of power transmitted during the fault, and hence the critical clearing angle. Increase in fault severity results in decrease of the value of K. In terms of decreasing severity, the fault conditions are given below:

1. Three-phase fault
2. Double line-to-ground fault
3. Line-to-line fault
4. Single line-to-ground fault

The effect of grounding impedance in the system neutrals is to decrease the severity. While the three-phase fault is the most severe one, it is the least likely one to occur. For complete reliability a system should be designed for transient stability for three-phase faults at the worst locations. However, reliability may sometimes be sacrificed from the economic standpoint.

Step-by-Step Method of Solution for Transient Stability Problem:

It is desirable to obtain swing curves (plots of δ versus t) for all the machines of interest in a large system for various clearing times. The most severe type of fault for which protection against loss of stability is desired will be considered for the calculations, and also the fault will be assumed to occur in a position for which the least transfer of power takes place from the machine. The swing curve is usually plotted over a period long enough to determine whether the angle δ will reach a maximum and start to decrease. It is the relative angle between machines that is important for our consideration, and in order to be stable, the angular difference between any two of all machines must decrease after the switching operation. The analysis based on the examination of the first swing of the machine angles is sometimes known as the first swing transient stability analysis. While this criterion is sufficient to indicate stability in most of the cases, caution needs to be exercised in a few cases to make sure that δ will not increase again without returning to a low value. It may then be necessary to plot the swing curve over a longer period of time, while taking into account various variables on an actual system. Another criterion may also be applied in such cases for the system to be transiently stable; the system must be in a steady-state stable condition for the system configuration and operating conditions after the transient disturbance and removal of all faulted circuits.

The solution of the system of second-order differential swing equations of the type of Eq. (6.3.13) is desired. A number of different methods are available for the numerical evaluation of such equations through step-by-step computations for small increments of time. The advent of the digital computer has made more elaborate and involved methods of computation practical. In order to gain a basic understanding of the step-by-step solution procedure, we shall consider a hand-calculating simpler procedure* in which the accelerating power curve as well as the angular velocity curve have been approximated as stepped waveforms for convenience as shown in Figure 6.3.5. The ends of time intervals of Δt are denoted by points (n - 2), (n - 1), and (n) etc., while the midpoints of time intervals are indicated by $(n - \frac{3}{2})$ and $(n - \frac{1}{2})$, etc. The angular velocity ω^* that is in excess over the synchronous angular velocity is shown in Figure 6.3.5 as a function of time. The steps are so chosen that the accelerating power P_a remains constant between midpoints of the time intervals, while ω^* has a constant value during the time interval Δt as calculated for the midpoint. The change of speed between any two midpoints caused by the constant accelerating power is given by:

$$\Delta\omega^* = \omega^*_{n-\frac{1}{2}} - \omega^*_{n-\frac{3}{2}} = (p^2\delta)\ \Delta t = \frac{P_{a(n-1)}}{M}\ \Delta t \qquad (6.3.21)$$

The change in δ during the (n - 1) and n^{th} intervals may then

*See Stevenson, Jr., W. D., Elements of Power System Analysis, 3rd Ed., Ch. 14, McGraw-Hill Book Company, New York, 1975.

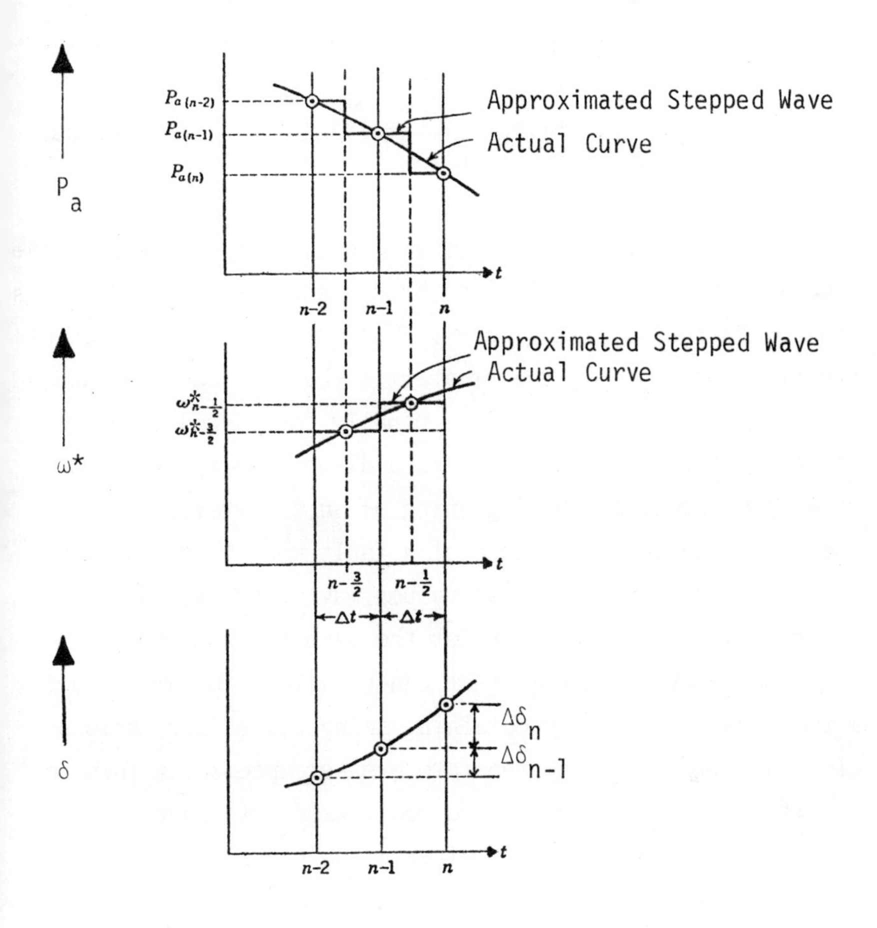

Figure 6.3.5. Approximations for the step-by-step method of solution for transient-stability problem.

be calculated as follows:

$$\Delta\delta_{n-1} = \delta_{n-1} - \delta_{n-2} = \Delta t\ \omega^*_{n-\frac{3}{2}} \tag{6.3.22}$$

and

$$\Delta\delta_n = \delta_n - \delta_{n-1} = \Delta t\ \omega^*_{n-\frac{1}{2}} \tag{6.3.23}$$

Eliminating ω^* from the above equations, one can obtain

$$\Delta\delta_n = \Delta\delta_{n-1} + \frac{(\Delta t)^2}{M} P_{a(n-1)} \tag{6.3.24}$$

which interrelates the changes in δ during any two successive time intervals. The solution through successive steps yields the points for the swing curve. The smaller the duration of the time interval Δt, the greater is the accuracy obtained; an interval of 0.05 seconds is usually considered to be satisfactory. Whenever there is a discontinuity in the accelerating power at the beginning of an interval, such as the case with the occurrence of a fault when t is equal to zero, the average of the two values may be treated as the constant accelerating power for the sake of calculations.

The duration of time that can be allowed before clearing a fault can be found by obtaining swing curves for various clearing times. The appropriate breaker speeds may then be specified, while noting the commonly used interrupting times as 8,5,3, or 2 cycles after a fault occurs.

Digital-Computer Methods for Transient Stability Study:

The second-order differential equation (6.3.13) can be written as two simultaneous first-order equations:

$$\frac{d\delta}{dt} = \omega - 2\pi f \tag{6.3.25}$$

and

$$\frac{d\omega}{dt} = \frac{1}{M} (P_m - P_e) \tag{6.3.26}$$

in which $(2\pi f)$ represents the rated synchronous speed in radians per second. For an m-machine system problem where each machine is represented in a simplified manner by a voltage of constant magnitude back of transient reactance, it is necessary to solve 2m simultaneous first-order differential equations given by

$$\left.\begin{aligned} \frac{d\delta_i}{dt} &= \omega_i - 2\pi f \\ \frac{d\omega_i}{dt} &= \frac{1}{M_i}(P_{mi} - P_{ei}) \end{aligned}\right\} \quad i = 1,2,\ldots,m \qquad (6.3.27)$$

where the internal voltage angles δ_i are measured from a common synchronously rotating reference axis. If no governor action is considered, P_{mi} will remain constant. When the effects of saliency and the changes in field-flux linkages are to be included in machine representations, the following 3m simultaneous first-order differential equations have to be solved:

$$\left.\begin{aligned} \frac{d\delta_i}{dt} &= \omega_i - 2\pi f \\ \frac{d\delta_i}{dt} &= \frac{1}{M}(P_{mi} - P_{ei}) \\ \frac{dE'_{qi}}{dt} &= \frac{1}{T'_{doi}}(E_{fdi} - E_i) \end{aligned}\right\} \quad i = 1,2,\ldots,m \qquad (6.3.28)$$

If the effects of the exciter control system are not included, E_{fdi} will remain constant.

In transient stability studies a load flow calculation is made first to obtain system conditions prior to the disturbance. The network is composed of system buses, transmission lines, and transformers; besides, equivalent circuits for machines and static impedances or admittances to ground for loads are included. The reader is assumed to be familiar with the load flow solution methods[*] such as Gauss-Seidel or Newton-Raphson iterative method. After the load flow calculation, the impedance or admittance matrix of the network needs to be modified to reflect the changes in the representation of the network. The operating characteristics of synchronous and induction machines are described by sets of differential equations. A transient stability analysis is performed by combining a solution of the algebraic equations describing the network with the numerical solution of the differential equations. While the application of Liapunov functions to power system stability studies[**] is considered to be outside the scope of this book, the modified Euler and Runge-Kutta methods[*] are briefly presented here for the solution of the differential equations in transient stability studies.

In thc application of the modified Euler method, for the simplified machine representation, the first estimates of the internal voltage angles and machine speeds at time $(t + \Delta t)$ are obtained from the following:

$$\left.\begin{aligned} \delta_i^{(1)}\Big|_{(t + \quad t)} &= \delta_i\Big|_{(t)} + \frac{d\delta_i}{dt}\Big|_{(t)} \Delta t \\ \omega_i^{(1)}\Big|_{(t + \quad t)} &= \omega_i\Big|_{(t)} + \frac{d\omega_i}{dt}\Big|_{(t)} \Delta t \end{aligned}\right\} i = 1,2,\ldots,m \tag{6.3.29}$$

[*] See Stagg, G.W. and El-Abiad, A.H., Computer Methods in Power System Analysis, Chs. 7-10, McGraw-Hill Book Company, New York, 1968.

[**] See Appendix E for details on Stability Criteria (E-7).

where the derivatives are evaluated from Eq. (6.3.27). Then the first estimates of voltages behind machine impedances and machine powers at time $(t + \Delta t)$ are computed. The second estimates are obtained by evaluating the derivatives at time $(t + \Delta t)$:

$$\delta_i^{(2)}\Big|_{(t + \Delta t)} = \delta_i\Big|_{(t)} + \frac{\left.\frac{d\delta_i}{dt}\right|_{(t)} + \left.\frac{d\delta_i}{dt}\right|_{(t + \Delta t)}}{2} \Delta t$$

$$\omega_i^{(2)}\Big|_{(t + \Delta t)} = \omega_i\Big|_{(t)} + \frac{\left.\frac{d\omega_i}{dt}\right|_{(t)} + \left.\frac{d\omega_i}{dt}\right|_{(t + \Delta t)}}{2} \Delta t \qquad (6.3.30)$$

$$i = 1,2,\ldots,m$$

where

$$\left.\begin{aligned} \left.\frac{d\delta_i}{dt}\right|_{(t + \Delta t)} &= \omega_i^{(1)}\Big|_{(t + \Delta t)} - 2\pi f \\ \left.\frac{d\omega_i}{dt}\right|_{(t + \Delta t)} &= \frac{1}{M_i}\left\{P_{mi} - P_{ei}^{(1)}\Big|_{(t + \Delta t)}\right\} \end{aligned}\right\} \qquad (6.3.31)$$

The sequence of steps for transient analysis by the modified Euler method and the load flow calculations is shown in the flowchart given in Figure 6.3.6.

The flowchart for transient calculations using Runge-Kutta method with a fourth-order approximation is given in Figure 6.3.7. For the simplified machine representation, the changes in the internal voltage angles and machine speeds are obtained as

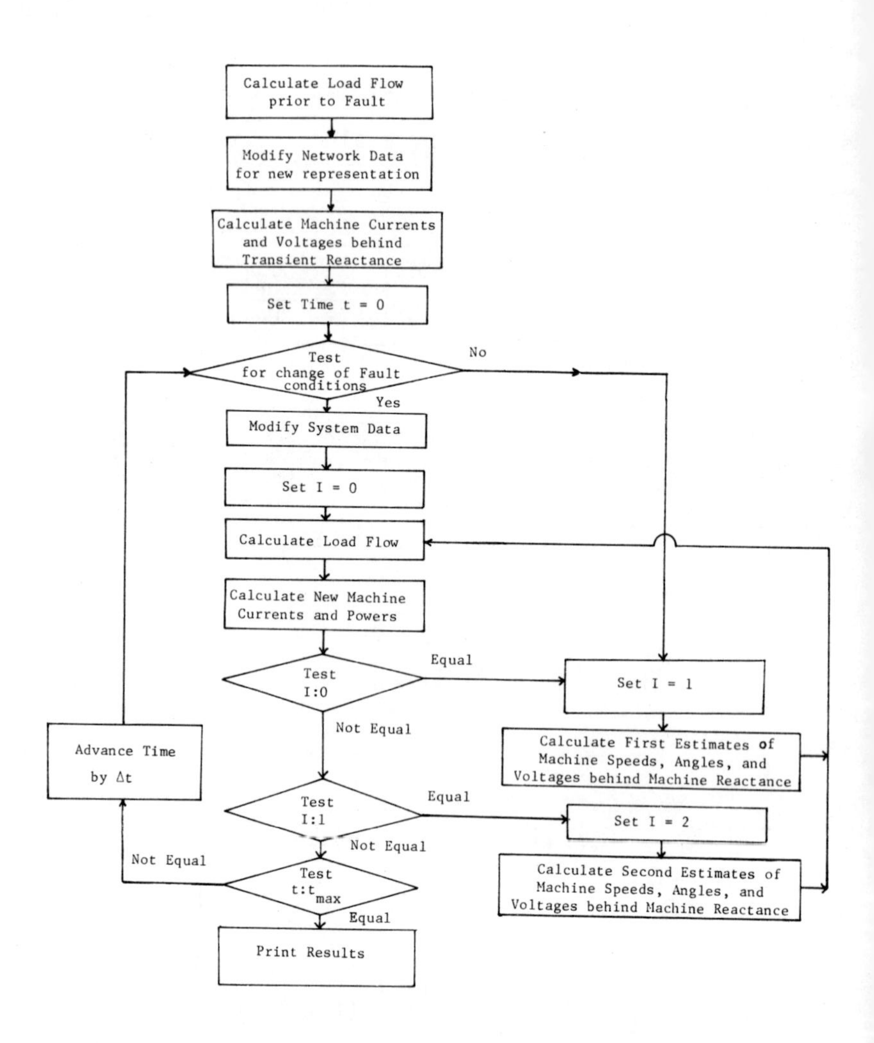

Figure 6.3.6. Flow chart for transient calculations with the application of modified Euler method.

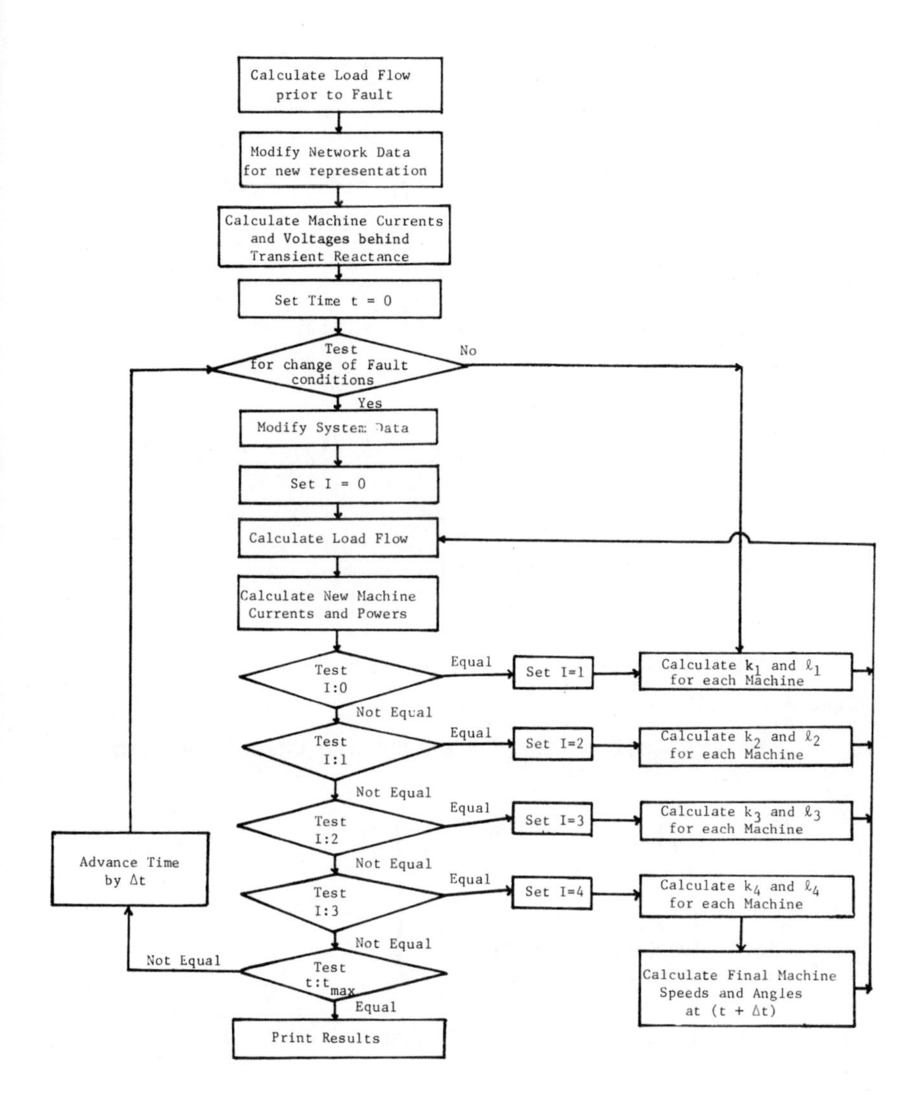

Figure 6.3.7. Flow chart for transient calculations using Runge-Kutta method with a fourth-order approximation.

$$\delta_i\big|_{(t+\Delta t)} = \delta_i\big|_{(t)} + \Delta\delta_i\big|_{(t+\Delta t)} =$$

$$= \delta_i\big|_{(t)} + \frac{1}{6}(k_{1i} + 2k_{2i} + 2k_{3i} + k_{4i})$$

$$\omega_i\big|_{(t+\Delta t)} = \omega_i\big|_{(t)} + \Delta\omega_i\big|_{(t+\Delta t)} =$$

$$= \omega_i\big|_{(t)} + \frac{1}{6}(\ell_{1i} + 2\ell_{2i} + 2\ell_{3i} + \ell_{4i})$$

$$i = 1,2,\ldots,m \tag{6.3.32}$$

where the k's and ℓ's are the changes in δ_i and ω_i respectively.

The first estimates of changes are calculated from

$$\left.\begin{aligned} k_{1i} &= [\omega_i\big|_{(t)} - 2\pi f]\,\Delta t \\ \ell_{1i} &= \frac{1}{M_i}[P_{mi} - P_{ei}\big|_{(t)}]\,\Delta t \end{aligned}\right\} i = 1,2,\ldots,m \tag{6.3.33}$$

The second set of estimates of changes in δ_i and ω_i are obtained from

$$\left.\begin{aligned} k_{2i} &= \{[\omega_i\big|_{(t)} + \frac{\ell_{1i}}{2}] - 2\pi f\}\,\Delta t \\ \ell_{2i} &= \frac{1}{M_i}[P_{mi} - P_{ei}^{(1)}]\,\Delta t \end{aligned}\right\} i = 1,2,\ldots,m \tag{6.3.34}$$

where $p_{ei}^{(1)}$ represent the machine powers corresponding to the internal voltage angles of $\{\delta_i|_{(t)} + (k_{1i}/2)\}$.

The third set of estimates are computed as

$$\left.\begin{aligned} k_{3i} &= \{[\omega_i|_{(t)} + \frac{\ell_{2i}}{2}] - 2\pi f\}\ \Delta t \\ \ell_{3i} &= \frac{1}{M_i}\ [P_{mi} - P_{ei}^{(2)}]\ \Delta t \end{aligned}\right\}\quad i = 1,2,\ldots,m \tag{6.3.35}$$

where $p_{ei}^{(2)}$ are obtained from a second solution of the network equations with the internal voltage angles of $\{\delta_i|_{(t)} + k_{2i}/2\}$.

The fourth set of estimates are calculated from

$$\left.\begin{aligned} k_{4i} &= \{[\omega_i|_{(t)} + \ell_{3i}] - 2\pi f\}\ \Delta t \\ \ell_{4i} &= \frac{1}{M_i}\ [P_{mi} - P_{ei}^{(3)}]\ \Delta t \end{aligned}\right\}\quad i = 1,2,\ldots,m \tag{6.3.36}$$

in which $p_{ei}^{(3)}$ are evaluated from a third solution of the network equations with internal voltage angles of $[\delta_i|_{(t)} + k_{3i}]$. The final estimates of the internal voltage angles and machine speeds at time $(t + \Delta t)$ are simply obtained by substituting the k's and ℓ's into Eq. (6.3.32). The solution progresses as per the flowchart shown in Figure 6.3.7 and the process is repeated until the time t equals the maximum time t_{max}.

Factors Affecting Transient Stability:

It is important for the power system designer and operating engineer to understand the causes of instability in a system, so that appropriate means for its prevention may be reliably employed from economic and technical viewpoints. Some of the factors which emphasize stability considerations are more remote plant siting, larger plant ratings, larger unit sizes, fewer transmission circuits, more heavily loaded transmission, increased pool interchanges, increased emphasis on reliability, and increased severity of various disturbance criteria as well as contingencies. Most of these are just manifestations of the problems of environmental considerations or economic pressures or both.

The direct effect of generator reactances, inertias, and generator internal voltage or the system voltage on stability can easily be seen from Eqs. (6.3.8), (6.3.14) and (6.3.19). Although an increase in M offers a means of increasing stability, it has not been used for economic reasons. It is necessary to keep the overall reactance within bounds, if a reasonable degree of stability is to be preserved. The overall system reactance is to be considered as consisting of the generators, transformers, transmission lines, and possible compensation. Future turbine-generator unit designs, particularly for the larger ratings, are expected to have greater generator reactances and lower unit inertias. Reduced short-time overload capability may also have an indirect effect on stability. Another form of overload limit from the viewpoint of the generator may be the permissible negative-phase-sequence current. Underexcited operation of turbo-generators (and high reactive-power demand) may also be additional features to be reckoned with. The control action of high-response, optimally-stabilized excitation systems and turbine valves as well as the reduction in circuit-breaker opening and reclosing times may be effectively used to increase the limiting permissible reactance to a certain extent. Excitation control alone cannot, however, be expected to completely eliminate the effects of increased generator reactances on stability.

One approach to designing power systems in order to deal successfully with unusual contingencies that endanger stability is to compensate the transmission system with series capacitors and to construct additional transmission lines to cope with the stability problem. Compensation for line reactance by series capacitors is often economical for increasing stability. However, some of the series capacitor application considerations that are to be borne in mind are transient voltages across capacitors, sparkover gap settings, re-insertion time, capacitor current rating for parallel line switching, ferro-resonance phenomena, effects on breaker recovery voltage and relaying, effects on untransposed lines and load sharing with parallel lines, turbo-generator self excitation and transient torques. The last two have been responsible for a certain amount of difficulty recently and these will be discussed later under Section 6.5 in some detail. It is also a common practice to increase the number of parallel lines between any two points in order to achieve not only reduced overall reactance but also increased power transferred during a fault.

Since circuit breakers must operate to clear the fault before the critical clearing time in order to maintain stability, it is easy to see that high-speed circuit breakers on power systems lead to improved stability. The critical clearing time itself may be increased by raising P_{max} of Eq. (6.3.19), or by stability control techniques. It can be shown* that the critical clearing time for faults in a fixed system is given by

$$\text{critical clearing time} \simeq \sqrt{\frac{H}{(x_d' + x_t)(x_d + x_t)}} \qquad (6.3.37)$$

where H is the inertia constant, x_d' is the generator transient reactance, x_d is the synchronous reactance, and x_t is the transformer reactance. In cases where a large load is suddenly removed or clearing a fault is delayed, a breaking resistor at or near the generator bus may be connected in order to compensate for some of the load reductions and thereby reduce the acceleration. Reducing the severity of the fault to that of a single line-to-ground fault through independent pole circuit breaker operation is an effective means of meeting three-phase breaker failure fault clearing system design criterion. It is expected that breaker failure will only very rarely involve more than one pole. Independent pole switching should especially be considered if back-up clearing times with back-up relay systems in case of stuck breakers are a problem. Another feature that is commonly employed is single-pole switching of the type which would trip only the faulted phase when a single line-to-ground fault occurs.

Automatic load shedding is sometimes done through the application of under-frequency relays during severe emergencies which result in insufficient generation to meet the load demands. Hand-in-hand with a load shedding program, a convenient means must be provided for restoring the load

---

*See Concordia, C. and Brown, P. C., "Effects of Trends in Large Steam Turbine Driven Generator Parameters on Power System Stability", IEEE Trans. 71-TP74-PWR, pp. 2211-2218, 1971.

after the system returns to a normal state. The use of under-frequency relays to reduce overload by disconnecting selected load circuits provides a means of achieving power balance in isolated power systems.

Generation shedding is also employed sometimes in response to line faults. However, fast repositioning of turbine valves provides an alternative to generation shedding that can, in part, accomplish the same purpose while avoiding problems attendant on the temporary outage of a generating unit. The sensing of the difference between mechanical input and electrical output of a generator due to a fault initiates the closing of a turbine valve to reduce the power input. Fast turbine valve control, which is a more recent development, can improve unit stability by increasing the critical fault clearing time.

There are a number of dc-transmission lines operating these days in association with ac-transmission system, and these bring about several distinctive operating situations. The dc-lines can have significant effects on ac-system stability. The very rapid control of power possible on dc-transmission can provide strong damping to disturbances within the power system. Substitution of a dynamically responsive dc-line for one of two parallel ac-lines may provide a marked improvement in the transfer capability from the stability criterion.

A summary of the methods used for improving transient stability is given below:

1. Increase in the system voltage.
2. Reduction of overall reactance by appropriately choosing individual system components, providing series compensation, and adding more parallel transmission lines.

3. Provision of a strong transmission network with various interconnections and additional transmission lines.
4. Control of high-response and optimally-stabilized excitation systems.
5. Application of high-speed circuit breakers, including reclosing breakers.
6. Use of breaking resistor in case of sudden loss of a large load.
7. Independent pole switching of circuit breaker.
8. Single-pole switching which would trip only the faulted phase.
9. Load shedding during periods of insufficient generation.
10. Generation shedding in response to line faults.
11. Fast turbine valve control.
12. Provision of a dc-transmission line.

The counter-measures taken against sudden, undesired changes in the network should match the disturbance with respect to location, amount, duration, and dependence on angular difference. A proper combination of measures accomplished rapidly under suitable control can improve system stability to a large extent. Various means of improving power system stability should be viewed as a coordinated system instead of as individual measures applied locally. Methods such as resistor breaking, load shedding, generation shedding, and switching of series capacitors on the electrical side have to be coordinated with fast valving and bypassing of steam or water on the mechanical side. Coordination and cooperation amid different ownerships of interconnected power systems is required in economical design and reliable operation. Although a central control by a digital computer may be conceived for coordination of various measures in the future, much can be accomplished by local analog-type or digital-type devices also for the time being.

6.4 Damper Windings

As indicated in Section 1.1 and Figure 1.1.7, the damper windings may be connected or non-connected. Complete or connected dampers are electrically superior and are more commonly used, while the incomplete or non-connected, open dampers may be preferred for mechanical reasons and high centrifugal forces on the unsupported connections between poles. Double-deck damper windings, similar to the double squirrel-cage windings for induction-motor rotors, are sometimes used. Field collars consisting of heavy, short-circuited copper loops placed around the pole core near the outer end are also occasionally used either as substitutes for damper windings or together with a single damper winding. Even though damper windings are not employed on turbo-generators, the solid steel rotor cores of such machines produce the same effects as dampers.

Amortisseur windings affect some of the synchronous machine constants discussed in Section 3.1. Subtransient reactances x''_d and x''_q are decreased by damper windings. A connected damper winding decreases x''_q more and makes it more nearly equal to x''_d than the non-connected damper winding. The negative-sequence reactance x_2, usually taken as the average value of x''_d and x''_q, is also decreased consequently. The negative-sequence resistance as well as the subtransient time constants are also naturally affected by the damper windings.

The damper action in synchronous machines can be approximately analyzed and qualitatively explained by induction-motor theory, as an amortisseur winding is similar to the squirrel-cage rotor winding and the armature winding of a synchronous machine is like the stator winding of an induction motor. It will be assumed that the reader is familiar with the induction-motor steady-state theory, which assumes constant slip and a symmetrical rotor with only one set of windings, with no difference between the direct- and quadrature-axis circuits. A synchronous machine, however, has an unsymmetrical rotor with a field winding on only one axis, even with the effect of damper windings in both axes. Also the slip of the machine varies during the disturbances treated in stability studies. Nevertheless, induction-machine theory is considered adequate for the calculation of negative-sequence braking.

Based on the well-known equivalent T-circuit of induction motor for positive phase sequence, the familiar torque-speed characteristic of an induction machine is shown in Figure 6.4.1 for three operating regions, namely motor, generator, and braking regions. The torque in synchronous watts is equal to the rotor input in watts, given by

$$T = I_r^2 \frac{R_r}{s} \tag{6.4.1}$$

in which I_r is the rotor current referred to stator, R_r is the rotor resistance referred to stator, and s is the slip. The range of slip $0 < s \leq 1$ corresponds to induction-motor operation. For $s < 0$, at speeds above synchronous, the machine operates as an induction generator. Both the internal mechanical power and torque developed by the machine become negative, and hence the torque is retarding. For

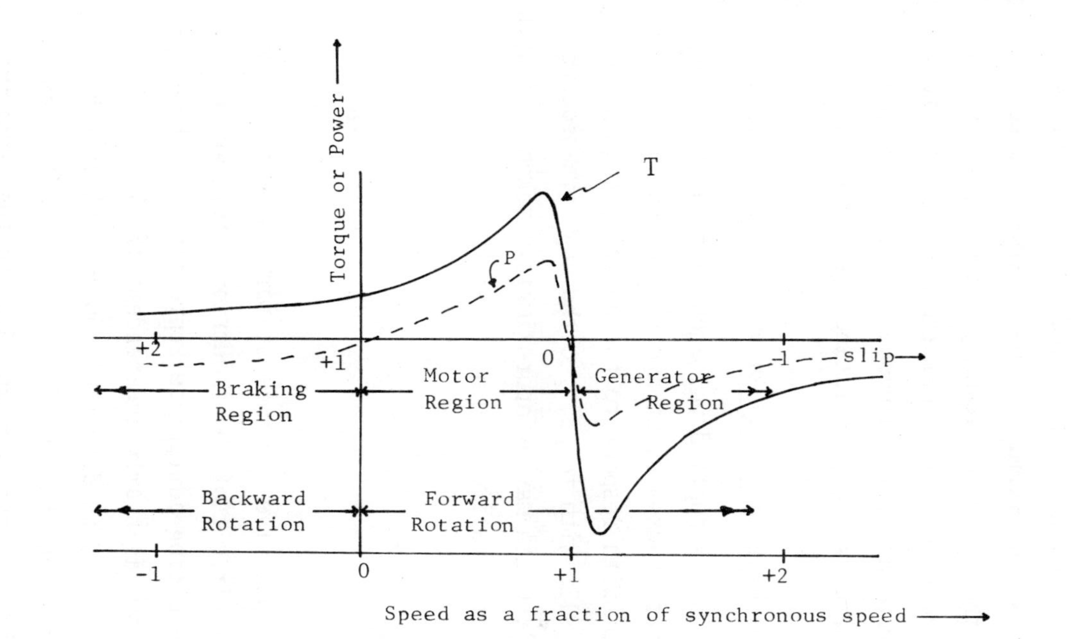

Figure 6.4.1. Induction machine torque-speed and power-speed characteristics for three operating regions.

(*Note:* T is the internal electromagnetic torque and P is the internal mechanical power developed.)

$s > 1$, the machine must be driven backward, against the direction of rotation of its magnetic field, by an external source of mechanical power capable of counteracting the internal torque. During the braking region ($s > 1$), the torque is positive while the mechanical power developed is negative; it is a braking torque, opposing the rotation.

The slip at which maximum torque occurs is directly proportional to the rotor resistance, and the maximum torque itself is independent of rotor resistance. In order to give high damping torque at small values of slip, the damper winding should have as low a resistance as possible.

Negative-Sequence Braking During an Unsymmetrical Fault:

Negative-sequence applied voltages cause an air-gap-flux wave to rotate backwards at synchronous speed. If the slip of the rotor with respect to the positive-sequence field is s, the slip with respect to the negative-sequence field is $(2 - s)$, which may be approximately taken as 2 with little error. The negative-sequence resistance r_2 can be seen to be

$$r_2 \simeq r_1 + \frac{R_r}{2} \qquad (6.4.2)$$

(where r_1 is the positive-sequence stator resistance) from the equivalent T-circuit of the induction motor, while neglecting the exciting admittance represented by the shunt branch. The torque is nearly independent of slip in the vicinity of $s = 2$, and is directly proportional to the mechanical power, since the speed is substantially constant. The mechanical power input corresponding to the braking torque is

$$P_b \simeq \frac{1}{2} I_{r2}^2 R_r \simeq I_{s2}^2 (r_2 - r_1) \tag{6.4.3}$$

where I_{r2} is the negative-sequence rotor current, R_r is the rotor resistance, and I_{s2} is the negative-sequence stator current. The negative-sequence armature current corresponding to an unsymmetrical fault can be calculated with the method of symmetrical components. From the known values of the positive- and negative-resistances of the machine, the negative-sequence braking power of synchronous machines can be computed. The damper winding should have high resistance, such that maximum torque occurs at or near s = 2, in order to give high negative-sequence braking torque.

Negative-sequence damping may be treated as a reduction of the input in the calculation of swing curves, since the negative-sequence braking torque has the same effect on the angular motion of the rotor as a decrease of mechanical input torque. The accelerating power, given earlier by Eq. (6.3.18), is then modified as

$$P_a = P_m - P_b - P_e \tag{6.4.4}$$

where P_m is the shaft power, P_b is the braking power, and P_e is the positive-sequence electric power developed for a generator. Unless the amortisseur winding has a high resistance, the negative-sequence braking torque may be neglected. It does not also come into the picture unless the fault is an unbalanced one.

D.C. Braking:

The d.c. components of armature current induce in the rotor circuits fundamental-frequency currents, which produce a braking torque similar to that caused by negative-sequence armature currents. The d.c. braking power may be approximately calculated from

$$P_b \simeq i_{dc}^2(r_2 - r_1) \tag{6.4.5}$$

where i_{dc} is the total instantaneous d.c. component of armature current. The d.c. braking power decreases the accelerating power and may hence be considered as all or part of P_b in Eq. (6.4.4). For types of fault other than three-phase and for faults not very near the armature terminals, i_{dc} decays so rapidly that its braking effect is negligible.

Positive-Sequence Damping:

While the negative-sequence braking results from the torque caused by the interaction of damper currents with the negative-sequence backward-rotating air-gap magnetic field, the positive-sequence damping is a consequence of the torque caused by interaction of amortisseur currents with the positive-sequence forward-rotating air-gap magnetic field. Positive-sequence damping is more effective after the fault clearance than during the fault. By absorbing energy from the oscillation, it may prevent a machine (which has survived the first swing) from going out of step on the second or subsequent swings. Its effect is almost always neglected in calculating power limits, as the damping is not great enough to increase significantly the power that can be carried through the first swing. The damping power of a synchronous machine connected to an infinite bus through an external series reactance x_e can be shown to be the following, based on the assumptions of negligible armature resistance, no resistance in field circuit, small values of slip, and damping action caused only by the damper windings: (See Park* and Dahl**.)

$$P_d = e^2 s\omega \left[\frac{(x_d' - x_d'')\, T_{do}''}{(x_e + x_d')^2} \sin^2 \delta + \right.$$

$$\left. + \frac{(x_q' - x_q'')\, T_{qo}''}{(x_e - x_q')^2} \cos^2 \delta \right] \tag{6.4.6}$$

in which e is the voltage of infinite bus and the notation for symbols is the same as introduced in earlier sections of the text. The slip in per-unit may be related to δ through the equation

$$s = \frac{1}{360f} \frac{d\delta}{dt} \tag{6.4.7}$$

where f is the frequency of the infinite bus in hertz, δ is in electrical degrees, and t is expressed in seconds. The effect of armature resistance is to decrease the damping torque. However, its effect is usually negligible in large machines with effective damper windings. Even though Eq. (6.4.6) is applicable for the case of one machine connected to an infinite bus, its use can be extended to a multi-machine system through successive approximations in the course of a point-by-point determination of swing curves.

---

* Park, R. H., "Two-Reaction Theory of Synchronous Machines", Pt. I, AIEE Trans., Vol. 48, pp. 716-727, July 1929; Pt. II, AIEE Trans., Vol. 52, pp. 352-354, June 1933.

** Dahl, O. G. C., Electric Power Circuits: Theory and Applications, Vol. II. Power System Stability, McGraw-Hill Book company, NY, 1938 (Ch. XIX: Damper Windings & Their Effect).

The damping power P_d may be treated as an additional component of positive-sequence electric output power of a generator. The accelerating power is thereby diminished as

$$P_a = P_m - P_b - P_d - P_e \tag{6.4.8}$$

In cases where the damping power P_d may be as large as 10% of the synchronous power P_e, it might effect the network conditions sufficiently to warrant taking it into account for the network solution. Then the usual representation of a synchronous machine, consisting of the transient reactance in series with the voltage behind transient reactance, may be modified by adding a parallel branch, consisting of a resistance and a source of power, which may be adjusted to have a power output equal to the calculated damping power at unity power factor. The assumption of unity power factor can be justified as the effect of rotor resistance at low values of slip is much greater than the induction effects.

Heating Effect of Rotor Currents:

Negative-sequence currents are induced in the rotor when a synchronous machine is subjected to an unbalanced short circuit. These currents are confined to paths near the surface of the rotor in view of their high frequency. In the case of the rotor of the salient pole generator, the use of a connected amortisseur winding provides a known path for the current. Without a connected damper winding, the circuits are much less well-defined. In turbine generators with solid steel rotor forgings, the negative phase sequence stator current generates double frequency currents in the surface of the rotor, the slot wedges (with the retaining rings acting as end connections between wedges), and to a smaller degree the field winding. The current distribution at the surface of the rotor corresponds to the current distribution in the rotor of a squirrel-cage induction motor. The currents flow axially over the length of the rotor and close circumferentially at the ends, with the same number of poles as in the stator winding.

The paths that the induced currents in the rotor surface take have a relatively high resistance, and considerable heat may be developed in some parts such as the slot wedges and the retaining rings, thereby causing damage to their mechanical properties, if the currents are sustained for more than few seconds. The complex nature of the three-dimensional nonlinear heat transfer and electromagnetic problems, which are inherently coupled together through material property-temperature effects, makes it necessary to rely heavily on full-scale machine testing if an understanding of negative-sequence heating is to be achieved. More comprehensive and sophisticated means for full-size generator unbalanced load testing and the introduction of innovative design features for control of temperatures during unbalanced faults have helped the generator designers to keep pace with the increased generator electromagnetic loadings accompanying the continuing growth in unit ratings. Nevertheless, provision for unbalanced fault capability continues to be one of the limiting generator design criteria.

Earlier experience with directly-cooled generators led to the revision of industry standards in 1935, when the required value of

$$I_2^2 t \left(= \int_0^t i_2^2 \, dt \right)$$

was changed from 30 to 10, I_2 being the negative-sequence current in per-unit and t the time in seconds. More recently, in order to keep pace with the growth of the largest unit ratings, a proposal for standards revision has been made* as shown in Figure 6.4.2. Table 6.4.1 presents the proposed standard for both types of machines: salient pole and cylindrical rotor. Substantial economic benefits are likely to result from reduced weights and physical dimensions of the new generator designs with I_2^2t capability below the present requirement of 10.

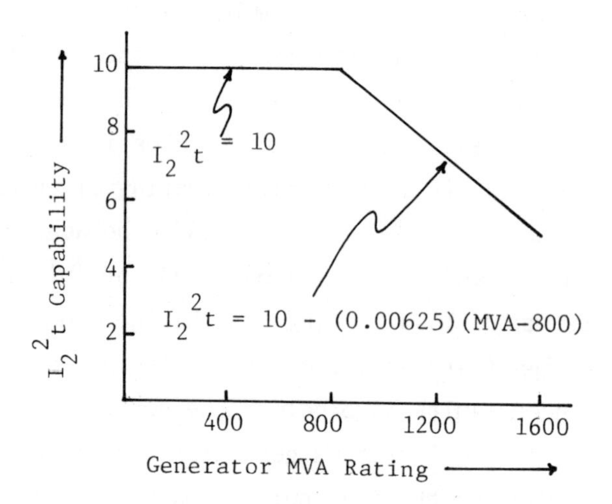

Figure 6.4.2. Proposed revised standard for unbalanced fault capability, I_2^2t, for directly-cooled generators as a function of generator rating.

---

*See IEEE Working Group Report, "A Standard for Generator Continuous Unbalanced Current Capability", IEEE Trans. 73-128-6, pp. 1547-1549, 1973.

Table 6.4.1

Continuous Unbalanced Current Capability

A generator shall be capable of withstanding, without injury, the effects of a continuous current unbalance corresponding to a negative phase sequence current of the values listed below, providing rated KVA is not exceeded, and the maximum current does not exceed 105 percent of rated in any phase. Negative phase sequence current is expressed in percent of rated stator current.

Type of Generator	Permissible I_2 (percent)
Salient Pole	
With connected amortisseur windings	10
With non-connected amortisseur windings	5
Continuous performance with non-connected amortisseur windings is not readily predictable. Therefore, for anticipated unbalanced conditions, machines with connected amortisseur windings should be specified.	
Cylindrical Rotor	
Indirectly cooled	10
Directly cooled - to 960 mva	8
961 - 1200 mva	6
1201 - 1500 mva	5
These values also express the negative phase sequence current capability at reduced generator KVA capabilities, in percent of the stator current corresponding to the reduced capability.	

Damper Winding Applications:

Amortisseur windings are provided on salient-pole synchronous machines for one or more of the following reasons:

(a) Starting synchronous motors as induction motors;

(b) Reducing hunting by low-resistance dampers in case of generators driven by reciprocating engines, motors driving loads of pulsating torque, and synchronous machines connected together by circuits having a high ratio of resistance to reactance;

(c) Damping rotor oscillations caused by short-circuits or switching operations;

(d) Aiding in synchronizing the generators by providing additional torque;

(e) Providing a braking torque on a generator during an unbalanced fault;

(f) Reducing overvoltages under certain short-circuit conditions, suppressing harmonics and balancing terminal voltages during unbalanced loading;

(g) Shielding the pole pieces from variations of flux and thereby preventing overheating.

6.5 Subsynchronous Resonance

Series capacitors are used quite extensively these days by some utilities to increase power transfer capability and system stability. The major alternatives to the application of series capacitors are additional transmission lines, higher voltages, d.c. transmission, and reduced reliability. All these alternatives, except the last, require greater capital investment and increase the environmental impact of transmission. In addition, there is increased emphasis on coal-fired generation where economic analysis, in general, favors mine-mouth plants connected to the load centers by series compensated transmission lines. For these reasons the continued use of series capacitors is highly desirable both from economic and environmental viewpoints, even though the phenomenon of subsynchronous resonance may result in an additional penalty on the use of series-capacitor compensation as a means of attaining power system stability.

Subsynchronous resonance encompasses the oscillatory attributes of electrical and mechanical variables associated with turbine-generators when coupled to a series capacitor compensated transmission system as shown in Figure 6.5.1. The oscillatory energy interchange between the system inductances and series capacitors may be lightly damped or even undamped. The terms subsynchronous and supersynchronous are used to denote frequencies below and above the synchronous frequency of the power system as seen by an observer on the generator rotor.

Since the inductive reactance goes down and the capacitive reactance goes up as the frequency is decreased, there is a natural frequency of the circuit at which the inductive reactance and capacitive reactance are equal in magnitude and opposite in sign. The resonance frequency, also known as the electrical subsynchronous frequency, at which the circuit reactance is zero and the current is limited only by the resistance component is given by

$$f_e = f\sqrt{\frac{X_C}{(\Sigma X_L)}} \tag{6.5.1}$$

The generator appears like a negative resistance at this frequency, or in other words, like a source of power. It is in fact acting like an induction generator at this frequency

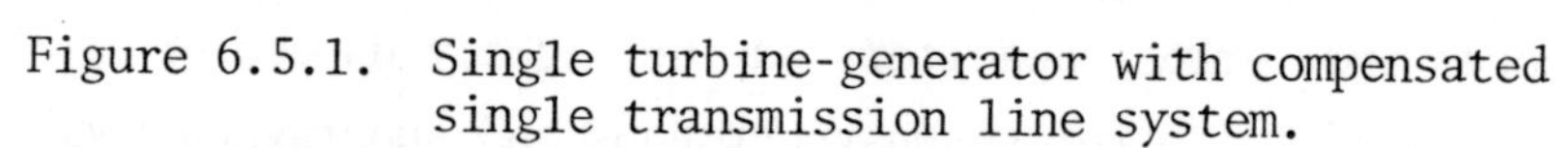

Figure 6.5.1. Single turbine-generator with compensated single transmission line system.

Notation for symbols:

M_n: Inertia of nth turbine-generator rotor element, in KW-Seconds/KVA.

K_{nm}: Stiffness of shaft between rotors n and m, in (KW/KVA)/radian.

D_{nm}: Viscous damper between rotors n and m, in KW-Seconds/KVA-radian.

D_n: Viscous damper, rotor n to synchronous reference frame in KW-Seconds/KVA-radian.

R: Resistance in per-unit

X: Inductive reactance in per-unit.

(*Note:* Subscripts A, T, G and L refer to armature, transformer, generator, and transmission line respectively.)

X_C: Capacitive compensation in per-unit.

$\Sigma X_L = X_G + X_T + X_L$

and is able to contribute the amount of energy at this natural frequency proportional to the value of its apparent negative resistance.

Electrical subsynchronous currents entering the generator terminals produce subsynchronous terminal-voltage components which sustain the currents. This effect is called self-excitation. There are two types of self-excitation, one involving resonance of the electrical system and the other involving both the electrical and mechanical systems of the turbine generator.

Self-excitation of the electrical system alone is caused by induction generator effect. The rotor resistance to subsynchronous current viewed from the armature terminals is negative because the rotor circuits are turning more rapidly than the rotating mmf caused by the subsynchronous armature currents. The phenomenon of self-excitation starts when the apparent negative resistance is equal (or greater) than the positive resistance of the rest of the circuit. At this point, there is no impedance in the circuit at all at this natural frequency and the current starts growing until it is limited by saturation or other effects which de-tune the circuit. This phenomenon must be avoided both because of the undesirable effects it has on the system, such as objectionable flickering of the lights at the resonant frequency, and also because of the very serious effect it has on the turbine generator.

As the rotor field flux overtakes the more slowly rotating subsynchronous mmf in the armature, it produces a subsynchronous torque having a forcing frequency which is the difference between the synchronous frequency and the electrical subsynchronous frequency. The subsynchronous electrical frequency and the forcing frequency are complimentary to each other because they add up to unity when expressed in per unit of synchronous frequency. A pulsating torque of frequency $(f - f_e)$ is applied to the rotor and stator of the generator.

When the forcing frequency is close to a torsional mode frequency of the shaft, the generator-rotor oscillation can build up to induce armature voltage components of subsynchronous and supersynchronous frequency. The subsynchronous frequency voltage is phased to sustain the subsynchronous torques. If the subsynchronous torque equals or exceeds the inherent mechanical damping of the rotating system, the system will become self-excited. This interplay between the electrical and mechanical networks is known as torsional interaction, due to which severe torque amplification can occur with resulting damage to the shaft of the machine.

Following a disturbance, the turbine generator-rotor masses will oscillate relative to one another at one or more of its natural frequencies depending on how it was disturbed. With mechanical oscillation at a critical frequency, the relative motion (amplitude and phase) of the individual turbine generator rotor elements is fixed and is called the mode shape of torsional motion, shown in Figure 6.5.2. For the mechanical system acting alone, with the absence of damping, the notion of mode shape is defined in this context. This mode shape, often displayed graphically, is an eigenvector of rotational displacement of the rotor inertial elements when the system is represented mathematically. The torsional modes are numbered sequentially according to mode frequency and number of phase reversals in the mode shape. Thus, Mode 1 has the lowest mode frequency and only one phase reversal in the mode shape, as can be seen from Figure 6.5.2. More generally, Mode n has the nth lowest frequency and a mode shape with n phase reversals. The maximum number of modes is one less than the number of

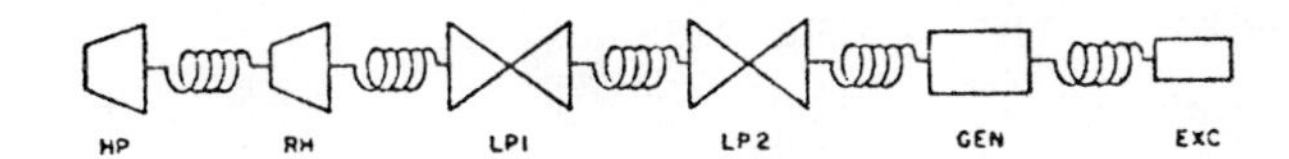

(a) Spring-Mass Model of Torsional System

<u>*Note:*</u> The turbine is tandem compound with high-pressure, reheat, and two low-pressure stages. In addition, the generator and exciter are on the same shaft system.

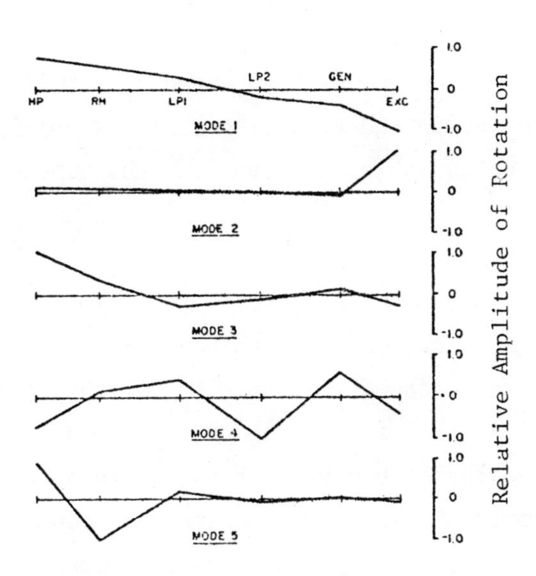

(b) Typical Mode Shapes of Torsional Mechanics

Figure 6.5.2.

inertial elements in the spring mass model.

Mathematical Models and Analysis:

Subsynchronous resonance with mechanical interaction can be traced through the electromechanical system by assuming a displacement at some point and tracing its effect. A displacement of the generator rotor produces an increment of terminal voltages and hence an increment of system current. The change in current will in turn produce a mechanical torque providing an incremenet of motion at the generator. In this circle of dependent relations, the responses may be regenerative or degenerative depending on the phase of the dependent to independent variables.

In view of the complexity of the problem, due to the limitations of space and scope of the textbook, it will only be attempted to provide an understanding of the significant factors and important variables involved in subsynchronous resonance phenomenon*. No detailed analysis with any particular method is pursued.

An analysis of turbo-generator self-excitation or transient shaft torques caused by disturbances in the electrical system requires mathematical modeling of the torsional mechanical system. A complete representation would require the solution of the elastic behavior of the whole turbine generator. It is only necessary to represent a mechanical system in the mathematical analysis with spring-coupled inertial elements that produce generator motion at

*See for additional information: "Analysis and Control of Subsynchronous Resonance", IEEE-76-CH1066-0-PWR, 1976. Figures 6.5.2-3 are adapted from "Proposed Terms and Definitions for Subsynchronous Resonance in Series Compensated Transmission Lines", pp. 55-58 of the above publication.

frequencies below the synchronous frequency, because sub-synchronous resonance involves only torsional mechanical frequencies below the synchronous frequency. A spring-mass modal mechanical model allows computation of rotor motion with mechanical and electrical torque as inputs. A separate model for each mode as shown in Figure 6.5.3 consisting of a single mass and spring tuned to the mode frequency, is normally used to represent the generator-rotor displacement. This model has the same energy storage as the real turbine generator rotor, at a critical frequency, for the same generator-rotor displacement. The mode spring-mass model is described in terms of its mode inertia, mode damping, mode spring constant, mode frequency, mode energy, etc.

The general mechanical response to a torque on the generator excites all modes of vibration of the mechanical system. The solution of the mechanical response is then a linear combination of the modal responses. These modal responses are to be obtained by the application of some proportion of the exciting torque to each of the mode spring-mass models. The multiplier, less than or equal to unity, is known as the mode transfer factor. The use of the mode shape vectors as weighting functions to combine linearly the modal responses allows the calculation of the actual response such as rotor displacement, speed, and shaft torque. The relationship between change in torque to change in speed through the mechanical systems is thus obtained through the use of modal analysis.

Next, the relationship between change in mechanical motion to torque through the electrical system is to be obtained. The change in electrical torque on the generator due to changes in mechanical motion is to be solved, while including the effects of series capacitors in the transmission

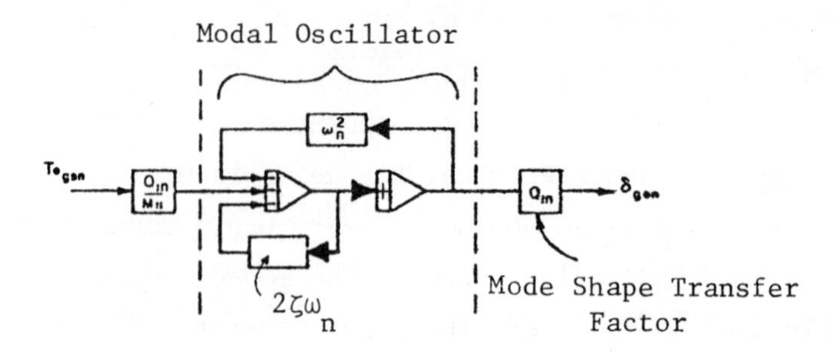

Figure 6.5.3. Modal spring-mass model.

$$\omega_n = \sqrt{\frac{2\pi f\ K_{nn}}{M_{nn}}} \quad ; \quad \zeta = \frac{\delta}{2\pi} \quad ; \quad \delta = \frac{1}{n}\ \ell_n \frac{A_o}{A_n}$$

Notation for symbols:

Q_{nm}: Transfer factor of the mth modal oscillator in the nth rotor motion.

K_{nn}: Modal stiffness mode n, in KW/KVA-radian.

M_{nn}: Inertia of nth modal oscillator alone or referenced to generator motion, in KW-Seconds/KVA. (Modal inertia at the nth natural frequency).

ω_n: n^{th} torsional mode angular frequency in radians/second.

ζ: Damping ratio, damping/critical damping, associated with turbine mechanical damping.

δ: Logarithmic decrement (the natural logarithm of the ratio of successive peaks of oscillation).

f: Power frequency (60) in hertz.

system. The currents in the well-known torque equation

$$T = \lambda_d i_q - \lambda_q i_d \tag{6.5.2}$$

are to be eliminated in favor of the mechanical variables (speed and angle) responsible for their production, while the flux linkages are expressed in terms of the machine terminal inputs using the operational impedances. The assumption of equal reactances in direct- and quadrature-axes, or no subtransient saliency, is usually allowed since the solutions required are in the frequency range of 10 to 50 Hz.

The difficulty with subsynchronous resonance has been the order of the problem even for the most simple case. Direct analytical solutions including the effects of torsional mechanical motion become almost impossible and one needs to take recourse to digital computer methods or hybrid computer simulation. Linear analysis methods are only valid for small perturbations at a given operating point. Transient performances can only be predicted through techniques which include significant nonlinear effects, such as torsional damping which varies with load and stress level.

Corrective Devices:

If series capacitors are to be applied to increase the power transfer capability from economic and environmental considerations, it is absolutely essential that the hazards of the subsynchronous resonance be avoided. Series capacitors can in fact be successfully applied and the associated risks can be avoided through a complete analysis of series capacitor application considerations including subsynchronous resonance and judicious application of corrective devices and operating procedures. Because of the very strong interactions among the various aspects of the phenomena of subsynchronous resonance and self-excitation, the problem should be treated as a whole for best results as another form of stability problem from the viewpoint of system planning considerations.

The theoretically ideal solution to the problem of subsynchronous resonance would be a generating-unit rotor with all elements tightly coupled, so that the rotor behaves as one piece, or at least a rotor that is critically damped so as not to have any torsional frequencies. But no one can see any prospect of this even in the remote future, and even then this would not eliminate completely the possibility of self-excitation of the electrical system. So one must take recourse to other possible solution methods.

You will recall that the phenomena of self-excitation started when the apparent negative rotor resistance equalled the positive circuit resistance. Also, the negative resistance is a function of the frequency in the rotor, while depending on other factors such as the degree of rotor saturation and presence of an amortisseur winding. In the first place, the apparent negative resistance of the generator can be reduced by adding an amortisseur winding to the machine. In addition, positive resistance can be added to the system, possibly at the point where the generator step-up transformer neutral is made up, while bypassing the 60-Hz component of current around the resistor by means of a filter circuit to minimize the losses at the power frequency. The next step is to be sure that the natural resonant frequency of the circuit is not at or near one of the torsional critical frequencies of the turbine-generator shaft system producing pulsating torques and leading to severe torque amplification. This latter criteria requires more than the simple network analysis, since all the important loops of the power system must be represented in all their various feasible switched conditions, including all of the possible variations of series capacitance due to the possibility of switching this equipment in steps since it is nearly always in the modular form. In the actual power system there may be several resonant frequencies and none of them should correspond to shaft torsional frequencies, regardless of whether or not the apparent circuit resistance is positive at these frequencies.

Corrective devices to torsional interaction problem may include a static blocking filter (provided at the generator step-up transformer neutral) consisting of separate filters connected in series. Each section of the filter is made up of a high-Q parallel resonant circuit tuned to block electrical currents at a frequency corresponding to one of the torsional modes, with negligible increase in resistance to the 60-Hz load current. Additional damping may be provided through a supplementary excitation damper control by injecting a properly phased sinusoidal signal (derived from the rotor motion) into the voltage regulator. The dominant effect of the filter for transient torque control is to produce multiple network series resonant frequencies detuned away from the corresponding shaft model frequencies. The effect of detuning is to minimize the build-up of shaft torques while the introduction of multiple mode electrical torques serves to spread the stored mechanical energy more uniformly among the shaft sections.

An alternative such as dynamic-filter device, which generates a voltage in series with the generator of equal magnitude and opposite in phase to that produced by an oscillating motion of the rotor at any of the frequencies corresponding to torsional resonant frequencies, may be considered for the torsional interaction problem.

Other schemes such as capacitor dual gap flashing may be utilized to reduce transient torques on the turbine-generator shafts to the level that shaft fracture may not occur for any single transient incident. The dual gap flashing scheme provides for gap flashing at about 2.2 times rated current with the gaps reset to the 3.5 level for successful reinsertion. Following reinsertion, the gaps are set again to the 2.2 level after a constant time delay introduced to allow the swing current to be damped below the level that would reflash the gaps.

The resistors, sometimes used along with the capacitors for damping transient oscillations following capacitor re-insertion, could also be used for additional transient torque control, provided additional thermal capacity is allowed. When subsynchronous current is detected, the resistor could be switched across the capacitors.

The subsynchronous over-current relay, combining a negative-sequence current network and suitable bypass as well as rejection filters to achieve a relaying output for current components of 20-40 Hz range, may be used to provide automatic relay protection of the generating units (connected to a series compensated power system) for sustained subsynchronous oscillations. Such a relay might be considered as a solution to the transient torque problem if such transients were rare and the consequential damage tolerable. Other relays based on sensing the rotor motion may be provided to detect subsynchronous oscillations and initiate turbine-generator trip-out.

In general, such a method of solution is to be preferred that does not put limitations on operation, that does not depend primarily on control or relay action, that does not require operator judgment or operational procedures to avoid certain possible network conditions, and that does not reduce the effectiveness of the series compensation in its function of improving power system transient stability.

The comparisons of the costs and benefits of series and shunt compensation to strengthen an a-c network and improve stability (or d-c transmission or d-c ties as an alternative) should be made for each particular case. Series compensation tends to be favored in the case of long-distance power transmission from a generating station to a relatively very large system, and for direct ties between very large systems where a substantial power transfer capability is needed. Shunt compensation tends to be favored in the case of long-distance power transmission from a generating station to a load area having little or no generating capacity, and for widespread relatively weak power systems where interconnecting ties may have to be made through several transmission systems, possibly belonging to different utility companies, in series. With a possible added penalty on series capacitors, and with the introduction of new static reactive power compensators which may be able to compete favorably with the synchronous condenser, shunt compensation deserves a new look in order to supply or absorb reactive power as a means of maintaining voltage at appropriate points in the network.

Problems

6-1. In order to study the effects of changing excitations

in two paralleled identical generators, consider them to be initially adjusted to share the active and reactive loads equally. Without touching the prime-mover throttles, and without changing the terminal voltage, load current, and load power factor, the excitation of generator 1 is increased so as to supply all the reactive KVA. Show by means of a phasor diagram the operating conditions of generator 2. Comment on the effects of change of excitation on the terminal voltage and reactive-KVA distribution. (*Hint:* For simplicity, you may assume round-rotor machines and neglect resistance drops.)

6-2. A salient-pole generator, initially operating at no load and at rated terminal voltage, is subjected to a three-phase short-circuit at its terminals. The machine parameters are given below:

$x_d = 1.05$ p.u. ; $x'_d = 0.29$ p.u. ;

$x_q = x'_q = 0.69$ p.u.; $T'_{do} = 5.0$ sec.

(a) Calculate and plot excitation voltage E as well as quadrature-axis transient voltage E'_q as functions of time for each of the following conditions:

(i) Constant exciter voltage E_{fd};

(ii) Linear exciter voltage build-up at the rate of 2.5 units per second.

(b) Under constant exciter voltage conditions, let the three-phase short-circuit at the machine terminals be cleared 0.2 seconds later.

Calculate and plot E_q' as a function of time.

6-3. A salient-pole generator is delivering rated current at 0.91 lagging power factor to an infinite bus having a voltage of 1.00 per-unit. The machine parameters are given below:

$$x_d = 1.15 \quad ; \quad x_q = 0.75 \quad ; \quad x_d' = 0.37 \quad ; \quad x_q' = 0.75$$

H = 2.5 KW-sec./KVA ; Rated frequency = 60 Hz.

A three-phase short-circuit, occurring at the terminals of the generator, is cleared in 0.20 seconds without disconnecting the generator from the bus. Compute and plot generator swing curves with each of the following assumptions:

(i) Constant generator field flux linkage;

(ii) Field decrement taken into account, with T_{do}' = 5.0 seconds.

6-4. A generator is connected to an infinitely large power system as shown in the figure below, with reactance values given in per-unit:

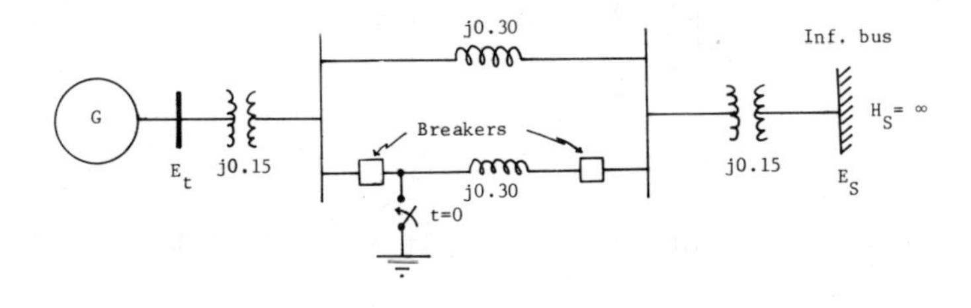

The generator is initially operating at 0.90 per-unit real power and at unity power factor, with the general terminal bus voltage of 1.0 per-unit. The machine parameters are given as:

$x_d' = 0.25$ p.u.; $x_d = 1.95$ p.u.; $H = 2.5$ KW/sec./KVA

A three-phase fault occurs at $t = 0$. During the fault, assume that P_e is 0.15 per-unit and is a constant. The fault is removed by opening the two breakers simultaneously.

(a) Plot the required torque-angle curves and find graphically the critical switching angle. Neglect transient saliency.

(b) Sketch the torque-angle curves and indicate the course traversed by the machine when the fault is cleared at the critical switching angle. Comment on how speed and acceleration change during the transient.

6-5. A generator is connected to an infinitely large power system as shown in the figure below, with reactance values given in per-unit:

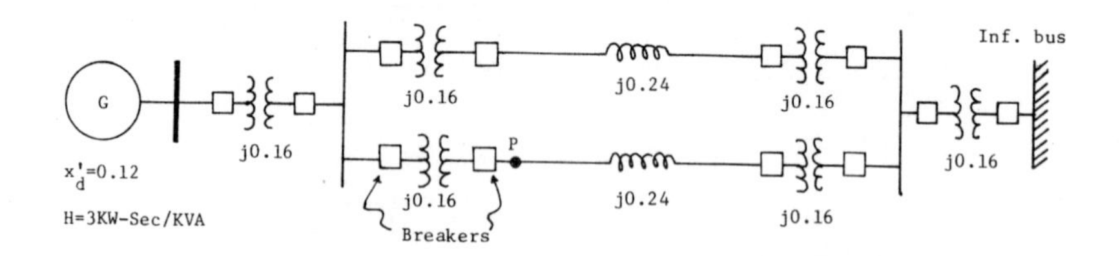

Breakers adjacent to a fault on both sides are arranged to clear simultaneously.

(a) Consider a three-phase fault at the point P, when the generator is delivering 1.0 per-unit power. Assuming that the voltage behind transient reactance is 1.25 p.u. for the generator and the voltage at the infinite bus is 1.0 p.u., calculate the critical clearing angle in electrical degrees for the generator.

(b) Plot swing curves corresponding to the following cases:

(i) fault cleared after 3 cycles.

(ii) fault cleared after 4.5 cycles.

(iii) fault not cleared (i.e., a sustained fault).

(*Hint:* Choose the time step of 0.05 seconds; follow the step-by-step solution of the swing curve; and plot the swing curve over a period of 0.7 seconds for the first two cases and over a period of 0.25 seconds for the third case.)

(c) Comment on whether a 5-cycle or 8-cycle breaker would be satisfactory for the application.

6-6. Considering Eq. (6.4.4) for the accelerating power, the term P_e represents only the positive-sequence power output; P_b is the breaking power; and P_m is the shaft power. Negative-sequence electric power does not appear in Eq. (6.4.4). Justify Eq. (6.4.4) through physical reasoning and balance-sheet of power, explaining how the mechanical power input at the shaft is consumed.

6-7. Consider the machine of Problem 6-3 with the same pre-fault conditions, the fault and the time of clearing. Additional machine parameters are given below:

$x''_d = 0.24$ p.u. ; $x''_q = 0.34$ p.u.

$T'_{do} = 5$ sec.; $T''_{do} = 0.035$ sec.: $T''_{qo} = 0.035$ sec.

Calculate and plot the swing curve, while taking into account both field decrement and damping. Compare this curve with those obtained in the solution of Problem 6-3.

6-8. Find the natural resonance frequency f_e and the slip S for the following configurations:

(a) A generator is connected to an infinitely large power system as shown in the figure below. The transmission line reactance of 1.0 p.u. is being compensated to 0.4 p.u. by the addition of 0.6 p.u. of series capacitance, which is often referred to as 60% line compensation.

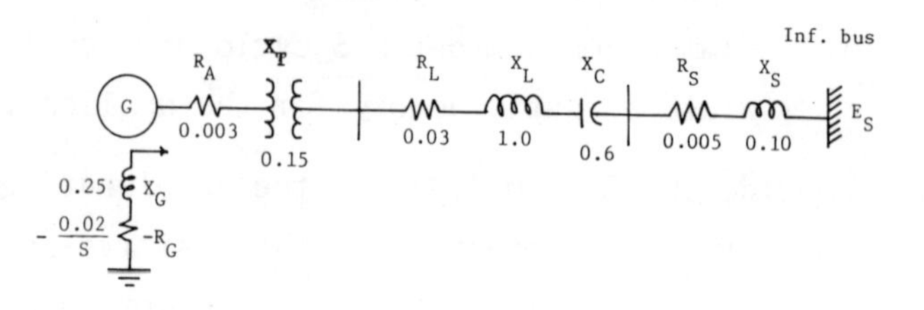

(b) A mine-mouth station is connected to a large system through two 500-KV lines with 75% compensation, as shown below:

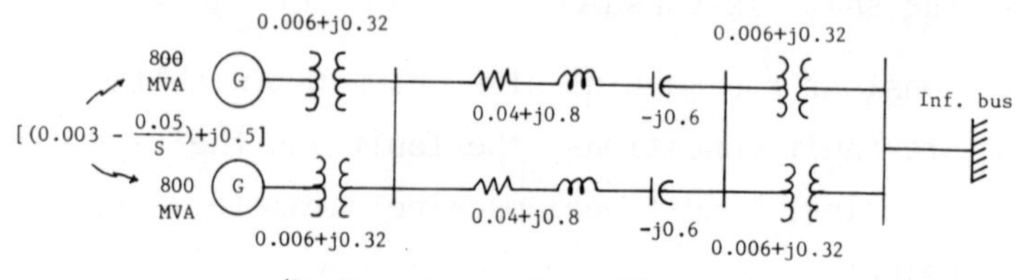

(Impedances in per unit on 1600 MVA base)

Consider also the event in which both lines are in service and one machine is off the line.

6-9. The differential equations relating the general motion of a multi-mass mechanical system, such as a turbine-generator, to a rotational force represent n second-order differential equations coupled to one another due to the off-diagonal elements of the stiffness matrix. These can be uncoupled through the introduction of a new coordinate system, and the transformation would yield the following:

$$[M_m][\ddot{q}] + [K_m][q] + [D_m][\dot{q}] = [T_m]$$

The diagonalized matrices $[M_m]$, $[K_m]$, and $[D_m]$ are referred to as the modal inertia, stiffness and damping matrices. The matrices $[\ddot{q}]$, $[\dot{q}]$, $[q]$ and $[T_m]$ are column matrices. Taking one equation as representative, one has a nonhomogeneous differential equation of one variable:

$$M \frac{d^2q}{dt^2} + D \frac{dq}{dt} + Kq = T$$

We are interested in the frequency response representation to define part of a linear control system relating an output q for an input T under sinusoidal conditions.

Develop an analog computer representation of the control path for the typical differential equation given above.

[*Note:* This problem would provide the basis for the

representation of the modal oscillator in the modal spring-mass model of Figure 6.5.3 in the text.]

6-10. A reduction of the equations of the electrical system, analyzing a change in the electrical torque on a generator due to changes in mechanical motion, is desired while including the effects of series capacitors in the transmission system. Consider the following set of equations relating to the generator and transmission system:

$$e_d = p[-L_d(p)i_d + G(p)e_{fd}] - [-L_q(p)i_q]\,p\delta - ri_d - V_d$$

$$e_q = p[-L_q(p)i_q] + [-L_d(p)i_d + G(p)e_{fd}]\,p\delta - ri_q - V_q$$

$$pV_d = (i_d/C) + V_q p\delta$$

$$pV_q = (i_q/C) - V_d p\delta$$

in which the voltages e_d and e_q are those of the infinite bus, and the voltages V_d and V_q are the voltage drops across the capacitors. Neglecting transient saliency, allowing $L_d(p) = L_q(p)$, and introducing

$$e_d = E_p \sin\delta + E_Q \cos\delta$$

$$e_q = E_p \cos\delta - E_Q \sin\delta$$

linearize the equations and obtain the torque equation for ΔT_e in linearized form as a function of δ.

[*Hint:* Let

$$\frac{1}{L_d(p)\ C} = \omega_n^2 \quad ,$$

where ω_n represents the oscillatory natural frequency. Recognizing that the frequency of interest from the viewpoint of subsynchronous resonance is in the vicinity of $(1 - \omega_n)$, center the interest on $p = j(1 - \omega_n)$ and neglect terms of insignificance in aid of simplification. It would then be possible to express in the form

$$\Delta T_e = K_1\ \Delta\delta + K_2\ \Delta p\delta]$$

CHAPTER VII

EXCITATION SYSTEMS

7.1 Types of Excitation Systems

The source of field current for the excitation of a synchronous machine is an excitation system that includes the exciter, regulator, and manual control. The modern excitation control system of a large synchronous machine is a feedback control system*. Figure 7.1.1 illustrates the essential elements of an automatic control system, while Figure 7.1.2 gives the general layout of the components included in an excitation system. The components commonly used in modern excitation control systems are shown in Table 7.1.1. Figures 7.1.3 through 7.1.11 show the actual configurations of the principal excitation systems currently being supplied by North American manufacturers. The use of solid-state components and the creation of new excitation systems prompted the requirement for new definitions**.

The main exciter could be one of the following:

(a) DC Generator-Commutator Exciter, whose energy is derived from a d-c generator with its commutator and brushes. The exciter may be driven by a motor, prime mover, or the shaft of the synchronous machine.

---

* See Appendix E for details on Block Diagrams, Transient Response, Frequency Response, and Stability Criteria.

** IEEE Committee Report, "Proposed Excitation System Definitions for Synchronous Machines", IEEE Trans. on PA&S, Vol. PAS-88, No. 8, August 1969, pp. 1248-1258. [Figures 7.1.1 through 7.1.11 and Table 7.1.1 are adapted from the above IEEE Committee Report.]

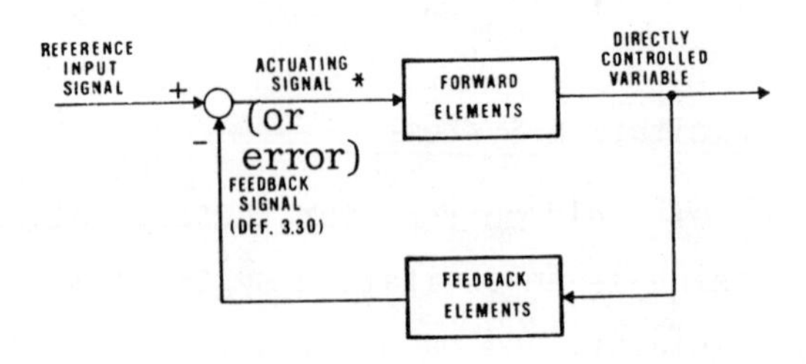

Figure 7.1.1. Essential elements of an automatic feedback control system.

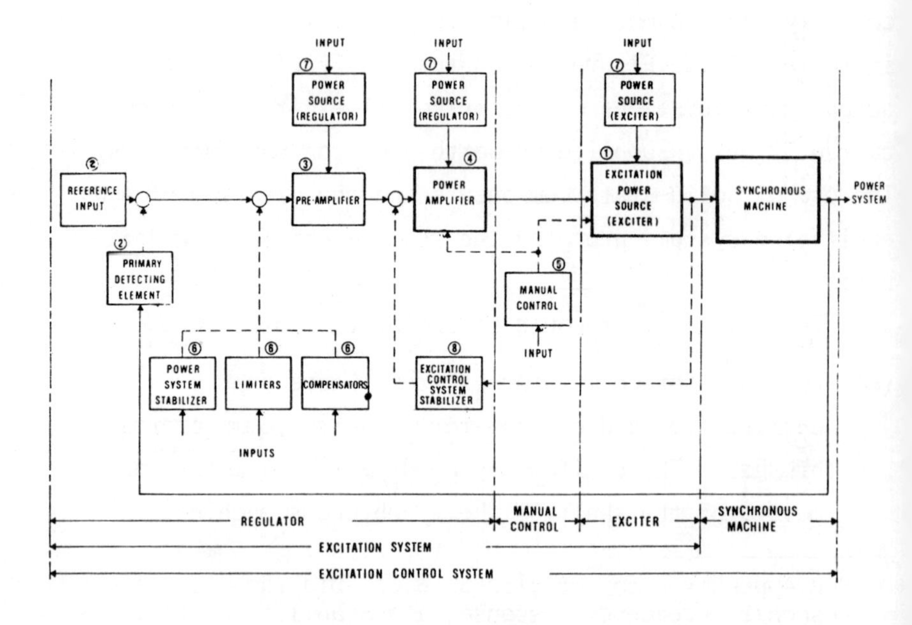

Figure 7.1.2. General layout of the components included in an excitation system.

[*Note:* The Numerals on this diagram refer to the column of Table 7.1.1.]

Table 7.1.1

Components Commonly Used in Excitation Control Systems

(1) TYPE OF EXCITER	(2) PRIMARY DET. ELEMENT & REF. INPUT	(3) PRE-AMPLIFIER	(4) POWER AMPLIFIER	(5) MANUAL CONTROL	(6) SIGNAL MODIFIERS	(7) POWER SOURCES: REGULATOR	(7) POWER SOURCES: EXCITER	(8) EXCITATION SYSTEM STABILIZERS	SYSTEM DIAGRAM REFERENCE
D.C. GENERATOR-COMMUTATOR EXCITER	SEE NOTE (2)	SEE NOTE (3)	ROTATING, MAGNETIC, THYRISTOR	SELF-EXCITED OR SEPARATELY EXCITED EXCITER	SEE NOTE (6)	M-G SET	M-G SET, SYNCHRONOUS MACHINE SHAFT	SEE NOTE (8)	FIGURES 7.1.3 7.1.4
ALTERNATOR-RECTIFIER EXCITER			ROTATING, THYRISTOR	COMPENSATED INPUT TO POWER AMPLIFIER, SELF-EXCITED FIELD VOLTAGE REGULATOR		SYNCHRONOUS MACHINE SHAFT, M-G SET, ALTERNATOR OUTPUT	SYNCHRONOUS MACHINE SHAFT		FIGURES 7.1.5 7.1.6
ALTERNATOR-RECTIFIER (CONTROLLED) EXCITER			THYRISTOR	EXCITER OUTPUT VOLTAGE REGULATOR		ALTERNATOR OUTPUT	SYNCHRONOUS MACHINE SHAFT		FIGURE 7.1.7
COMPOUND-RECTIFIER EXCITER			MAGNETIC, THYRISTOR	SELF-EXCITED		SYNCHRONOUS MACHINE TERMINALS	SYNCHRONOUS MACHINE TERMINALS		FIGURE 7.1.8
COMPOUND-RECTIFIER EXCITER PLUS POTENTIAL-SOURCE-RECTIFIER EXCITER			THYRISTOR	COMPENSATED INPUT TO POWER AMPLIFIER		SYNCHRONOUS MACHINE TERMINALS	SYNCHRONOUS MACHINE TERMINALS		FIGURE 7.1.9
POTENTIAL-SOURCE-RECTIFIER (CONTROLLED) EXCITER			THYRISTOR	EXCITER OUTPUT VOLTAGE REGULATOR, COMPENSATED INPUT TO POWER AMPLIFIER		SYNCHRONOUS MACHINE TERMINALS	SYNCHRONOUS MACHINE TERMINALS		FIGURES 7.1.10 7.1.11

NOTE (2) – PRIMARY DETECTING ELEMENT AND REFERENCE INPUT – CAN CONSIST OF MANY TYPES OF CIRCUITS ON ANY SYSTEM INCLUDING: DIFFERENTIAL AMPLIFIER, AMP-TURN COMPARISON, INTERSECTING IMPEDANCE AND BRIDGE CIRCUITS

NOTE (3) – PRE-AMPLIFIER – CONSISTS OF ALL TYPES BUT ON NEWER SYSTEMS IS USUALLY A SOLID-STATE AMPLIFIER.

NOTE (6) – SIGNAL MODIFIERS – A) AUXILIARY INPUTS – REACTIVE AND ACTIVE CURRENT COMPENSATORS; SYSTEM STABILIZING SIGNALS PROPORTIONAL TO POWER, FREQUENCY, SPEED, ETC.
B) LIMITERS – MAXIMUM EXCITATION; MINIMUM EXCITATION; MAXIMUM VOLTS HERTZ

NOTE (8) – EXCITATION CONTROL SYSTEM STABILIZERS – CAN CONSIST OF ALL TYPES FROM SERIES LEAD-LAG TO RATE FEEDBACK AROUND ANY ELEMENT OR GROUP OF ELEMENTS OF THE SYSTEM

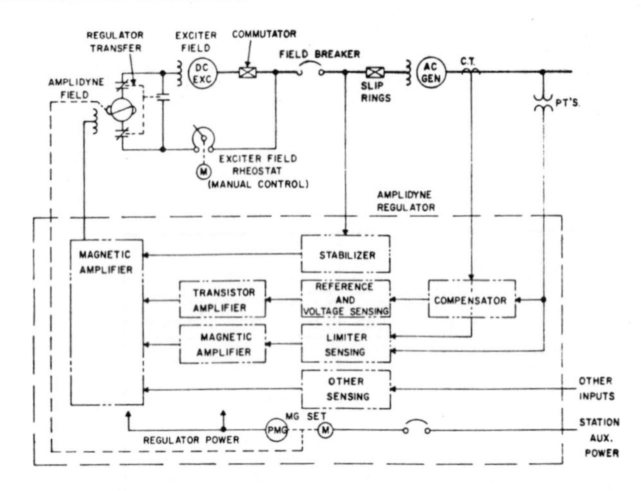

Figure 7.1.3. Excitation control systems with d-c generator-commutator exciter.

Example: General Electric type NA 143 amplidyne system

Reference: G. S. Chambers, A. S. Rubenstein, and M. Temoshok "Recent developments in amplidyne regulator excitation systems for large generators", AIEE Trans. (PA&S) Vol. 80, pp. 1066-1072, 1961 (February 1962 Sec.).

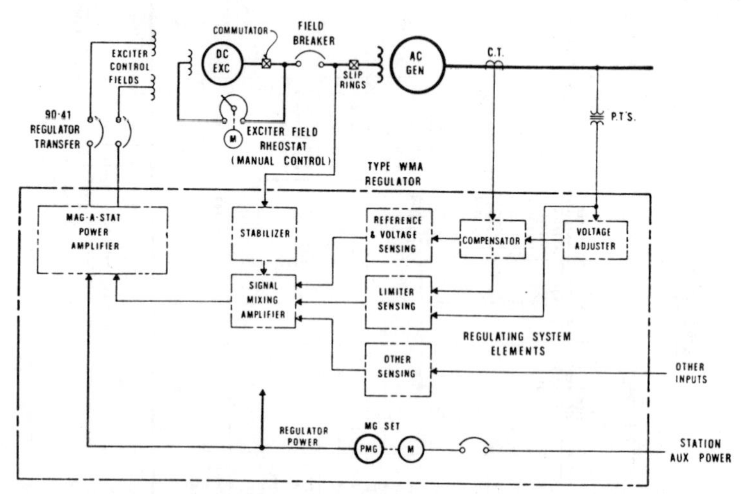

Figure 7.1.4. Excitation control system with d-c generator-commutator exciter.

Example: Westinghouse type WMA MagA-Stat system.

Reference: P. O. Bobo, J. T. Carleton, and W. F. Horton, "A new regulator and excitation system", AIEE Trans. (PA&S), Vol. 72, pp. 175-183, April 1953.

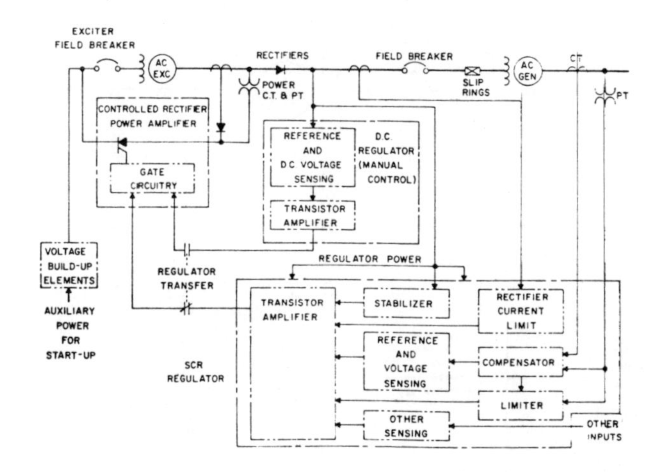

Figure 7.1.5. Excitation control system with alternator-rectifier exciter employing stationary non-controlled rectifiers.
Example: General Electric Alterrex excitation system.
Reference: H. C. Barnes, J. A. Oliver, A. S. Rubenstein, and M. Temoshok, "Alternator-rectifier exciter for Cardinal plant", IEEE Trans. PA&S, Vol. PAS-87, pp. 1189-1198, April 1968.

(b) Alternator-Rectifier Exciter, whose energy is derived from an alternator and converted to d-c by rectifiers. The power rectifiers may be either controlled or non-controlled; they may be either stationary or rotating with the alternator shaft. The alternator may be driven by a motor, prime mover, or the shaft of the synchronous machine.

(c) Compound-Rectifier Exciter, whose energy is derived from the currents and potentials of the a-c terminals of the synchronous machine and converted to d-c by rectifiers. The exciter includes the power transformers (current and potential), power reactors, and power rectifiers which may be either noncontrolled or controlled, including gate circuitry.

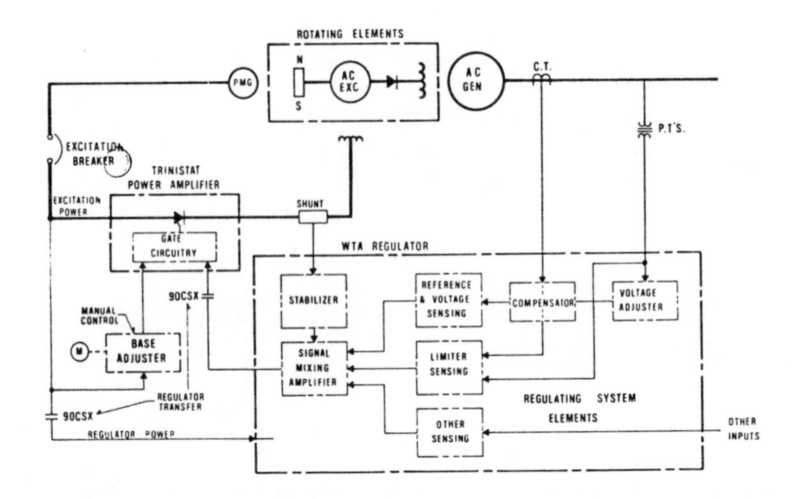

Figure 7.1.6. Excitation control system with alternator-rectifier exciter employing rotating non-controlled rectifiers.

Example: Westinghouse type WTA-Brushless excitation system.

Reference: E. C. Whitney, D. B. Hoover, and P. O. Bobo, "An electric utility brushless excitation system", AIEE Trans. (PA&S), Vol. 78, pp. 1821-1824, 1959 (February 1960 sec.); E. H. Meyers and P. O. Bobo, "Brushless excitation system", Proc. 1966 SWIEEECO.

(d) Potential-Source-Rectifier Exciter, whose energy is derived from a stationary a-c potential source and converted to d-c by rectifiers. The exciter includes the power potential transformers, where used, and power rectifiers which may be either controlled or non-controlled, including gate circuitry.

A synchronous machine regulator couples the output variables of the synchronous machine to the input of the exciter through feedback and forward controlling elements for the purpose of regulating the synchronous machine output

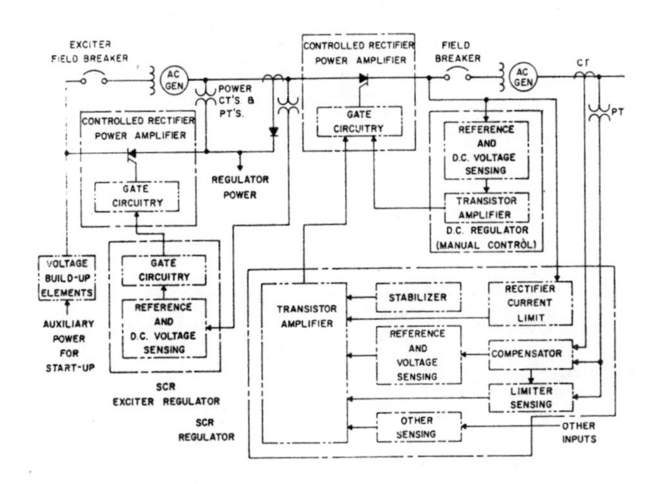

Figure 7.1.7. Excitation control system with alternator-SCR exciter system.
Example: General Electric Althyrex excitation system.
Reference: A. S. Rubenstein and M. Temoshok, "Excitation systems-designs and practices in the United States", presented at Association des Ingenieurs Electriciens de l'Institute Electro-technique Montefiore, A.I.M., Liege, Belgium, May 1966.

variables. The regulator generally consists of an error detector, preamplifier, stabilizers, auxiliary inputs, and limiters. A regulator may be continuously acting or non-continuously acting; the former type initiates a corrective action for a sustained infinitesimal change in the controlled variable, while the latter one requires a sustained finite change in the controlled variable to initiate corrective action. Regulators in which the regulating function is accomplished by mechanically varying a resistance are known as the rheostatic-type regulators.

Power system stabilizers provide additional input to the regulator to improve power system dynamic performance. Quantities such as shaft speed, frequency, synchronous

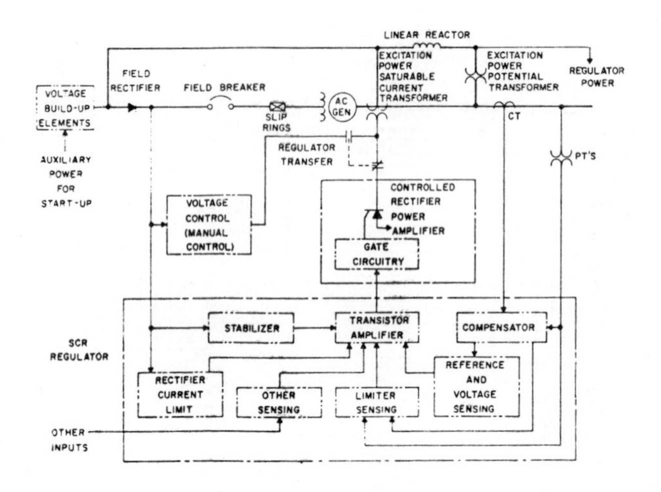

Figure 7.1.8. Excitation control system with compound-rectifier exciter.
Example: General Electric SCTP static excitation system.
Reference: L. M. Domeratzky, A. S. Rubenstein, and M. Temoshok, "A static excitation system for industrial and utility steam turbine-generators", AIEE Trans. (PA&S), Vol. 80, pp. 1072-1077, 1961 (February 1962 sec.); L. J. Lane, D. F. Rogers and P. A. Vance, "Design and tests of a static excitation system for industrial and utility steam turbine-generators", AIEE Trans. (PA&S), Vol. 80, pp. 1077-1085, 1961 (February 1962 sec.).

machine electrical power may be used as input to the power system stabilizer. Other signal modifiers include limiters and compensators. Limiters act to limit a variable by modifying or replacing the function of the primary detector element when predetermined conditions have been reached. Examples include maximum excitation, minimum excitation, and maximum volts per hertz. Compensators act to compensate for the effect of a variable by modifying the function of the primary detecting element. Examples include reactive current compensator, active current compensator, and line drop

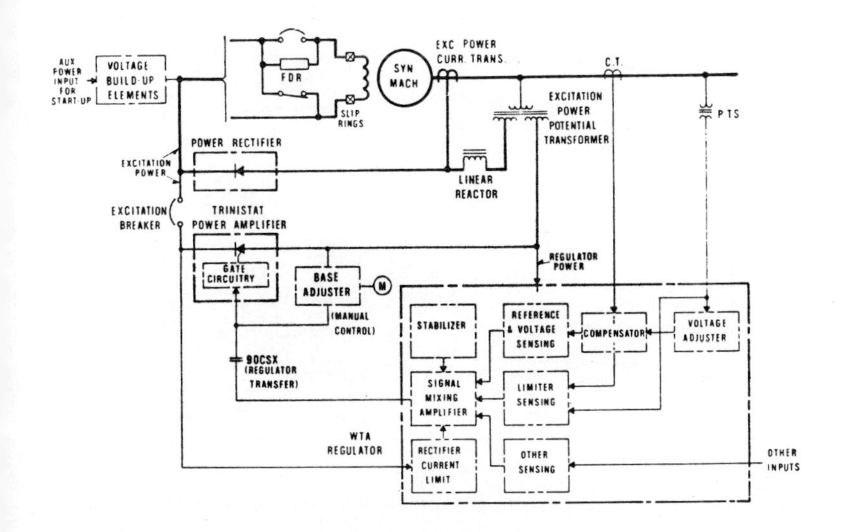

Figure 7.1.9. Excitation control system with compound-rectifier plus potential-source-rectifier exciter.
Example: Westinghouse type WTA-PCV static excitation system.
Reference: C. H. Lee and F. W. Keay, "A new excitation system and a method of analysing voltage response", 1964 IEEE Int'l Conv. Rec., Vol. 12, Pt. 3, pp. 5-14.

compensators. To obtain reactive current sharing among generators operating in parallel, generator voltage regulators are equipped with reactive compensators. Reactive droop compensation, which is more common, creates a droop in generator voltage proportional to reactive current and equivalent to that which would be produced by the insertion of a reactor between the generator terminals and the paralleling point. In cases where droop in generator voltage is not desired, reactive differential compensation is used by a series differential connection of the various generator current transformer secondaries and reactive compensators.

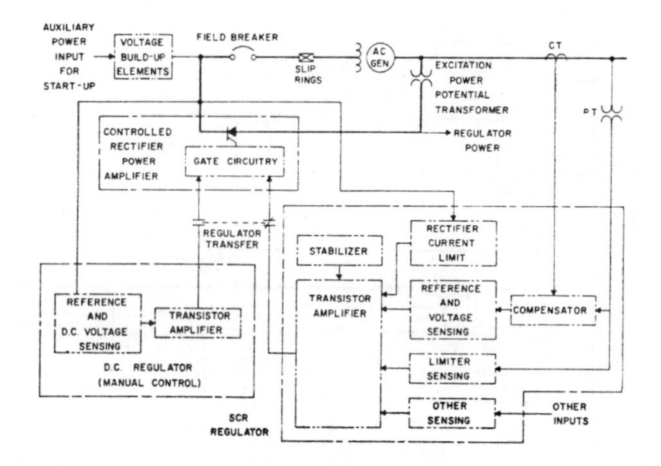

Figure 7.1.10. Excitation control system with potential-source-rectifier exciter.
Example: General Electric SCR static excitation system.

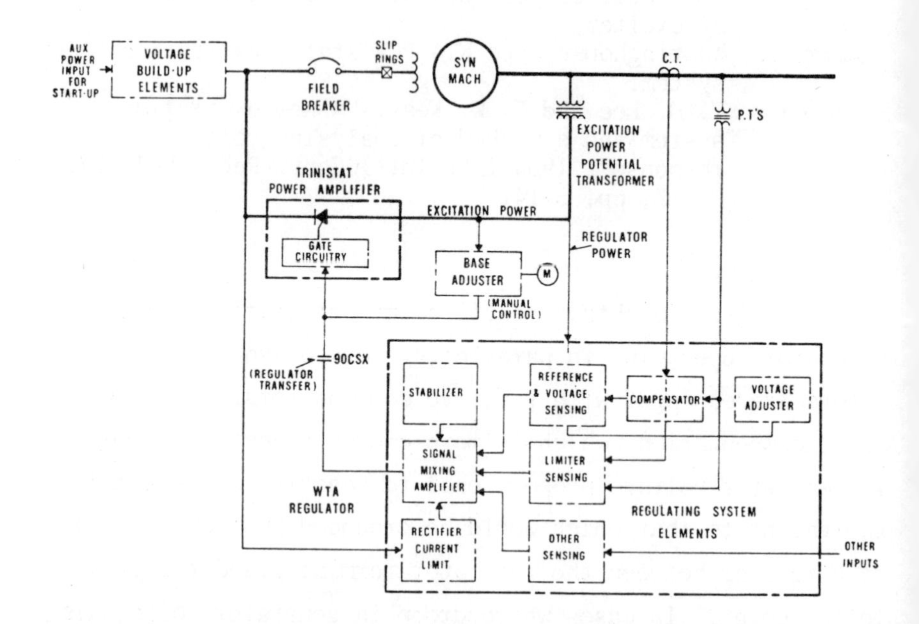

Figure 7.1.11. Excitation control system with potential-source-rectifier exciter.
Example: Westinghouse type WTA-Trinistat static excitation system.

The difference current for any generator from the common series current produces a compensating voltage in the input to the particular generator voltage regulator which acts to modify the generator excitation to reduce its differential reactive current to zero. Line drop compensators introduce within the regulator input circuit a voltage equivalent to the impedance drop and thereby modify generator voltage. The voltage drops of the resistive and reactive portions of the impedance are obtained respectively by active and reactive compensators.

The excitation control system stabilizer modifies the forward signal by either series or feedback compensation to improve the dynamic performance of the excitation control system, by eliminating undesired oscillations and overshoot of the regulated voltage.

The exciter control system provides the proper field voltage to maintain a desired system voltage. An important characteristic of such a control system is its ability to respond rapidly to voltage deviations during both normal and abnormal or emergency system operation. While many different types of exciter control systems are employed in practice, the basic principle of operation may be described in the following way: The voltage deviation signal is amplified to produce the signal required to change the exciter field current, which in turn produces a change in exciter output voltage thereby resulting in a new level of excitation for the synchronous generator.

7.2 Exciter Response

The excitation system voltage-time response curve is shown by the solid line in Figure 7.2.1. The straight line

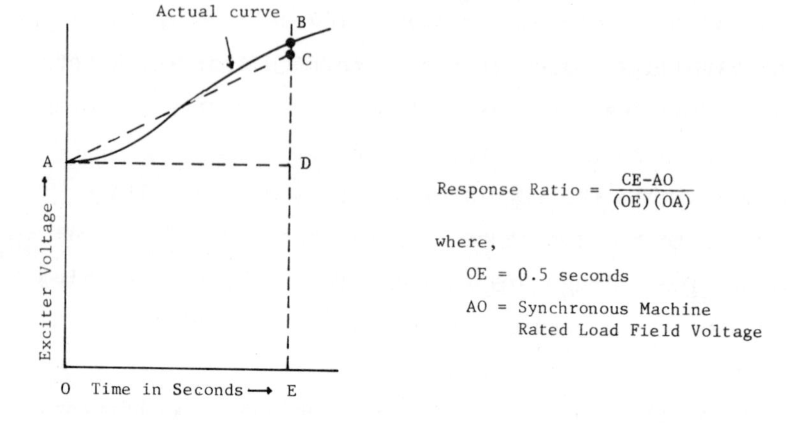

Figure 7.2.1. Exciter or synchronous machine excitation system voltage response.

AC is drawn in such a way that the area ACD is equal to the area ABD enclosed by the actual curve. The starting point A for determining the rate of voltage change is given by the initial value OA of the excitation system voltage-time response curve. OA is the synchronous machine rated load field voltage. The excitation system voltage response is given by the rate of increase or decrease of the excitation system output voltage, i.e., the slope determined by $\frac{CE - AO}{OE}$ in Figure 7.2.1. The time in seconds for the excitation voltage to reach 95 percent of ceiling voltage is known as the excitation system voltage response time. An exciting system having an excitation system voltage response time of 0.1 second or less is to be understood as a high initial response excitation system. The excitation system voltage response ratio is obtained as $\frac{CE - AO}{(OE)(OA)}$, where OE is taken to be 0.5 second, unless otherwise specified. This ratio usually applies only to the increase in excitation system voltage, because the build-up action is required during and immediately after the presence of a fault in order to improve system stability. The time of 0.5 seconds is chosen because the time during which quick-response excitation is applied in practice is of this order of magnitude. The 0.1-second time interval is chosen as a practical value for any high-response, fast-acting system. Similar definitions of voltage response and response ratio can be applied to the excitation system major components such as the exciter and regulator. The main exciter response ratio, also known as nominal exciter response, is determined with no load on the exciter (and at rated speed for a rotating exciter), because change of voltage caused by load is not great and also because both test and calculation are simplified immensely by the assumption of no load.

Calculation of Exciter Response:

The exciter response is usually shown by curves of exciter armature voltage as a function of time. Such curves can be determined quite easily and accurately by test procedures. However, calculating methods are needed for building new exciters or for making alterations in the existing exciters, and may be divided as follows depending on the technique used for solving the resultant differential equations with variable coefficients:

(a) Formal Integration
(b) Graphical Integration
(c) Point-by-Point Calculation
(d) Computer Methods (analog or digital).

Since separate excitation gives faster exciter response than self-excitation and also it has an advantage over self-excitation in the possibility of decreasing the time constant of the field circuit by external series resistance, we shall consider the calculation of response of a separately excited exciter, the circuit diagram of which is shown in Figure 7.2.2. The voltage build-up for the unloaded case will now be analyzed.

The loop equation around the field circuit of the main exciter is given by

$$N \frac{d\phi_f}{dt} + Ri = E \tag{7.2.1}$$

where N is the number of field turns in series, ϕ_f is the flux in webers per pole, R is the field-circuit resistance, i is the field current, and E is the pilot-exciter armature voltage. The effect of eddy currents induced in solid parts of the magnetic circuit is neglected here. E may be taken as a constant for the case of a flat-compounded pilot exciter; or correction may be applied suitably to take care of under-compounded or shunt-wound effect. Assuming the speed of the exciter to be a constant, the armature voltage e of the main exciter is directly proportional to the armature flux ϕ_a crossing the air-gap and linking the armature conductors:

$$e = k\phi_a \tag{7.2.2}$$

in which the proportionality constant k is given by

$$k = \frac{Znp}{60a} \tag{7.2.3}$$

where Z represents total number of armature conductors, n is the speed of rotation of exciter in rpm, p is the number of

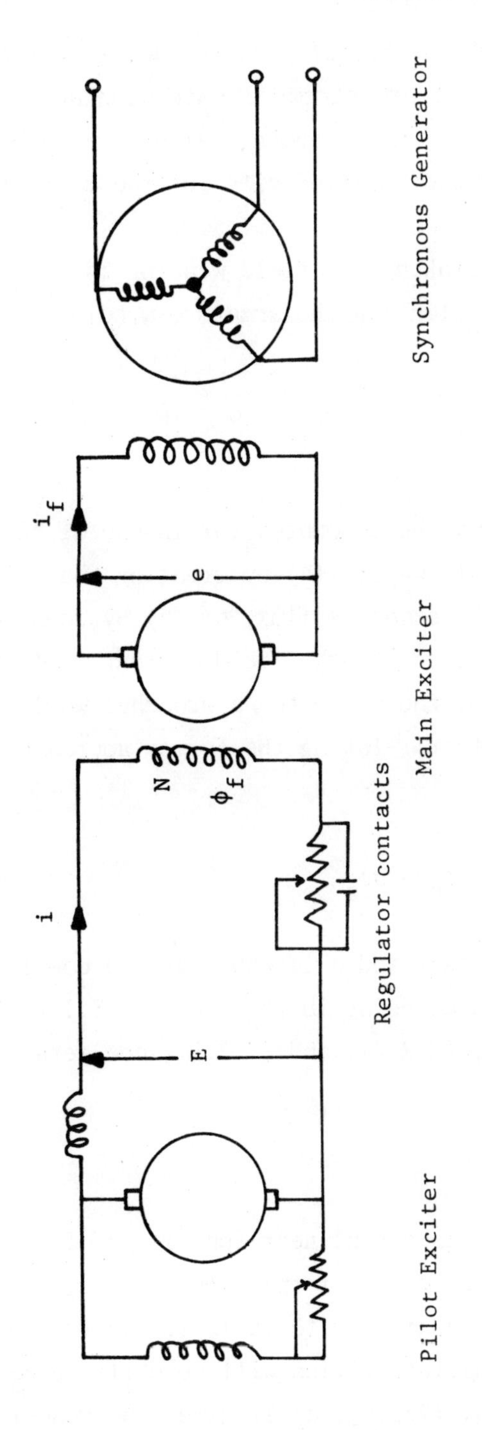

Figure 7.2.2. Separately excited exciter of a synchronous generator.

poles, and a is the number of parallel paths of the armature (which is equal to 2 for a simplex wave winding, and p for a simplex lap winding; the foregoing values are to be multiplied by the degree of multiplicity for the case of multiplex windings).

The flux ϕ_f linking the field winding is given by the sum of the flux ϕ_a linking the armature winding and the leakage flux ϕ_ℓ:

$$\phi_f = \phi_a + \phi_\ell \tag{7.2.4}$$

The leakage flux may be accounted for in one of the two possible approximate ways: (a) by treating that it is proportional to the armature flux, or (b) by considering that it is proportional to the field current. If there were no saturation, the above two approaches would be equivalent to one another. Taking the former approach, one has

$$\phi_f = (1 + k_1)\,\phi_a = \sigma\phi_a \tag{7.2.5}$$

where k_1 is a constant and σ is known as the coefficient of dispersion, its value being in the range of 1.1 to 1.2. From Eqs. (7.2.1), (7.2.2), and (7.2.5), one gets

$$\frac{N\sigma}{k}\frac{de}{dt} + Ri = E \tag{7.2.6}$$

Recognizing that e is a nonlinear function of i, as given by the magnetization curve of the main exciter, Eq. (7.2.6) is a differential equation with variable coefficients. The method of graphical integration will be utilized here for the solution of Eq. (7.2.6), as it gives the reader a better physical picture of the happenings.

Referring to Figure 7.2.3(a), let us consider the armature voltage and field current building up from a point on the magnetization curve having coordinates i_1 and e_1 to a point having coordinates i_2 and e_2. Equation (7.2.6) is to be rearranged so as to express time t as an integral. Multiplying Eq. (7.2.6) by e_2/E, one gets

$$\frac{N\sigma}{k}\frac{e_2}{E}\frac{de}{dt} = e_2 - \frac{Re_2}{E} i = e_2 - \frac{e_2}{i_2} i \tag{7.2.7}$$

which may be expressed as

$$T\frac{de}{dt} = \Delta e \tag{7.2.8}$$

where T stands for $\left(\frac{N\sigma}{k}\frac{e_2}{E}\right)$ and has dimensions of time. Solving for t, one obtains

$$t = T\int_{e_1}^{e} \frac{de}{\Delta e} \tag{7.2.9}$$

which gives the time required for the voltage to build up from its initial value e_1 to any value e. The corresponding Δe for a point having coordinates i and e is shown in Figure 7.2.3(a). By plotting $\frac{1}{\Delta e}$ as a function of e as in Figure 7.2.3(b), and measuring the area between the curve and the e-axis between the lower limit e_1 and a running upper limit e, one can evaluate the integral graphically and obtain a plot of e as a function of time as illustrated in Figure 7.2.3(c).

The retarding effect of eddy currents in the iron may be approximately taken care of by increasing T by an empirical

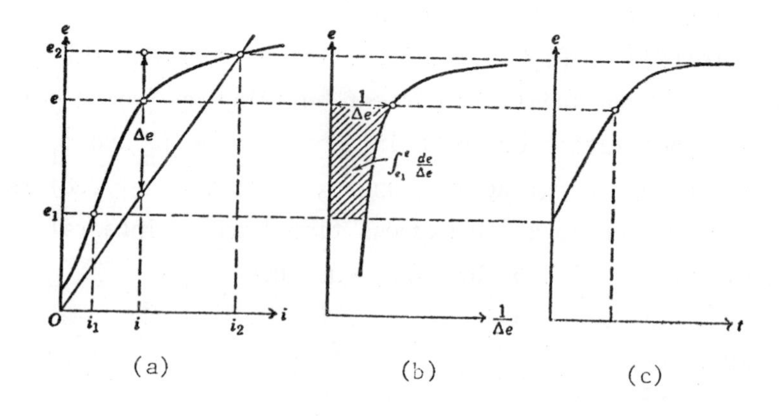

Figure 7.2.3. Graphical integration method for the voltage build-up of a separately excited exciter.

amount such as 5%. The effect of load will be to reduce Δe due to the effects of armature resistance (including brush drop), armature inductance, and armature reaction, and hence to increase the time of build-up. Besides reducing the response by a few percent, the effect of load is also to reduce the ceiling voltage by a small percent. The assumption that the voltage-time curve for no-load is valid also for the load condition causes little error for the case of separately excited exciters.

In the point-by-point calculation method, Eq. (7.2.6) will be rearranged as

$$\Delta t = \frac{(N\sigma/k)\ \Delta e}{(E - Ri)} \tag{7.2.10}$$

by replacing $\frac{de}{dt}$ by $\frac{\Delta e}{\Delta t}$ in terms of finite increments. The time Δt required for the armature voltage to change by an assumed increment of Δe may be computed from Eq. (7.2.10), while making use of the average value for i during the interval. Starting at zero time with the known initial value of voltage, the voltage-time curve can be plotted from the increments of voltage and time*.

7.3 Excitation System Modeling

Until the 1960's, transient stability studies generally considered no control system reactance and generators were represented as constant voltage behind transient reactance. The availability of large digital computers permitted more rapid and economical computations. When a detailed analysis of system response is required or the period of interest extends beyond one second, it is important to include the effects of the exciter and governor system. It is no longer necessary to assume a constant voltage behind a reactance and more realistic as well as accurate machine representations are now available.

We shall now consider control system representations of the various excitation systems currently available from the viewpoint of system studies. A convenient form of representing a control system is a block diagram** that relates the input and output variables of the principal components of the system through transfer functions. After considering different types of excitation systems (now in service in the United States and Canada, and contemplated

---

* See Kimbark, E. W., Power System Stability: Synchronous Machines, Ch. XIII, Dover Publications, Inc., New York, 1956/1968.
[Figure 7.2.3 is adapted from the above Reference.]

** See Appendix E for details.

for the immediate future), an IEEE Committee* has defined four excitation system types (Figures 7.3.1 to 7.3.4) to be used in computer representations. These should be adequate to represent all modern systems; the symbols used are explained directly below the figures.

In the development of the excitation system block diagrams, the per-unit system is usually chosen for convenience so that one per-unit generator voltage is defined as rated voltage, and one per-unit exciter output voltage is that voltage required to produce rated generator voltage on the generator air-gap line.

Type 1 Excitation System Representation - Continuing Acting Regulator and Exciter:

The type 1 excitation system is representative of most continuously acting systems with rotating exciters such as:

Allis Chalmers --	Regulex Regulator
General Electric --	Amplidyne Regulator (Figure 7.1.3) Alterrex (Figure 7.1.5) Alterrex-Thyristor (Figure 7.1.7)
Westinghouse --	Mag-A-Stat Regulator (Figure 7.1.4) Brushless System (Figure 7.1.6) Rototrol Silverstat Regulator TRA Regulator

Referring to Figure 7.3.1, V_T is the generator terminal

*IEEE Committee Report, "Computer Representation of Excitation Systems", IEEE Trans. on PA&S, Vol. PAS-87, No. 6, June 1968, pp. 1460-1464.
[Figures 7.3.1 through 7.3.5 and Table 7.3.1 are adapted from the above IEEE Committee Report.]

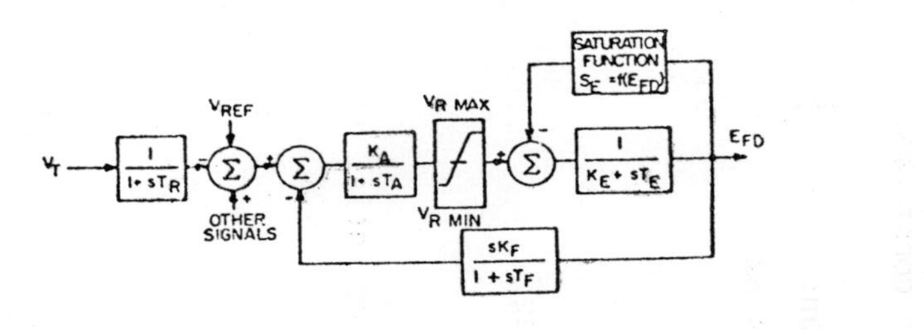

Figure 7.3.1. Type 1 excitation system representation, continuously acting regulator and exciter.

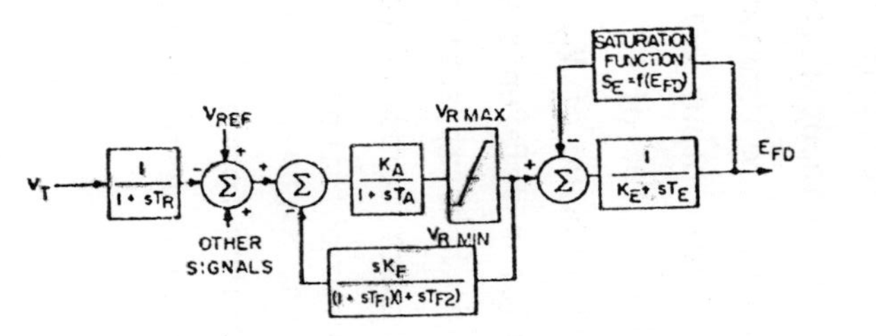

Figure 7.3.2 Type 2 excitation system representation, rotation rectifier system.

(SEE NEXT PAGE FOR NOMENCLATURE.)

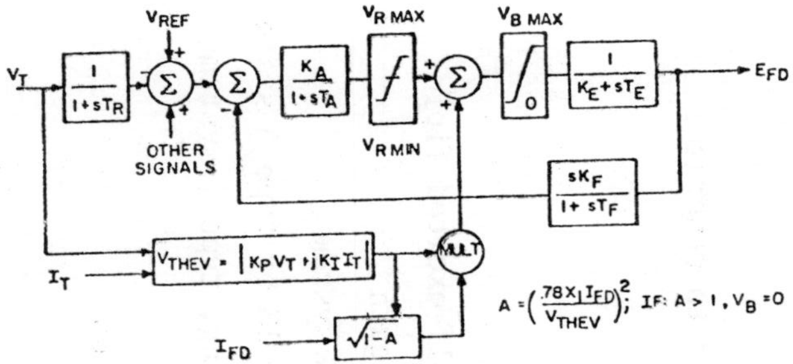

Figure 7.3.3. Type 3 excitation system representation, static with terminal potential and current supplies.

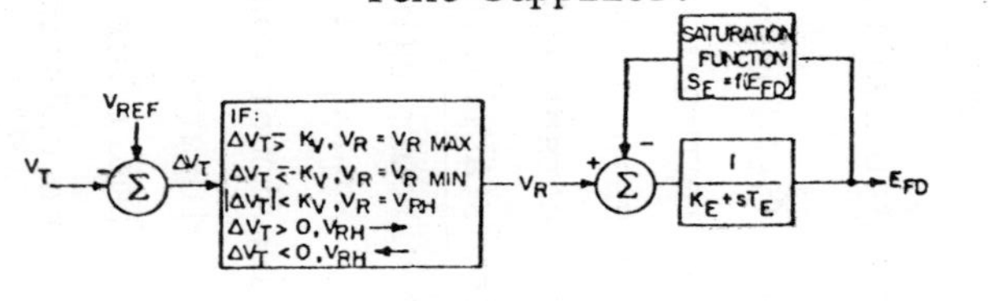

Figure 7.3.4. Type 4 excitation system representation, noncontinuously acting regulator.

[Note: V_{RH} limited between V_{Rmin} & V_{Rmax}; Time constant of rheostat travel T_{RH}.]

Nomenclature

E_{FD}	exciter output voltage (applied to generator field)
I_{FD}	generator field current
I_T	generator terminal current
K_A	regulator gain
K_B	exciter constant related to self-excited field
K_F	regulator stabilizing circuit gain
K_I	current circuit gain of type 3 system
K_P	potential circuit gain of type 1S or type 3 system
K_V	fast raise/lower contact setting, type 4 system
S_E	exciter saturation function
T_A	regulator amplifier time constant
T_E	exciter time constant
T_F	regulator stabilizing circuit time constant
T_{F1}, T_{F2}	regulator stabilizing circuit time constants (rotating rectifier system)
T_R	regulator input filter time constant
T_{RH}	rheostat time constant, type 4 system
V_R	regulator output voltage
$V_{R\,max}$	maximum value of V_R
$V_{R\,min}$	minimum value of V_R
V_{REF}	regulator reference voltage setting
V_{RH}	field rheostat setting
V_T	generator terminal voltage
V_{THEV}	voltage obtained by vector sum of potential and current signals, type 3 system
ΔV_T	generator terminal voltage error
S	Laplace transform variable

voltage applied to the regulator input; T_R is a very small time constant representing regulator input filtering. The first summing point determines the voltage error input to the regulator amplifier. The second summing point combines this with the excitation major damping loop signal, which is provided by the feedback transfer function $SK_F/(1 + ST_F)$ from exciter output voltage E_{FD}. The main regulator transfer function is characterized by a gain K_A and a time constant T_A. Following this, the maximum and minimum regulator limits are imposed so that large input error signals will not produce a regulator output exceeding practical limits. The exciter output voltage or the generator field voltage E_{FD} is multiplied by a nonlinear saturation function and subtracted from the regulator output signal. The resultant is applied to the exciter transfer function $1/(K_E + ST_E)$.

The exciter saturation function S_E represents the increase in exciter excitation requirements because of saturation. The calculation of a particular value of S_E is illustrated in Figure 7.3.5, in which the quantities A and B are defined as the exciter excitation to produce the output voltage on the constant-resistance load-saturation curve and air-gap line respectively.

An interrelation amid exciter ceiling voltage $E_{FD\ max}$, regulator ceiling voltage $V_{R\ max}$, exciter saturation factor S_E, and the exciter constant K_E is to be satisfied:

$$V_{R\ max} - (K_E + S_{E\ max})\ E_{FD\ max} = 0 \qquad (7.3.1)$$

K_E is usually specified, and the specification of any two of the remaining leads to establish the third. Under steady-state conditions, the following must be satisfied:

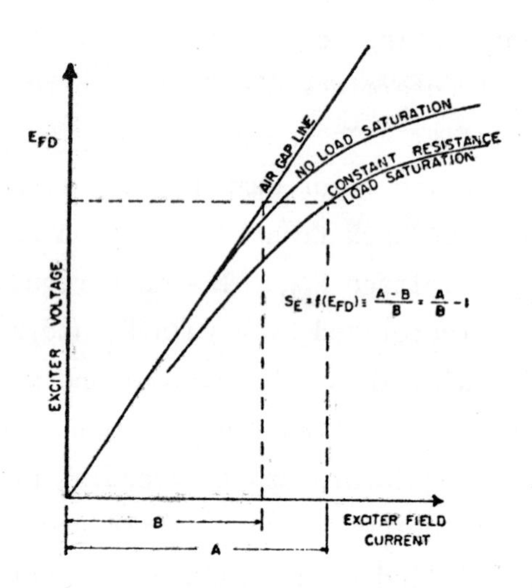

Figure 7.3.5. Exciter saturation curves showing procedure for calculating the saturation function S_E.

$$V_R - (K_E + S_E)\, E_{FD} = 0 \quad , \quad E_{FD\,min} \leq E_{FD} \leq E_{FD\,max} \tag{7.3.2}$$

It may be pointed out here that the sign of K_E is negative for a self-excited shunt field. For some systems employing noncommutating-type exciters, the minimum value of E_{FD} is zero and cannot be negative.

Table 7.3.1 gives typical excitation system constants of modern 2-pole steam-turbine generators.

Type 1S - Controlled-Rectifier Systems with Terminal Potential Supply only:

Excitation systems such as those shown in Figures 7.1.10 and 7.1.11 employing an excitation source from terminal voltage with controlled rectifiers only form a special case

Table 7.3.1

Typical Constants of Excitation Systems in Operation on 2-Pole Steam-Turbine Generators (Excitation system voltage response ratio - 0.5 p.u.)

Symbol	Self-Excited Exciters, Commutator or Silicon Diode, with Amplidyne Voltage Regulators	Self-Excited Commutator Exciter with Mag-A-Stat Voltage Regulator	Rotating Rectifier Exciter with Static Voltage Regulator
T_R	0.0 to 0.06	0.0	0.0
K_A	25* to 50*	400.0	400.0
T_A	0.06 to 0.20	0.05	0.02
$V_{R\ max}$	1.0	3.5	7.3
$V_{R\ min}$	-1.0	-3.5	-7.3
K_F	0.01 to 0.08	0.04	0.03
T_F	0.35 to 1.0	1.0	1.0
K_E	-0.05	-0.17	1.0
T_E	0.5	0.95	0.8
$S_{E\ max}$**	0.267	0.95	0.86
$S_{E\ 0.75\ max}$***	0.074	0.22	0.50

* For generators with open-circuit field time constants greater than four seconds.

** Corresponds to $E_{FD\ max}$.

*** Corresponds to 0.75 $E_{FD\ max}$.

of type 1. The ceiling voltage is proportional to generator terminal voltage:

$$V_{R\ max} = K_P\ V_T \qquad (7.3.3)$$

and in general, the constants for this type of system are such that

$$K_E = 1.0 \quad , \quad T_E = 0 \quad , \quad \text{and } S_E = 0 \qquad (7.3.4)$$

Type 2 Excitation System - Rotating Rectifier System:

This type, shown in Figure 7.3.2, applies for generator units with major damping loop input from the regulator output. An additional time constant is introduced into the transfer function in order to compensate for the exciter that is not included within the damping loop. Other characteristics of type 2 systems are very similar to those of type 1. An example of this type is the Westinghouse brushless system put into service up to and including 1966.

Type 3 Excitation System - Static with Terminal Potential and Current Supplies:

Static systems such as those shown in Figures 7.1.8 and 7.1.9, in which generator terminal current is used with potential as the excitation source, fall under this category. Figure 7.3.3 gives the transfer functions making up a type 3 system. K_P is the coefficient of the shunt excitation supply proportional to terminal voltage, and K_I is the coefficient of the supply obtained from terminal current transformers. The multiplier (MULT of Figure 7.3.3) allows for the variation of self-excitation with change in the angular relation of field current I_{FD} and self-excitation voltage V_{THEV}. The regulator output after the limiter is combined with the signal representing the self-excitation from the generator terminals. The excitation system output is set to zero by the $V_{R\ max}$ limiter when the field current exceeds the excitation output current (i.e., when A > 1). Excess generator field current bypasses the excitation supply by flowing through the output rectifier. The excitation transformer is represented by the transfer function $1/(K_E + ST_E)$, while damping is provided by the beedback transfer function $sK_F/(1 + ST_F)$.

Type 4 Excitation System - Noncontinuously Acting:

In this type, shown in Figure 7.3.4, no major damping loop is represented. Examples include noncontinuously acting excitation systems such as the General Electric GFA4 regulator and Westinghouse BJ30 regulator, that were used prior to the development of the continuously acting excitation systems. In these, adjustment is made with a motor-operated rheostat for small magnitudes of voltage error. Larger errors cause resistors to be quickly shorted or inserted, and a strong forcing signal is applied to the exciter.

Referring to Figure 7.3.4, different regulator modes are specified depending upon the magnitude of voltage error ΔV_T. If voltage error is larger than the fast raise/lower contact setting K_V (typically 5 percent), $V_{R\ max}$ or $V_{R\ min}$ is applied to the exciter depending upon the sign of the voltage error. For a voltage error less than K_V, the exciter input is the same as the rheostat setting V_{RH}, which is adjusted up or down depending upon the sign of the error.

T_{RH} is the time constant allowing for the slow adjustment of the exciter field voltage.

A Note on the Regulator Input Signals:

In addition to the terminal voltage, other signals such as accelerating power, speed, frequency, and rate of change of terminal voltage are employed to provide positive damping of power-system oscillations to improve generator stability and damp the line oscillations. Such signals are added at the voltage reference summing point to the terminal voltage error, as shown in Figures 7.3.1 through 7.3.4 as other signals. As indicated in Section 7.1, such a stabilizing signal is introduced through a transfer function providing gain adjustment and a lead-lag comepnsation for phase shifting. Depending upon the individual case under consideration, specific representation needs to be made to account for such stabilizing signals.

7.4 Field Excitation in Relation to Machine and System Operation

The excitation system affects stability under both transient and steady conditions. Either in a two-machine system or in a multi-machine system, based on the knowledge gained from Sections 6.2 and 6.3, it can be seen that raising the internal voltages of the machines increases the stability limits. Also, it is generally true that the higher the speed of exciter response, the greater is the power limit. Even with short fault duration in view of the standard practice of high-speed clearing of faults, the excitation system can aid transient stability appreciably by lessening the decrease of flux linkages or causing an increase of linkages. The faster the excitation system responds to correct the voltage, the more effective it will be in improving stability. A voltage stabilizer can be used to insure that stability will be maintained on subsequent swings, provided the machine survives the first swing of its rotor in the case of a sustained fault.

Quick-response excitation systems raise the steady-state stability limit; the limit is approximately equal to that calculated on the assumption of constant voltage behind a reactance intermediate between saturated synchronous reactance and transient reactance. That is to say that quick-response excitation system counteracts approximately half of the armature reaction. The recent trend in building large turbo-generators of lower short-circuit ratio inherently implies lower steady-state stability, the margin of which can be increased with the use of suitable excitation system. The excitation system should have a voltage regulator with no dead band in order to be effective in increasing the steady-state stability limit. The gain in steady-state stability limit ascribable to the excitation system is small if the external impedance is large. Properly designed excitation systems, in addition to raising stability limits, improve the quality of electric power service by reducing the voltage disturbance caused by such events as a short circuit, the opening of a transmission line, swinging of machines, or the disconnection of a generator.

Let us now take a look at the generator capabilities* within which the machine may be operated, and the effects of

---

*See Farnham, S. B., and Swarthout, R. W., "Field excitation in relation to machine and system operation", AIEE Trans. Paper 53-387, November 1953. [Figures 7.4.1 through 7.4.6 are adapted from the above Reference.]

field excitation on the individual machine as well as on the system of which that machine is a part. As previously indicated in Section 2.2, a synchronous machine can be operated in any quadrant of Figure 7.4.1, in which point A is plotted corresponding to the name-plate rated conditions for an assumed typical 0.85 power factor generator. We shall limit our discussion primarily for generator operation and as such restrict our attention to the right-hand or positive kilowatt side of Figure 7.4.1. Since it is usual that generators be suitable for delivering rated kilovolt-amperes at unity power factor corresponding to point B in Figure 7.4.1, one can draw the arc AB with its center at 0 and radius equal to rated armature current.

Starting with rated terminal voltage e and rated armature current i (each equal to 1.0 per unit), Figure 7.4.2 gives a typical simplified generator phasor diagram corresponding to a power-factor angle of ϕ. E is the internal or generated voltage corresponding to the rated terminal conditions. It would also be equal to the terminal voltage if full load were removed without making any change in field current.

Dividing the three sides of the phasor triangle by x_d and reorienting the diagram to suit the kilowatt-kilovar coordinates of Figure 7.4.3, triangle OAC is constructed in which OA represents the rated armature current, OC corresponds to the short-circuit ratio ($1.0/x_d$), and CA represents the rated full-load field current. Then with C as center and CA as radius, the arc AD is drawn representing the locus of the rated field current, thereby closing off the top of the area within which the machine may be operated. Thus the output of the typical machine is limited by field heating from D to A and by armature heating from A to B.

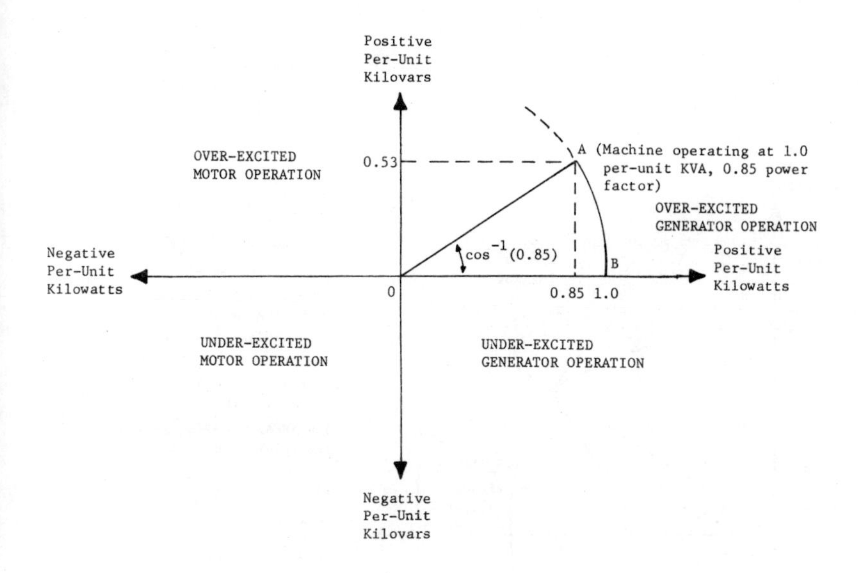

Figure 7.4.1. Synchronous machine operating modes.

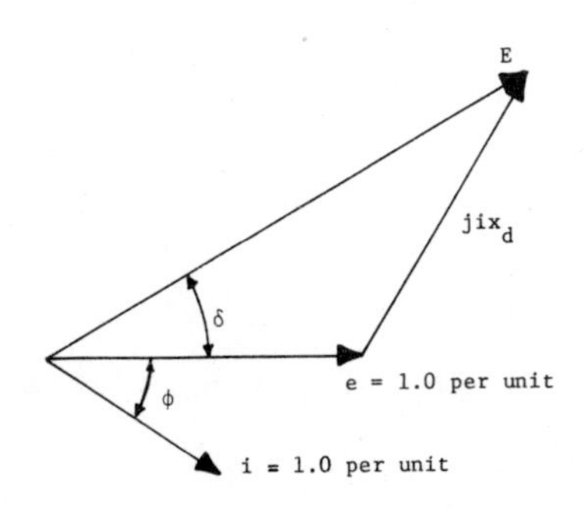

Figure 7.4.2. Typical simplified generator phasor diagram.

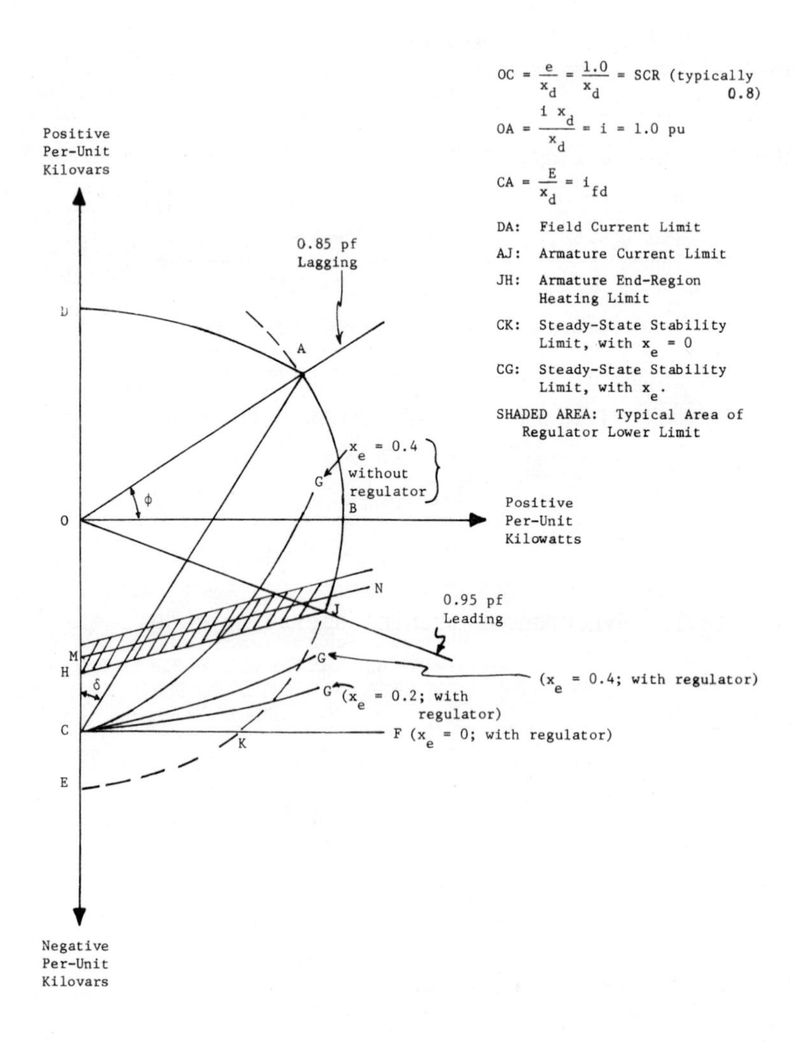

Figure 7.4.3. Generator capability limits.

The vertical intercept OD readily gives the maximum permissible per-unit kilovars to scale, which is a measure of the capability as a synchronous condenser.

Corresponding to the under-excited generator operation in the fourth quadrant, since this is a region of low field current, one might be tempted to extend the armature current arc AB all the way around to E. This could be wrong from the viewpoint of system stability as well as localized heating in the machine iron. Modern generators can be operated successfully in the under-excited region down to a line such as HJ in Figure 7.4.3, where the point H is typically at 60-percent rated KVA with zero power factor leading and the point J is typically at rated KVA with 0.95 power factor leading. It should be pointed out here that these limits may not apply to older machines and as such it may become necessary to investigate the capabilities of each such machine in question. The line such as HJ will then represent the end-region heating limit. While this limit applies to steam-turbine generators in particular, it is not a limitation of water-wheel generators because of their generally different construction.

The reason for the problem of end-region heating stems from the armature reaction end-leakage flux at both ends of the stator. While the main flux in the body of the stator is parallel to the laminations, the end-leakage flux enters and leaves the ends of the stator in a direction essentially perpendicular to the laminations as shown in Figure 7.4.4. Since core losses are typically 100 times greater for perpendicular flux than for flux parallel to the laminations, the effect of laminations in reducing eddy currents caused by the end-leakage flux is diminished. Hence, considerable additional heat is generated in only a relatively small

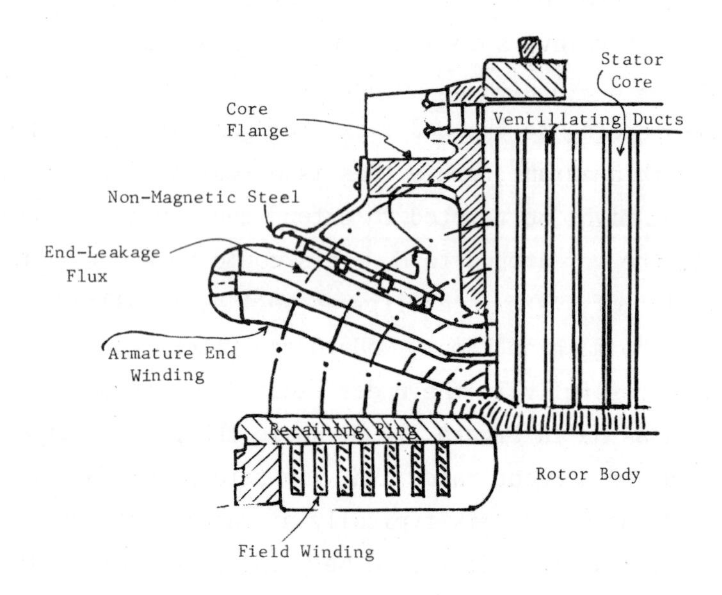

Figure 7.4.4. Sectional view of a modern turbine-generator end region, indicating the end-leakage flux.

volume of material, leading to dangerously high local temperatures within only a matter of minutes. Normal values of field current keep the retaining ring saturated, so that only a relatively small amount of armature end-leakage flux traverses. However, when the field excitation is reduced corresponidng to the operation of the machine in the region of unity and leading power factor, the retaining ring is no longer saturated and allows an increase in armature end-leakage flux. With the use of nonmagnetic materials for the retaining rings and parts of the stator-end structure, and with the use of magnetic shields to control the flux paths, and also by changing the end structure configuration so as to reduce the leakage flux and modify the flux paths not to be at right angles to the plane of stator laminations, the armature end-leakage flux and the resultant heating can be reduced.

If the machine under consideration is connected to an infinitely large system through negligibly small impedance, then the stability limit may be represented by a line CF in Figure 7.4.3 passing through the point on the negative vertical axis representing the short-circuit ratio. Operation along the arc EK is impossible as the machine will not remain in synchronism with the system, even if it did not exceed any heating limitation. Since the machine operates in practice through impedance representing transformers, lines, and the paralleled value of the impedances of all the other machines on the system, the resultant external impedance is typically about 0.2 to 0.4 per unit based on the individual machine rating. The effect of this external impedance x_e is to bend upward the straight line CF to some position such as CG. Holding the terminal voltage constant at 1.0 per-unit, it can be shown that CG is the arc of a circle with its center on the reactive axis at

$$\left[\frac{x_d + x_e}{2x_d x_e} - \frac{1}{x_d}\right]$$

and a radius of

$$\left[\frac{x_d + x_e}{2x_d x_e}\right]$$

Thus the part of the boundary dictated by the steady-state stability can be established. This limit, as determined above, is slightly conservative in view of approximating saturation and neglecting saliency. With no voltage regulator, the curve CG of Figure 7.4.3 would be more bent upwards as shown in the diagram. The role of the automatic voltage regulator in permitting the widest possible flexibility of machine operation within the boundaries established can be appreciated.

While the transient stability is also affected to some degree by the machine excitation, it would appear to be a very extreme case where the system transient stability is critically dependent on the field excitation, since many other factors such as type and location of the fault, operating times of relays and clearing times of circuit breakers, system grounding, machine inertias, and automatic reclosing play a dominant part in the transient stability considerations.

The permissible operating region is bounded by the limits determined by the field heating, the armature heating, the armature end-region heating, and the steady-state stability. As can be seen from Figure 7.4.3, an increase of the generator load, without corresponding adjustment of the field current, pushes the power factor of the machine toward the leading (underexcited) region, and an increase of excitation causes the generator power factor to lag more. The more lagging the initial load power factor is, the greater is the total load the machine can carry before running into one of the limits.

Figure 7.4.5 shows the pull-out power as a function of the power factor of the initial load, corresponding to different initial constant per-unit kilowatt loads. Margins in the permissible power increment are small in the higher load regions at 0.8 power-factor leading. Any practical concern over the steady-state stability can be dismissed with an automatic voltage regulator, since the regulator will hold the machine excitation at a proper level. The

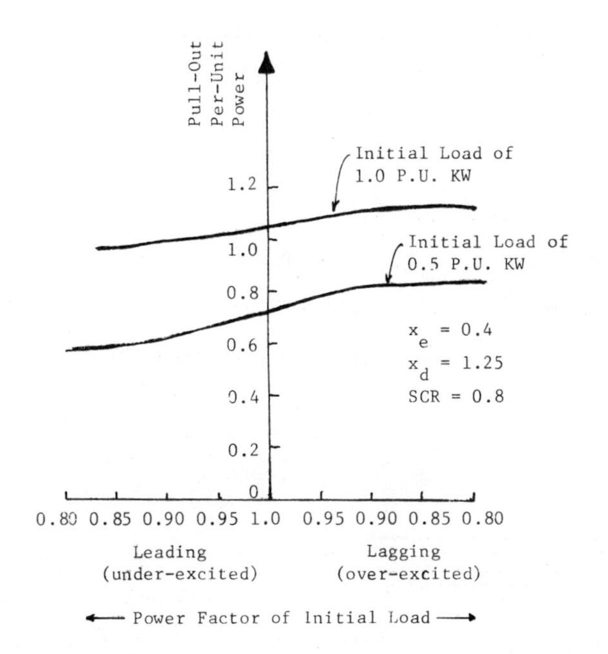

Figure 7.4.5. Pullout power vs. power factor at constant kilowatt loads.

limits, which may be ample at rated terminal voltage, become materially reduced at lower values of terminal voltage, as can be seen from Figure 7.4.6. In view of this, some modern voltage regulators are equipped with a lower excitation limit which automatically comes into play below a predetermined level of excitation to prevent loss of synchronism. Considering the end-region heating limit (or the steady-state stability limit, as determined for the particular machine and its associated system), the region within which the lower excitation limit is typically set to come into operation is shown by the shaded area of Figure 7.4.3. The characteristic of the device, given by a straight line such as MN, may be located anywhere within the shaded area

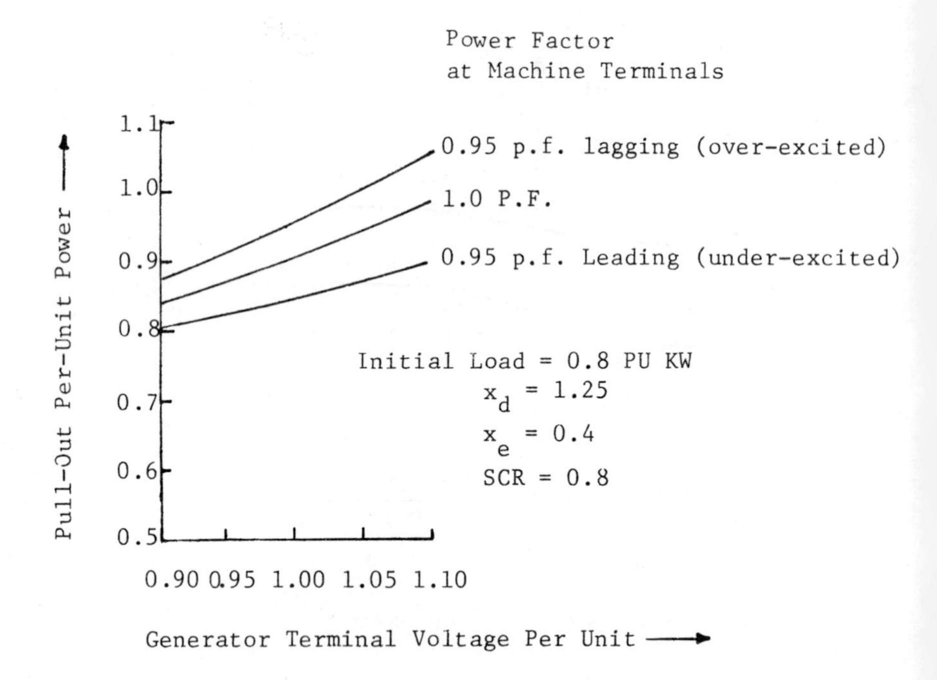

Figure 7.4.6. Pull-out power vs. generator terminal voltage (for assumed initial load of 0.8 per-unit KW).

through independent adjustments of its slope and point of intersection with the vertical axis. The lower excitation limit is a standard feature of all amplidyne-type automatic voltage regulators, but is not adaptable to rheostatic-type regulators.

When a field lead is broken or when the field breaker is inadvertently tripped, complete loss of excitation may occur in which case loss of synchronism with the system will result, although the machine may continue to run as an induction generator producing kilowatts. The machine could be damaged due to heating caused by large eddy currents flowing in the surface of the rotor. Also, the system may not be able to withstand successfully the large kilovar load suddenly imposed on it. Especially, prior to the disturbance, if the machine was delivering kilovars to the system in an overexcited mode of operation, the net change from the viewpoint of the system could be very serious indeed. The system voltage may drop and other machines may lose synchronism, unless they are adequate to meet the demand of kilovars and their excitation controls provide immediate response to the need. Since the machine itself may be damaged and instability may occur leading to a major system shutdown, it becomes desirable to provide a protective relay that can isolate the machine from the system in the event of loss of excitation.

Problems

7-1. (a) A transfer function is given by the voltage ratio

$$G(s) = \frac{V_o(s)}{V_e(s)} = \frac{400}{1 + 0.85s}$$

and is represented in a block diagram as follows:

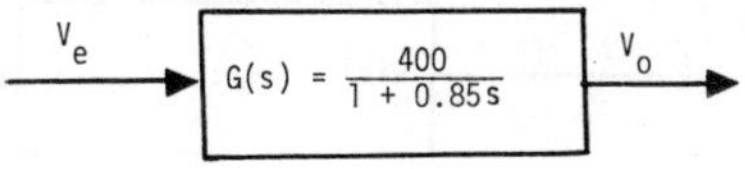

V_e → $G(s) = \frac{400}{1 + 0.85s}$ → V_o

Show that the circuits given below are capable of yielding the above transfer function, while filling the details left out.

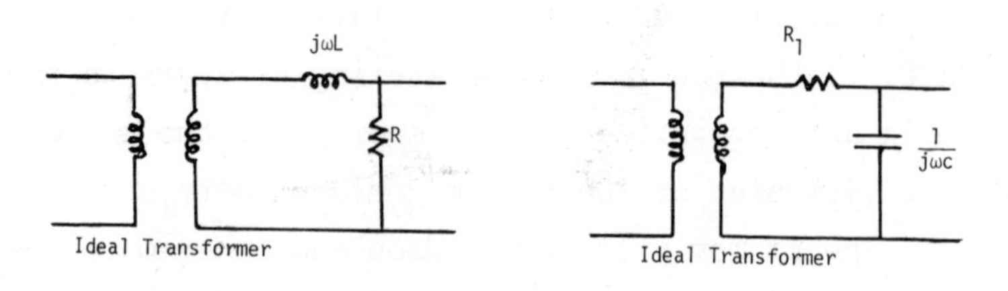

(b) Next consider the transfer function

$$H(s) = \frac{V_f}{V_o} = \frac{0.03s}{1 + s}$$

and find the corresponding equivalent circuits that can produce the transfer function.

(c) Now consider a block diagram of a simple feedback control system given below with parameters that might correspond to a static excitation system:

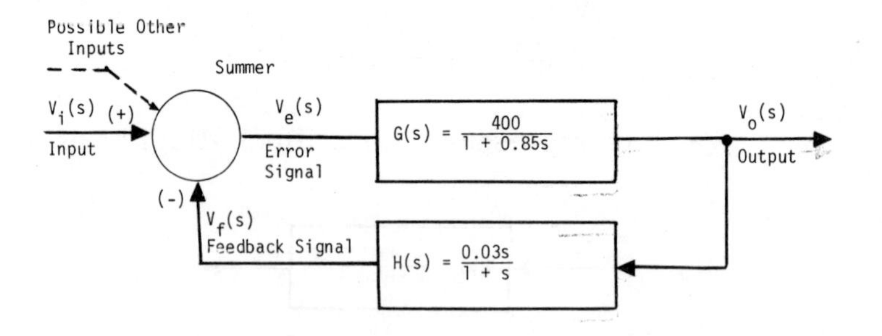

Determine the open-loop and closed-loop transfer functions.

(d) The transient response of a circuit or system is described, usually by means of a graph, showing how the magnitude of the output signal of the circuit or system varies with time for any given input signal, usually a step function or an impulse function. Obtain the time (or transient) response of the output of the closed loop feedback control system (given in part c) for a unit step input in analytical form, and sketch the same.

(e) Make an s-plane plot of the closed-loop transfer function.

(f) The frequency response is given by the variation of the output signal as a function of the frequency of a sinusoidal input signal. The Bode diagram is one of the methods for describing the frequency response of a circuit. Obtain the Bode plots for closed-loop as well as open-loop transfer functions.

[*Note:* See Appendix E for details on Block Diagrams, Transient Response, Frequency Response, and Stability Criteria.]

7-2. In Section 7.2 of the text, for the calculation of the exciter response of a separately excited exciter, a constant coefficient of dispersion has been assumed. Instead, assume constant leakage inductance and go through the entire formulation.

7-3. Let the magnetization curve be represented by an empirical equation:

$$e = \frac{ai}{b + i} + ci$$

Obtain a formal solution of Eq. (7.2.6) by substituting

the above into it and integrating.

7-4. The exciter with the following data is operating with no load, separately excited from a 250-volt source. The field circuit is initially open and later closed through such a resistance as to give an ultimate voltage of 275 volts. The magnetization characteristic of the exciter is given below. The time constant is to be increased by 5% to allow for the effect of eddy currents in the yoke. Plot the armature voltage and the field current, as a function of time.

Exciter Data:

Rated Voltage	250 Volts
Rated Output	100 KW
Rated Speed	1800 rpm
Number of Poles	6
Type of Armature Winding	Simplex Lap
Number of Armature Slots	108
Number of Conductors Per Slot	4
Coefficient of Dispersion	1.15
Shunt Wound	
Number of Field Turns Per Pole	1500
Number of Field Circuits	1
Field Resistance	14.6 ohms

Magnetization Characteristic of the Exciter:

Field Current (amperes)	1	2	3	4	5	7.5	10	13.2	15	20
Armature e.m.f. (volts)	50	100	150	200	220	250	270	275	280	290

7-5. Compute the nominal response of a 167 KW, 250 volt, 1200 rpm, six-pole, separately excited exciter with the following additional data. Assume constant leakage inductance and use the point-by-point method of calculation. For the effect of eddy currents, multiply the calculated nominal exciter response by 0.97.

Additional Data:

Number of Field Turns per Pole	1,450
Number of Parallel Paths of Field	3
Field Resistance per Path at 75°C	27 ohms
$k = e/\phi_a$ [See Eq. (7.2.2).]	10,080
Leakage Flux/Useful Flux on the Air-Gap Line	0.15
No-load Saturation Curve	See Table* below
When used with an a-c generator having a nominal collector-ring voltage	190 Volts

The following data applies to the external apparatus in the field circuit of the exciter:

External Resistance (for all paths)	11.6 ohms
Separate Excitation	
Voltage of Pilot Exciter	250 Volts

*Table:

Field Current per Circuit (amperes)	0.5	1.0	1.5	2.0	2.5	3.0	3.78
No-load Armature Voltage (Volts)	60	120	186	235	270	295	320

7-6. A block diagram is a convenient form of representing a control system, relating the input and output variables of the principal components of the system through transfer functions. Develop the block diagram for a simplified representation of a continuously acting exciter control system, for which the differential equations relating the input and output variables of the regulator, amplifier, exciter, and stabilizing loop, respectively, are given below:

$$\frac{dE^{\ddot{v}}}{dt} = \frac{1}{T_R}\left(E_s - E_t - E^{\ddot{v}}\right)$$

$$\frac{dE^{iii}}{dt} = \frac{1}{T_A}\left\{K_A\left(E^{\ddot{v}} + \frac{E_o^{iii}}{K_A} - E^{i\ddot{v}}\right) - E^{iii}\right\}$$

$$\frac{dE_{fd}}{dt} = \frac{1}{T_E}\left(E^{ii} - K_E\, E_{fd}\right)$$

$$\frac{dE^{i\ddot{v}}}{dt} = \frac{1}{T_F}\left\{K_F\,\frac{dE_{fd}}{dt} - E^{i\ddot{v}}\right\}$$

where E_s is the scheduled voltage in per-unit

E_t is the terminal voltage of the generator

E_o^{iii} is the output voltage of the amplifier in per-unit prior to the disturbance

T_R is the regulator time constant

K_A is the amplifier gain

T_A is the amplifier time constant

K_E is the exciter gain

T_E is the exciter time constant

K_F is the stabilizing loop gain,

T_F is the stabilizing loop time constant,

and

E^{ii}, E^{iii}, E^{iv}, and E^{vi} are the intermediate variables.

E^{ii} is given by

$$E^{ii} = E^{iii} - E^{vi}$$

where E^{vi} is read as a function of E_{fd} to take care of the demagnetizing effect due to saturation in the exciter:

$$E^{vi} = Ae^{BE_{fd}}$$

in which A and B are constants depending upon the exciter saturation characteristic.

7-7. Referring to Figure /.4.3, holding the terminal voltage constant at 1.0 per-unit, show that the steady-state stability limit CG is the arc of a circle with its center on the reactive axis at

$$\left[\frac{x_d + x_e}{2x_d x_e} - \frac{1}{x_d}\right] \text{ and a radius of } \left[\frac{x_d + x_e}{2x_d x_e}\right] .$$

7-8. A 350 KVA, 3.3 KV, 3-phase, 6-pole, 60 HZ, star-connected synchronous generator has a synchronous impedance per phase of (1.0 + j 10.0) ohms, and operates at 0.9 lagging power factor. Construct the circle diagram and hence determine the following:

(a) the internal (or excitation) voltage and the

load angle.

(b) the current and power factor when motoring at maximum gross torque.

(c) the current and torque when operating as a motor at 0.6 power-factor leading.

(d) the power-factor when running as a generator delivering 80 amperes, with the excitation adjusted to the same value as the terminal voltage.

7-9. Consider the case of a single machine connected to an infinite bus with voltage V through an external impedance $(r_e + jx_e)$. Neglecting amortisseur effects, armature resistance, armature $(p\lambda)$ terms, and saturation, the following relationships apply:

$$e_t^2 = e_d^2 + e_q^2$$

$$-e_d = \lambda_q = -x_q i_q$$

$$e_q = \lambda_d = E'_q - x'_d i_d$$

$$E_q = E'_q + (x_q - x'_d)\, i_d$$

$$T_e = E_q i_q$$

$$i_d = [E_q - V\cos\delta]\{(x_e + x_q)/[r_e^2 + (x_e + x_q)^2]\} + \\ - V\sin\delta\,\{r_e/[r_e^2 + (x_e + x_q)^2]\}$$

$$i_q = [E_q - V\cos\delta]\{r_e/[r_e^2 + (x_e + x_q)^2]\} + \\ + V\sin\delta\,\{(x_e + X_q)/[r_e^2 + (x_e + x_q)^2]\}$$

$$E'_q = x_{ad}\, I_{fd} - (x_d - x'_d)\, i_d$$

$$T'_{do}(dE'_q/dt) = E_{fd} - x_{ad}\, I_{fd}$$

$$T_m - T_e = M[d(p\delta)/dt]$$

(a) Preserving the basic variables Δe_t, $\Delta E'_q$, and $\Delta\delta$, express the equations in small oscillation form under small perturbations:

$$\Delta e_t = K_5\, \Delta\delta + K_6\, \Delta E'_q$$

$$\Delta E'_q = \frac{K_3\, \Delta E_{fd}}{1 + sT'_{do}\, K_3} - \frac{K_3 K_4\, \Delta\delta}{1 + sT'_{do}\, K_3}$$

$$\Delta T_e = K_1\, \Delta\delta + K_2\, \Delta E'_q$$

(b) For the special case of zero external resistance, find K_1 through K_6.

(c) Using the linearized small-perturbation relations, develop a block diagram relating the pertinent variables of electrical torque, speed, angle, terminal voltage, field voltage, and flux linkages. [It may be pointed out here that such block diagrams are utilized to explore the phenomena of stability of synchronous machines.]

CHAPTER VIII

SYNCHRONOUS MACHINE MODELING

Dynamics of power systems cover a wide spectrum of phenomena, electrical, electromechanical, and thermo-mechanical in nature. Dynamic problems may be classified under the major categories of

(a) electrical machine and system dynamics

(b) system governing and generation control

and

(c) prime-mover energy supply system dynamics and controls.

The time range of the study may vary from milliseconds to microseconds for studies of switching transients, through seconds for stability studies, to many minutes for load-frequency control and boiler response, including possible operator action. As the time range of the study increases, the components of the system that should be represented in greater detail shift more or less uniformly through transmission lines, generators, turbines, and boilers. Figure 8.1.1 shows the time ranges of various studies, with logarithmic time scale. The bars represent the time range of a class of study; while the left end of the bar indicates the fastest dynamics that are usually of importance, the right end of the bar indicates the time over which the study is generally carried out enabling useful conclusions to be drawn. The X near the left end gives an indication of the value (or range) of the time step customarily used for simulation.

Some of the major engineering aspects involving power

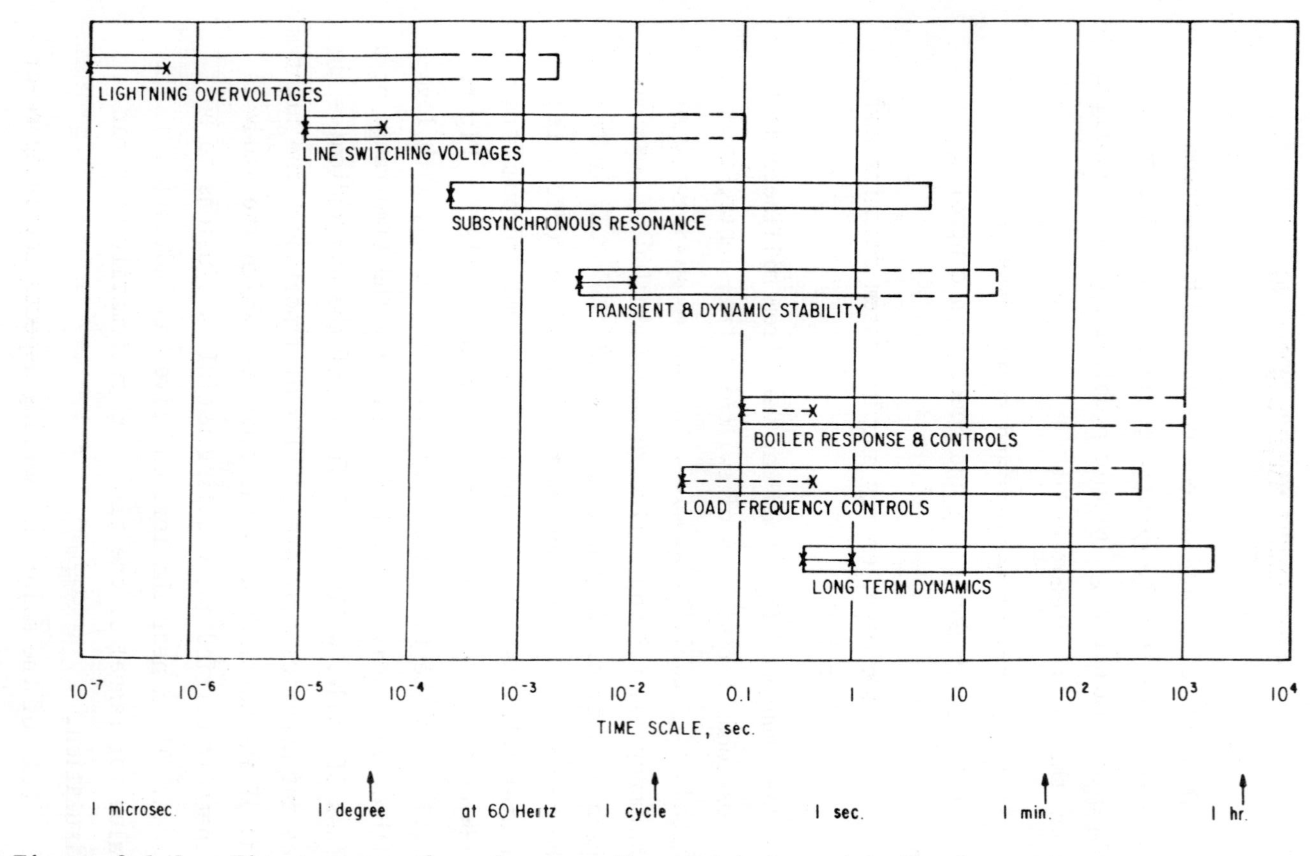

Figure 8.1.1. Time ranges of various studies, with logarithmic time scale.

system dynamics may require highly complex analysis of system behavior. Exploration of power system dynamic effects is critically a function of the engineers' ability to predict the system behavior through simulation. The less this ability, the more conservative must be the design criteria and the greater must be the margins allowed in designs. Most of the discussions on effects to be considered in power system dynamics become academic unless engineering solution methods can be made available to yield results within the constraints of time, cost, and human endurance which are part of every real-life situation.

The degree of detail of representation and the simplifying assumptions that can be made are a function of the particular problem on hand*. This fact places a great deal of importance on a thorough understanding of the fundamentals and orders of magnitude of the effects, so essential for the choice of the right model with the available data. It then becomes necessary to provide computational tools tailored to the task, with interactive computation and rapid access to results. It is equally important to interpret the results properly.

This chapter will be devoted to the modeling of the synchronous machine used as a generator in a large power system. In the case of three-phase systems (linear or nonlinear) with or without imbalances, for the study of switching surges, recovery voltages, energization transients, and short-circuits in which the duration of transients is short relative to the flux-decay time constants of generators,

---

*See "Symposium on Adequacy and Philosophy of Modeling: Dynamic System Performance", IEEE 75-CHO-970-A-PWR, 1975. [Figures 8.1.1 through 8.1.9 are adatped from the above Reference.]

the source is represented by the ideal sinusoidal balanced three-phase sources behind constant inductances in each phase with a constant or varying frequency; the source inductance is usually represented by x''_d. In order to take the generator flux effects into account, the source can be represented by the equations of the flux behavior in the d- and q-axes. Figure 8.1.2 shows the modeling details of representation of the generator through differential equations in the d- and q-axes, accounting for the transients in the stator and rotor currents and their effect on voltage producing fluxes. Such effects as subsynchronous resonance, harmonics generated by saturation, load rejection transients can be studied with this representation.

For the solution of the fundamental-frequency effects, the network differential equations can be reduced to algebraic equations, while preserving the time-varying nature of fluxes and rotor speed giving rise to varying magnitude and frequency of the generated voltages. As shown in Figure 8.1.3, the source is represented by equations of the flux behavior in the d- and q-axes, while the terminal voltage is obtained as the fundamental-frequency positive-sequence phasor by multiplying the flux with the speed. The armature and network transients are neglected. Such a modeling technique is applicable to the balanced three-phase systems for fundamental-frequency effects over several seconds following load rejections or other balanced network disturbances.

The effect of frequency variations in the generated voltages and in the network parameters is usually neglected in the stability studies. However, the behavior of the rotor flux will be accounted for the effect on synchronizing and damping torques. Assuming the speed to be at its rated

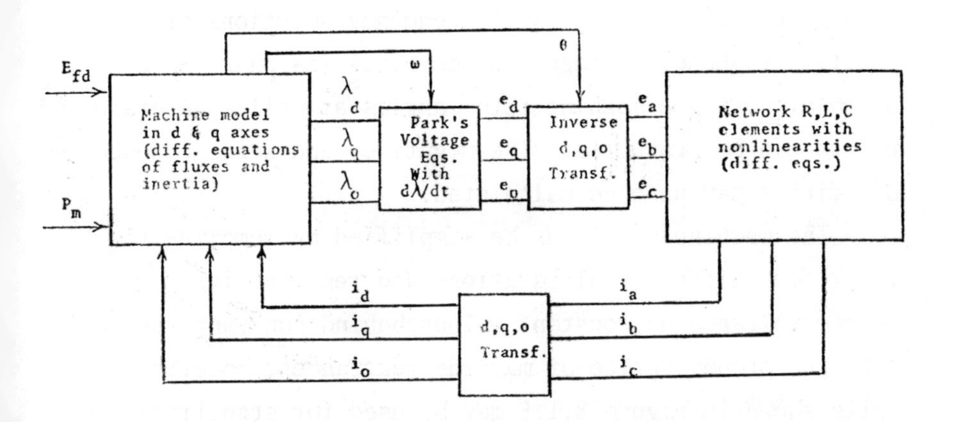

Figure 8.1.2. Modeling applicable to nonlinear 3-phase systems, with or without imbalances.

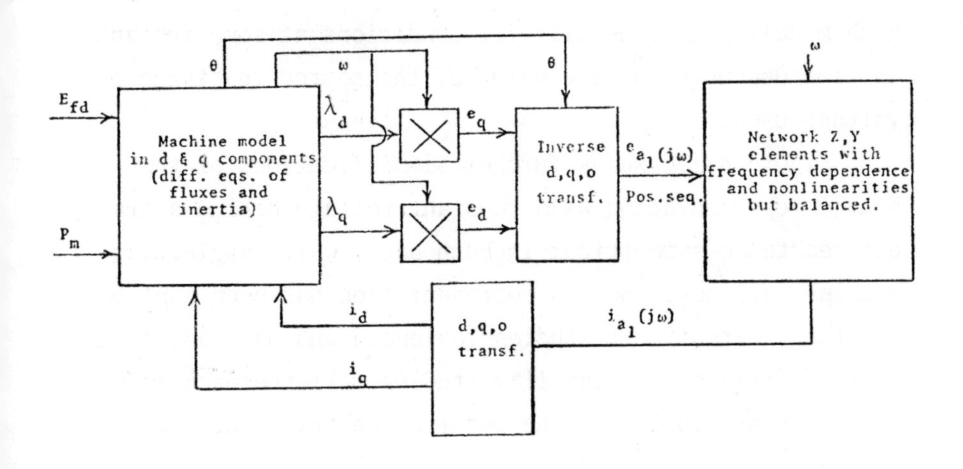

Figure 8.1.3. Modeling applicable to balanced 3-phase systems, for fundamental frequency effects over several seconds.

value, the voltage will be taken equal to the flux as shown in Figure 8.1.4. Fundamental-frequency solutions of positive-sequence voltages and currents (as phasors) can be obtained. The machine-rotor angles as well as average powers can be evaluated. The negative- and zero-sequence quantities may also be calculated.

The machine model can be simplified by removing the rotor flux differential equations and representing the source voltages as constant values behind constant reactances. With the proper choice of machine reactances, modeling details shown in Figure 8.1.5 may be used for stability calculations. Transient reactance values are used for solving swing equations and the solution approximates the conditions in the first half-second. Short-circuit currents (balanced and unbalanced) can be studied, obtaining fundamental-frequency solutions of the sequence currents and voltages. Such modeling is generally for conditions at some instant in time depending on the value of the source reactance and voltage used.

Figure 8.1.6 shows further simplifications of the source representation as a constant voltage behind a transient reactance symmetrical in both axes, while neglecting machine saliency. Such a representation has been applied for the short-circuit studies (balanced and unbalanced) as well as first-swing stability studies. If the machine swing equations are included, the rotor angle transients can be obtained.

Accuracy of load modeling, especially in system configurations where stability is a governing criterion, should be given adequate attention.

In widespread interconnected systems, analysis of any segment of the system additions must consider the effects

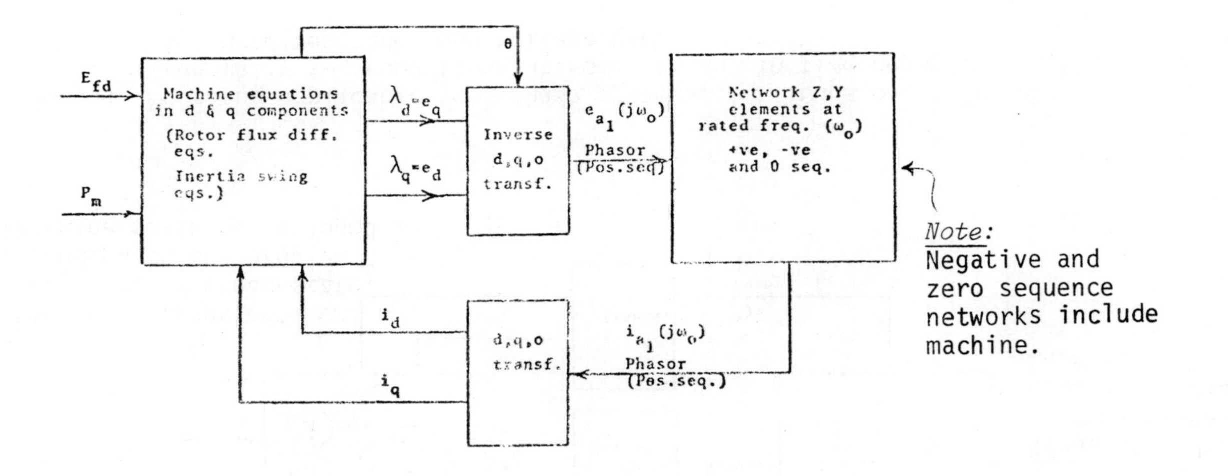

Figure 8.1.4. Modeling applicable to 3-phase systems, for fundamental-frequency effects over several seconds (for the study of stability phenomena).

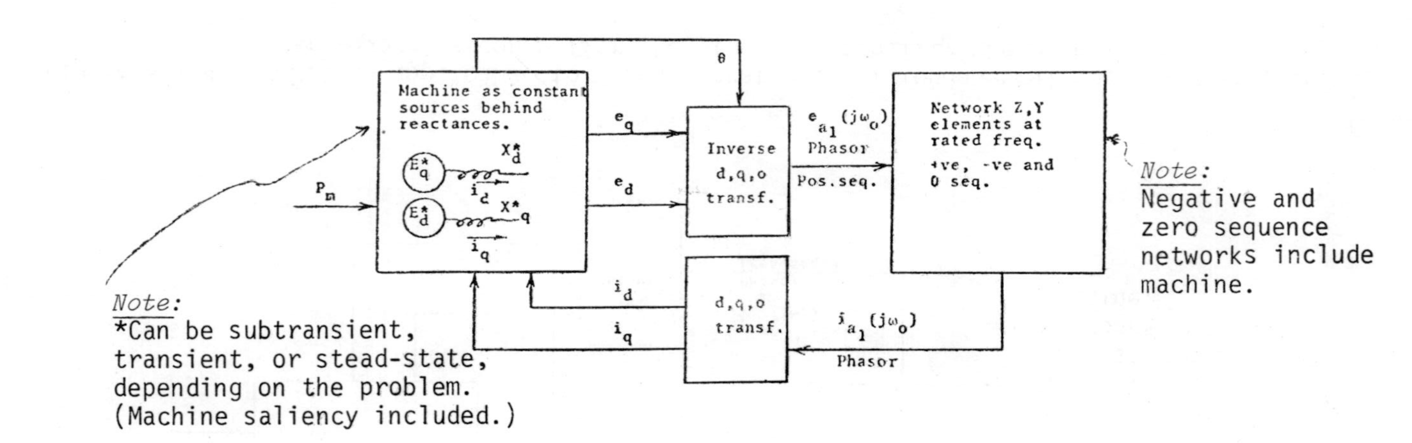

Figure 8.1.5. Modeling applicable to 3-phase systems, for fundamental-frequency effects (generally for conditions at some instant in time depending on the value of source reactance and voltage used).

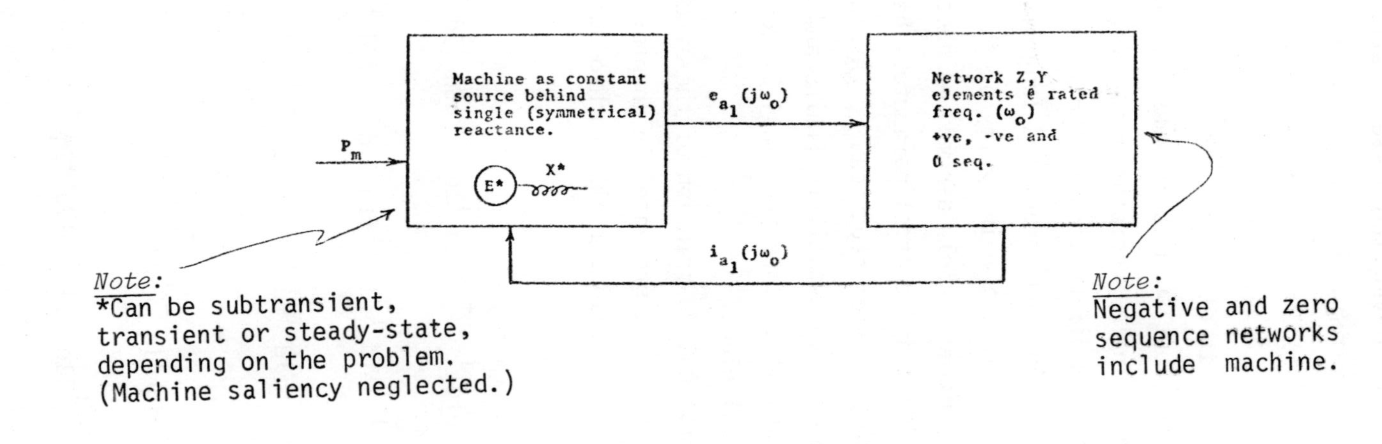

Figure 8.1.6. Modeling applicable for short-circuit studies (balanced and unbalanced) and first-swing stability studies.

throughout the whole interconnected system, particularly when the size of each system addition is a significant fraction of the overall system. However, large scale system simulations should not ordinarily be used to experiment with parameters or arrangements of a unit, such as the generator excitation and control; but such refinemenets should rather be studied first on a single-generator basis before incorporating the resultant modifications in the large system. For a proposed new station and associated long-distance transmission, in cases where the integrity of the remainder of the system can be either safely assumed or checked after the best generating unit and control parameters have been obtained, the single-generator study can be gainfully applied for the determination of the stability limits and of possible means to improve stability. Thus the small system study still has an essential place.

From the mathematical description of the synchronous machine in per-unit notation presented in Chapter 1, the set of general equations valid for all synchronous machines is given below:

$$\lambda_d = -L_d\, i_d + L_{afd}\, i_{fd} + L_{a1d}\, i_{1d} + L_{a2d}\, i_{2d} + \cdots$$

$$\lambda_{fd} = -L_{afd}\, i_d + L_{ffd}\, i_{fd} + L_{f1d}\, i_{1d} + L_{f2d}\, i_{2d} + \cdots$$

$$\lambda_{1d} = -L_{a1d}\, i_d + L_{f1d}\, i_{fd} + L_{11d}\, i_{1d} + L_{12d}\, i_{2d} + \cdots$$

$$\vdots \qquad \vdots \qquad \vdots \qquad \vdots$$

$$\lambda_q = -L_q\, i_q + L_{a1q}\, i_{1q} + L_{a2q}\, i_{2q} + \cdots$$

$$\lambda_{1q} = -L_{a1q}\, i_q + L_{11q}\, i_{1q} + L_{12q}\, i_{2q} + \cdots$$

$$\vdots \qquad \vdots \qquad \vdots \qquad \vdots$$

$$\vdots \qquad \vdots \qquad \vdots \qquad \vdots$$

$$p\lambda_d = e_d + ri_d + \lambda_q p\theta$$

$$p\lambda_{fd} = e_{fd} - r_{fd}\, i_{fd}$$

$$p\lambda_{1d} = -r_{11d}\, i_{1d} - r_{12d}\, i_{2d} - \cdots$$

$$\vdots \qquad \vdots \qquad \vdots \qquad \vdots$$

$$p\lambda_q = e_q + ri_q - \lambda_d p\theta$$

$$p\lambda_{1q} = -r_{11q}\, i_{1q} - r_{12q}\, i_{2q} - \cdots$$

$$\vdots \qquad \vdots \qquad \vdots \qquad \vdots$$

$$T_e = \lambda_d\, i_q - \lambda_q\, i_d$$

$$p^2\delta = \omega_o(T_m - T_e)/2H$$

Eq. Set (8.1.1)

In order to apply the above equations to a particular problem, they must be modified to use the available data and to represent the interconnections between the unit and the system. Let us consider a model which includes two rotor circuits per axis and solves for armature fluxes as a function of the field voltage and the armature currents. Eliminating the currents of two rotor circuits i_{1d} and i_{2q}, one can obtain the following equation-set for the particular model:

$$\lambda_d = \frac{E'_q(x''_d - x_\ell) - \lambda_{kd}(x'_d - x''_d)}{(x'_d - x_\ell)} - i_d x''_d$$

$$\lambda_q = \frac{E'_d(x''_q - x_\ell) - \lambda_{kq}(x'_q - x''_q)}{(x'_q - x_\ell)} - i_q x''_q$$

$$\frac{dE'_q}{dt} = \frac{1}{T'_{do}} (E_{fd} - x_{ad} I_{fd})$$

$$\frac{dE'_d}{dt} = \frac{1}{T'_{qo}} (-x_{aq} I_{1q})$$

$$\frac{d\lambda_{kd}}{dt} = \frac{1}{T''_{do}} (E'_q - \lambda_{kd})$$

$$\frac{d\lambda_{kq}}{dt} = \frac{1}{T''_{qo}} (E'_d - \lambda_{kq})$$

$$x_{ad} I_{fd} = \frac{(x'_d - x''_d)(x_d - x'_d)}{(x' - x_\ell)^2} \cdot$$

$$\cdot \left[E'_q - \lambda_{kd} + i_d \frac{(x'_d - x_\ell)(x''_d - x_\ell)}{(x'_d - x''_d)} \right] + F(E'_q)$$

$$x_{aq} I_{1q} = \frac{(x'_q - x''_q)(x_q - x'_q)}{(x'_q - x_\ell)^2} \cdot$$

$$\cdot \left[E'_d - \lambda_{kq} + i_q \frac{(x'_q - x_\ell)(x''_q - x_\ell)}{(x'_q - x''_q)} \right] + E'_d$$

$$e_d = p\lambda_d - \lambda_q p\theta - ri_d$$

$$e_q = p\lambda_q + \lambda_d p\theta - ri_q$$

$$T_e = \lambda_d i_q - \lambda_q i_d$$

$$p^2\delta = \omega_o(T_m - T_e)/2H \qquad \text{Eq. Set (8.1.2)}$$

The equation-set (8.1.2) may be represented in a block diagram shown in Figure 8.1.7; corresponding set of equivalent circuits is given in Figure 8.1.8. Means to account for nonlinearity due to saturation has been included in the particular model structure. It should be pointed out that there is a lack of consensus on the preferred modeling to be used for representing saturation effects on the machine dynamics. Further study needs to be carried out to understand what portions of the machine saturate most significantly under a particular load condition, how the model should be altered to represent the saturated machine, and what data should be gathered for each modeled generator to allow an economical yet accurate representation of the significant effects of saturation.

Lower-order models using less data would be more economical from the point of view of computation time. It is a common practice to leave the $p\lambda$-terms out of the stator voltage equations and assume the speed as rated, if the simulation is intended for the usual step-by-step time domain simulation of power system problems. From the viewpoint of consistency in modeling, if the $p\lambda$-terms of the machine are included, then the corresponding $p\lambda$-voltage terms will have to be included in all the transmission system equations, which make the simulation very cumbersome. Neglecting the $p\lambda$-terms implies not including the armature transients and resultant torques.

In cases where the interaction of the field and the

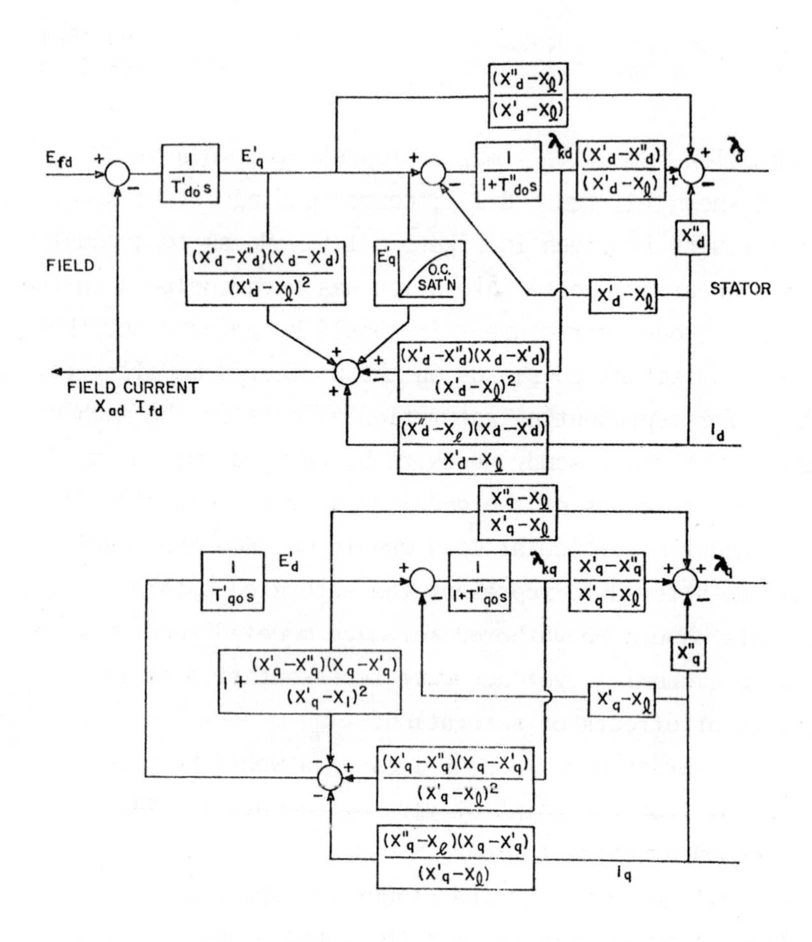

Figure 8.1.7. Block diagram of a particular model of a synchronous machine [modeled by Eq. Set (8.1.2)].

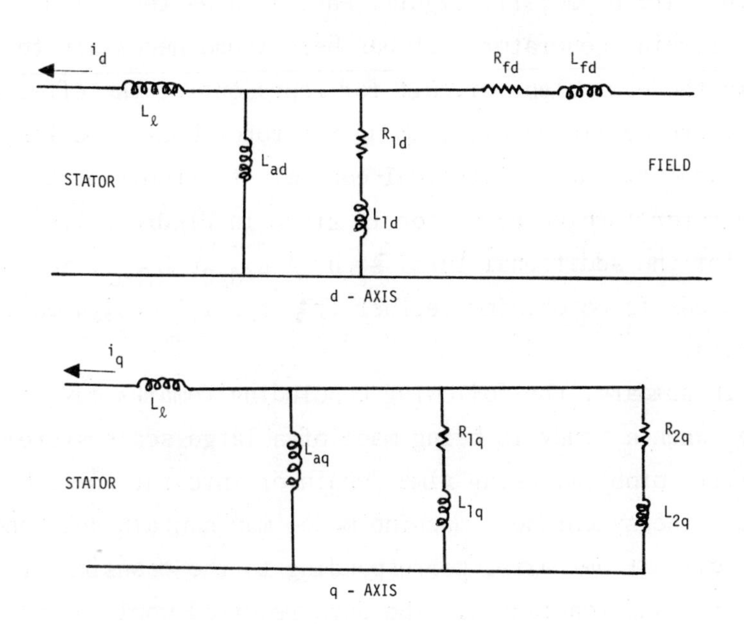

NOTE:

Data required for the model: x_ℓ, x_d, x'_d, x''_d, T'_{do}, T''_{do}, x_q, x'_q, x''_q, T'_{qo}, T''_{qo}.

Equivalent circuit elements: L_ℓ, L_{ad}, L_{fd}, L_{1d}, R_{fd}, R_{1d}, L_{aq}, L_{1q}, L_{2q}, R_{1q}, R_{2q}.

Figure 8.1.8. Equivalent circuits of a synchronous machine, corresponding to the Eq. Set (8.1.2).

armature are especially significant, such as the solid-rotor turbine generators, it has been found necessary to modify the equivalent circuit for representing the effects of differences in coupling among the rotor body circuits, field and armature. A typical equivalent circuit of a solid-rotor turbine generator is given in Figure 8.1.9. The data for the additional terms such as L_{fdk0}, L_{fkd1}, and R_{fkd} needs to be obtained either from special design analysis or tests.

In summary, the following concluding remarks may be made: When a study is being made of a large-scale system stability problem, using time domain or integration techniques, the synchronous machine model may contain only one rotor circuit per axis, corresponding to the transient time constants and reactances. The data required would be x_ℓ, x_d, x_d', T_{do}', x_q, x_q', and T_{qo}'. For critical cases, the model may include two rotor circuits per axis and the data needed will be x_ℓ, x_d, x_d', x_d'', T_{do}', T_{do}'', x_q, x_q', x_q'', T_{qo}', and T_{qo}''. The $p\lambda$-terms of the armature voltage equations would be left out.

In specialized studies of power system dynamic problems in which the effects of induced field and rotor body currents and/or the effects of harmonics or d-c offsets of armature currents are likely to be significant, the model should include the $p\lambda$- and $\lambda p\theta$-terms in the armature voltages, should include as many rotor circuits as may be deemed to be necessary within the available data, and should also include the mutual impedances between the field and d-axis rotor circuits, if data for these are available. Such studies may only be performed on a very small number of machines connected to the system at a time, while including the $p\lambda$-terms in the voltages of the transmission system connecting these

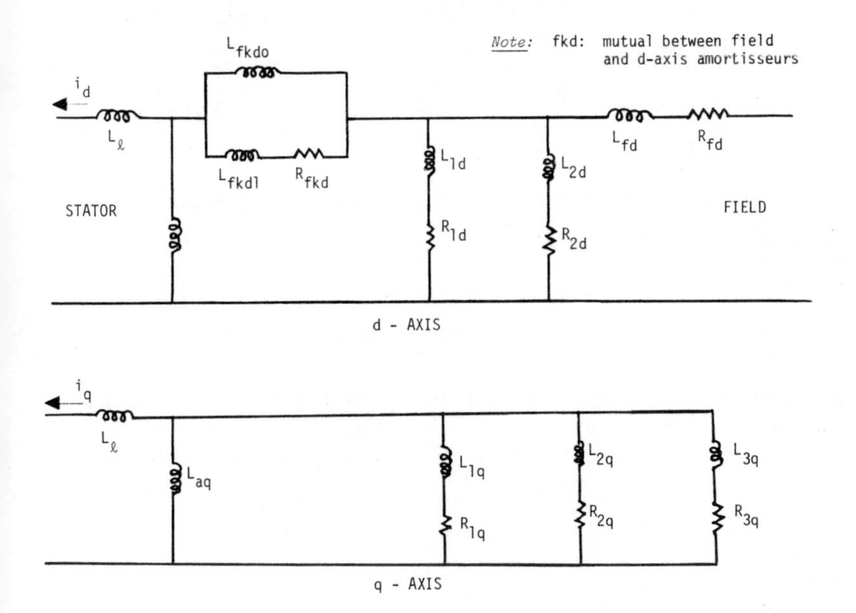

Figure 8.1.9. A typical equivalent circuit of a solid iron rotor turbine generator.

machines to loads or equivalents for large systems.

Depending on the nature of the problem at hand, time range involved, computation equipment, costs, and consistency of representation, appropriate models need to be chosen judiciously for the power system components and of their interconnections for the dynamic simulation of power systems.

Problems

8-1. (a) Starting from Eq. Set (8.1.1), obtain the Eq. Set (8.1.2) for the synchronous machine model which includes two rotor circuits per axis and solves for armature fluxes as a function of the field voltage and the armature currents.

(b) Justify the block diagram given in Figure 8.1.7 as represented by the Eq. Set (8.1.2).

8-2. Consider four models of a synchronous machine, based on the data shown in the following Table, which could be utilized for making large-scale power system stability studies.

Table

Parameters	Model I	Model II	Model III	Model IV
x''_d				x
x'_d	x	x	x	x
x_d		x	x	x
x'_q			x	x
x_q		x	x	x
T'_{do}		x	x	x
T''_{do}				x
T'_{qo}			x	x
T''_{qo}				x
x_ℓ				x
H	x	x	x	x

Each of the mathematical models used to represent a synchronous machine in a stability study consists of a set of simultaneous algebraic equations and a set of simultaneous differential equations. The phasor diagram associated with each representation is a way of describing algebraic equations which relate the internal and terminal conditions of the machine. Additional algebraic equations may exist in some cases. In all cases, the differential equations are to be expressed separately from the phasor diagram.

(a) Develop the phasor diagrams for each of the models and write down the equations not represented by the phasor diagram for each of the representations considered.
(b) State the simplifying assumptions on which each of the models is based.
(c) Discuss the limitations, advantages, and disadvantages of these models as a choice for making large-scale power system stability studies.

CHAPTER IX

GENERATOR PROTECTION

The generators are the most expensive pieces of equipment in the a.c. power system. The protection of generators involves the consideration of more possible abnormal operating conditions than the protection of any other system element. One should try to protect the generator against all the abnormal conditions and yet keep the protection simple and reliable. The choice must be carefully made since inadvertent operation of the relays is almost as serious as failure to operate. Unnecessary removal of a large generator may overload the rest of the system and cause power oscillations leading to possible disruption of the system; on the other hand, failure to clear a fault promptly may cause expensive damage to the generator. In order to provide higher standards of service with greater efficiency, centralized control is becoming more and more popular, requiring more automatic protective-relaying equipment to provide the protection that was formerly taken care of manually by the attendants. Only a brief treatment of the various aspects of the generator protection* is included here in order to make the text on synchronous machines more comprehensive.

Stator Protection:

Short-circuit forces, unbalanced currents, ventilation troubles, etc., may cause damage to the insulation, over-

*See for details the books by Mason and Warrington quoted under Bibliography.

voltage or overheating which in turn cause the breakdown of conductor insulation that may result in a fault between conductors or between a conductor and the iron core. The ground-fault current is usually limited by the impedance in the neutral of the generator which may be a resistance, a distribution transformer with resistance loading, a reactance, or a potential transformer. Modern practice is to use a resistance-loaded distribution transformer; to avoid the possibility of harmful high transient overvoltages, it has been found by test that the resistance of the resistor should not exceed

$$R_n = \frac{X_c}{3N^2} \text{ ohms} \tag{9.1.1}$$

where X_c is the capacitive reactance of the stator circuit to earth per phase and N is the turns-ratio of the transformer. A safe value of the resistance can be used which will limit the fault current to approximately 15 to 30 amperes.

(a) Phase and Ground Faults:

Percentage-differential relaying is generally applied for generators rated 1 MVA or higher. The protection is offered by comparing the two currents in the CT's at the two ends of each phase winding, which should theoretically be the same under normal conditions. The connections of the differential relays are shown schematically in Figure 9.1.1 for a wye-connected generator, and in Figure 9.1.2 for a combined generator-transformer arrangement. In machines below 10 MVA, time-overcurrent relays may replace the differential relays; but they should be monitored by an instantaneous under-voltage relay. For the cases where conventional

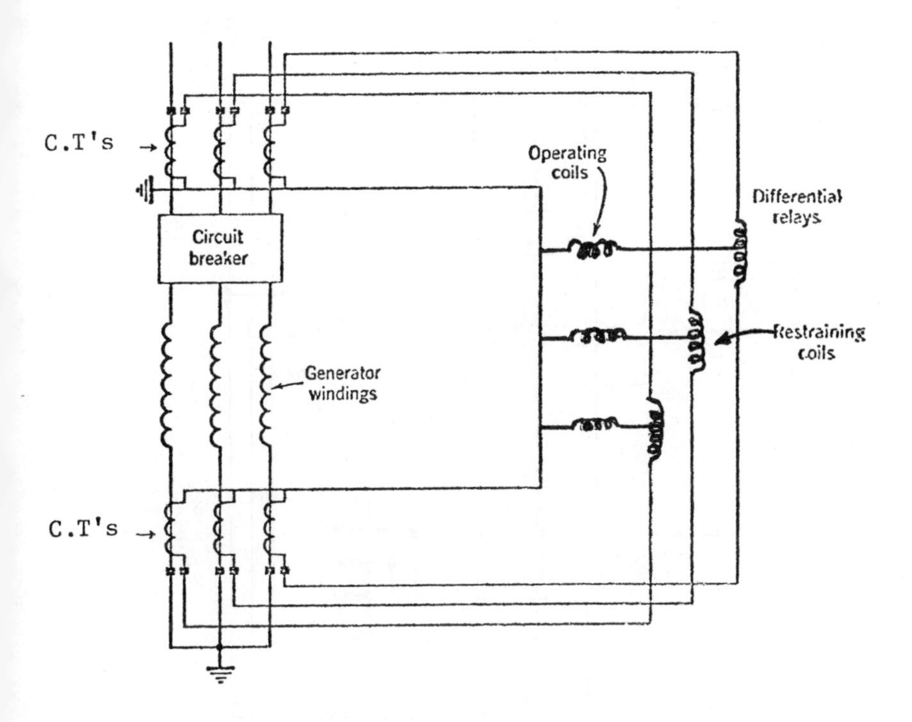

Figure 9.1.1. Schematic connections for percentage-differential relaying of a star-connected generator (for protection against stator phase and ground faults).

percentage-differential relaying equipment is not sensitive enough due to high impedance grounding, the stator ground-fault relaying can be made more sensitive by adding a current-polarized directional relay as illustrated in Figure 9.1.3. Figure 9.1.4 shows the preferred way to provide ground-fault protection for a generator that is operated as a unit with its power transformer.

(b) Stator Inter-Turn Faults:

Differential relaying shown in Figures 9.1.1 and 9.1.2 cannot be relied upon to detect inter-turn raults except

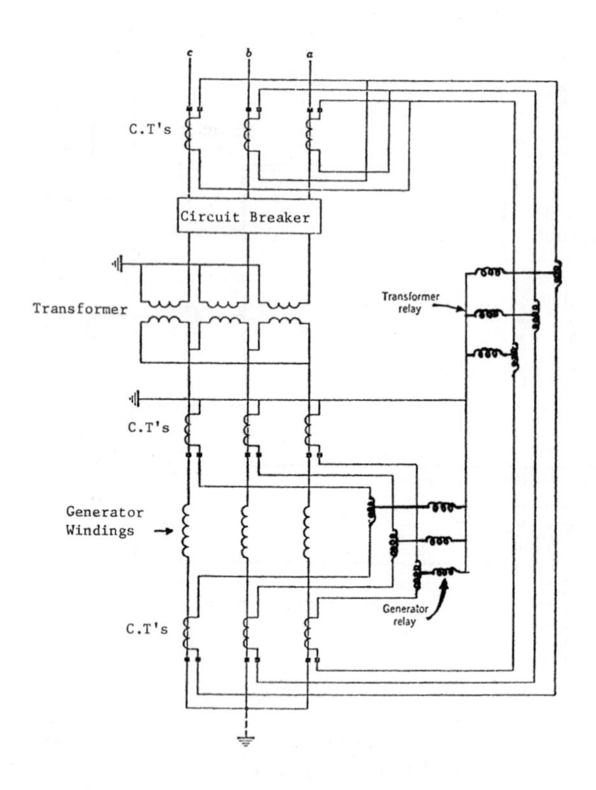

Figure 9.1.2. Schematic connections for percentage-differential relaying of a unit generator and transformer (for protection against phase and ground faults).

those between conductors of different phases which are in the same slot. A turn fault would have to burn through the major insulation to ground or to another phase before it could be detected. In large generators inter-turn fault protection is often unnecessary because there is only one turn per phase per slot and turn faults cannot occur without involving ground. The greatest value of this fault protection is for a generator with its neutral ungrounded, or grounded

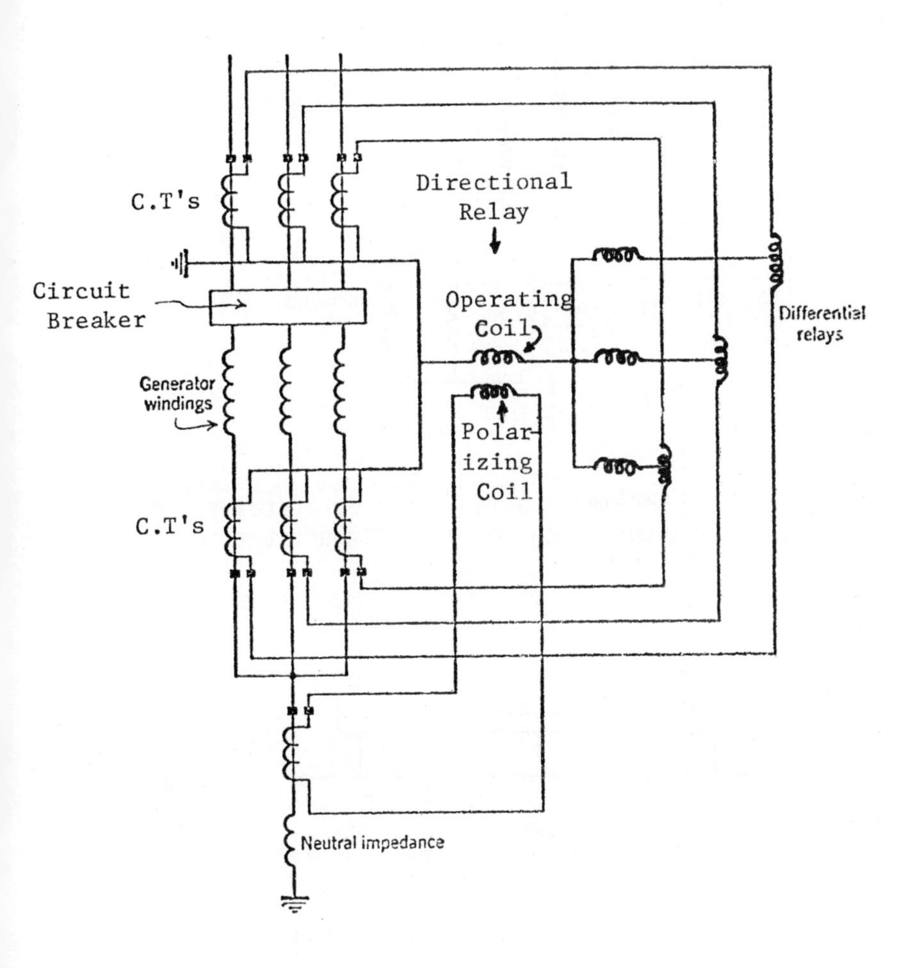

Figure 9.1.3. Sensitive relaying scheme for generator-stator ground-fault protection.

through a high impedance. Turn-fault protection has been devised for multi-circuit generators. With generators having parallel windings separately brought out to terminals, split-phase relaying illustrated in Figure 9.1.5 will detect faults between turns of the same winding. For generators without access to parallel windings, protection can be developed

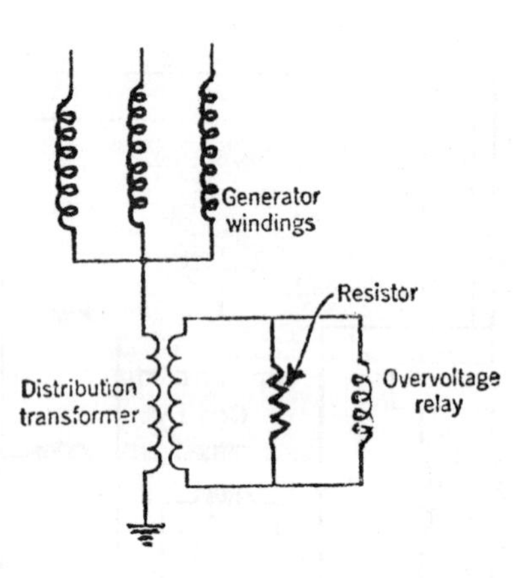

Figure 9.1.4. Schematic arrangement of stator ground-fault protection for a unit generator and transformer.

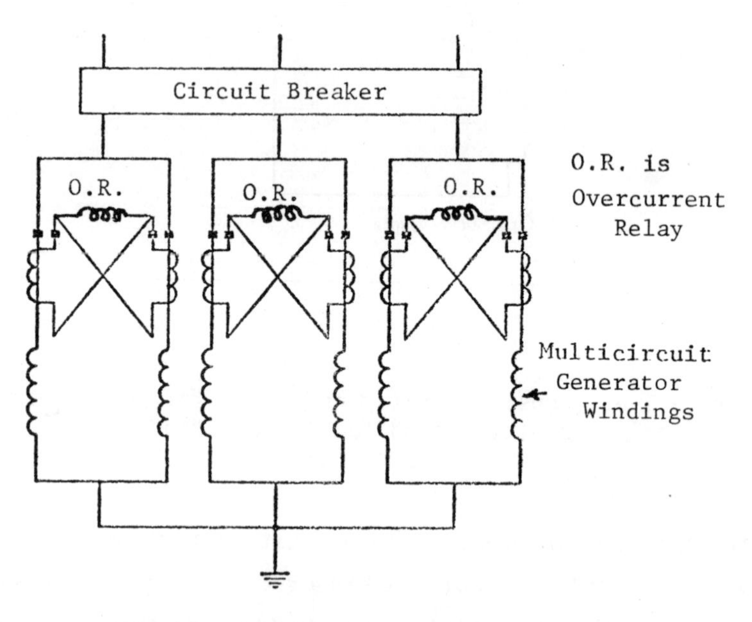

Figure 9.1.5. Schematic connections for split-phase relaying of a multi-circuit generator (for protection against stator inter-turn faults).

based on the zero-sequence component of voltage caused by the reduction of e.m.f. in the faulted phase. Since any dissymmetry of the stator currents produces a negative-sequence component (rotating at the same speed as the armature reaction field but in the reverse direction) which in turn induces a double-frequency current in the field circuit, a method based on this fact can also be relied upon for detecting turn-to-turn faults. It is possible to combine the split-phase and overall differential relaying for the generator without any sacrifice for sensitivity, as shown in Figure 9.1.6.

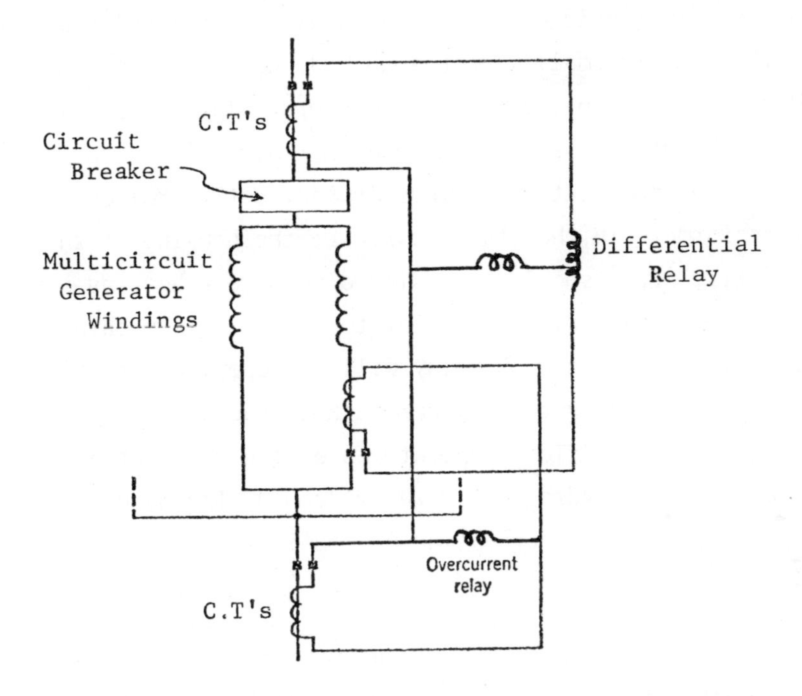

Figure 9.1.6. Sensitive relaying combining split-phase and differential protective schemes (shown for one phase only) (for protection against stator inter-turn, phase, and ground faults).

(c) Stator Overheating:

Ventilation failure, overloading, short-circuited laminations, and failure of core-bolt insulation may cause overheating of the stator. For machines rated 1 MVA or higher, the stator-overheating protection is generally provided. A common method is to embed temperature indicating devices (such as resistance temperature detectors, thermocouples, or thermistors) in the slots at different points in the winding; a selector switch checks each one in turn long enough to operate an alarm relay. Figure 9.1.7 shows one form of detector-operated relaying arrangement in which a Wheatstone-bridge circuit and a directional relay are utilized. In small machines that do not have temperature detectors, a replica-type temperature relay is used which has a bimetallic strip heated either directly by the current in one of the stator windings or indirectly from current transformers in the stator circuit; the housing of the bimetallic strip should be designed to have a heating and cooling characteristic similar to that of the machine. It is obvious that such a relay will not operate for failure of the cooling system. Other methods based on comparing the inlet and outlet temperatures of the ventilating medium (such as air, hydrogen, or water) could also be used for providing stator-overheating protection.

(d) Overvoltage Protection:

When a modern steam-driven turbine generator loses its load, the steam can be throttled before any great increase in speed has taken place; any overvoltage associated with overspeed will be controlled by the automatic voltage regulator. In hydroelectric generators, however, the water flow cannot be stopped or deflected so quickly and as such

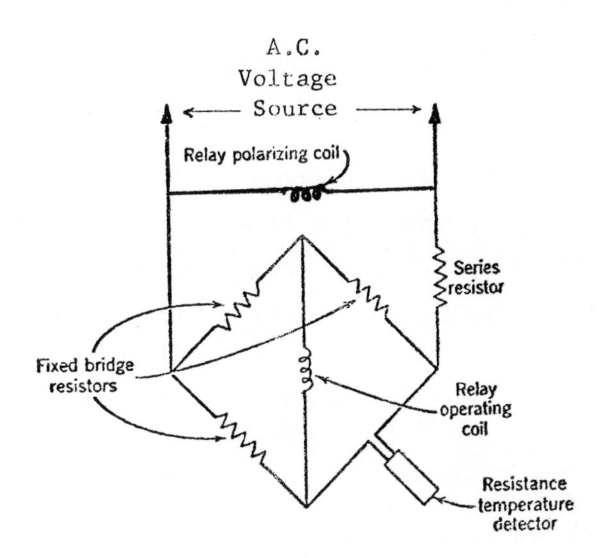

Figure 9.1.7. Relaying scheme with resistance temperature detector for protection against stator overheating.

overspeed can occur. In cases where the exciter is directly coupled to the machine, the voltage tends to go up nearly as the square of the speed. Hence overvoltage protection is needed for all hydroelectric or gas-turbine generators. It is not generally required with steam-turbo alternators.

The overvoltage protection is often provided by the voltage regulating equipment. If it is not, it should be provided by an overvoltage relay having two units, an instantaneous unit tripping on 25% (steam) or 40% (hydro) overvoltage, and an inverse time unit starting on 10% overvoltage. Both units must be compensated for the effect of varying frequency and must be energized from a potential transformer other than the one used for the automatic voltage regulator. The high-set relay would first insert additional resistance in the generator or exciter field circuit and, if the overvoltage persists, the low-set inverse time relay will shut down the generator.

(e) Stator Open Circuits:

Since open circuits are most unlikely in well-constructed machines, it is not a common practice to provide protective-relaying equipment specifically for open circuits. Negative phase sequence relaying that is used for protection against unbalanced phase currents contains a sensitive alarm unit that will alert the operator to the abnormal condition.

Rotor Protection:

Ground faults or open circuits may damage the rotor windings. Overheating due to unbalanced armature currents may cause structural damage to parts of the rotor.

(a) Ground Faults:

A single ground fault has no effect because field circuits are operated ungrounded. However, a single ground fault increases the potential of the whole field and exciter system when voltages are induced in the field by stator transients, and thereby increases the probability of a second ground occuring. A second ground fault will increase the current in a part of the winding, and bypass part of the field winding due to which air-gap fluxes may become sufficiently unbalanced, leading to unbalanced magnetic forces on opposite sides of the rotor, serious vibration problems and consequent damage. A second fault to earth may sometimes cause local heating due to which the rotor may be slowly distorted.

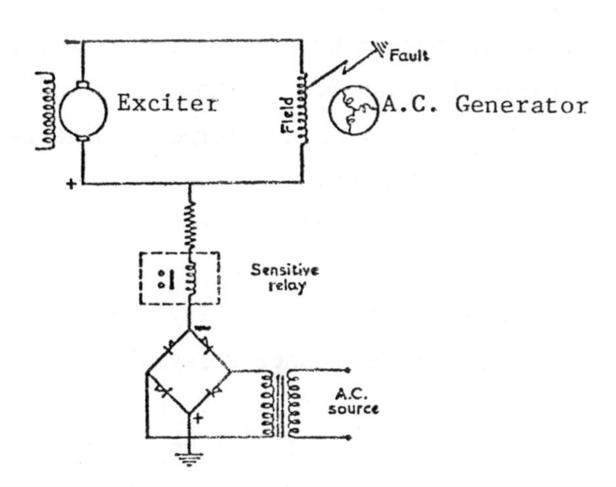

Figure 9.1.8. Generator-field earth-fault detection scheme.

A modern method of field ground-fault protection is shown in Figure 9.1.8. The field circuit is biased by a d.c. voltage which causes the current to flow through the relay if a ground fault occurs.

(b) Open Circuits and Loss of Excitation:

The relay to detect a rotor open circuit is the same one as is used for detecting the loss of field. Even though rotor open circuit is very rare, it must be promptly dealt with as otherwise the ensuing arc may cause damage.

When a generator loses its excitation, it speeds up slightly above synchronous speed and acts as an induction generator. Amortisseur windings normally provided on the salient-pole generators carry the induced rotor currents under such conditions. However, the rotor of a steam turbine generator will get overheated rather quickly from the induced currents flowing in the rotor iron, particularly at the ends of the rotor where the currents flow across the slots through the wedges and the retaining ring. Also, the wattless current drawn from the system by the machine as magnetizing current may overheat the stator. Furthermore, the system stability may be easily upset when a large machine is running out of step with the system, unless the generator is equipped with a quick-acting voltage regulator and connected to a stiff system.

The most reliable field failure relay is either a mho relay or a directional impedance relay with its characteristic in the negative reactance area. Figure 9.1.9 shows several loss-of-excitation characteristics and the operating characteristic of one type of loss-of-excitation relay on an R-X diagram. When the excitation is severely reduced or lost, the equivalent generator impedance traces a path from the first quadrant into a region of the fourth quadrant. The relay will operate to trip the field breaker and disconnect the generator from the system, when the generator first starts to slip poles. The characteristic of the relay is such that the relay is affected only by the loss of field and not by any other condition such as loss of synchronism.

(c) Unbalanced Stator Currents:

System conditions such as the open-circuiting of one phase of a line or the failure of one contact of a circuit breaker, an unbalanced fault near the station that is not cleared promptly, or a fault in the armature winding, may cause harmful unbalanced stator currents. The negative sequence component of these currents induces double-frequency currents in the rotor, leading to severe overheating and consequent damage to the structural parts of the rotor (such as the slot wedges and retaining rings) if the degree of unbalance is sufficiently large.

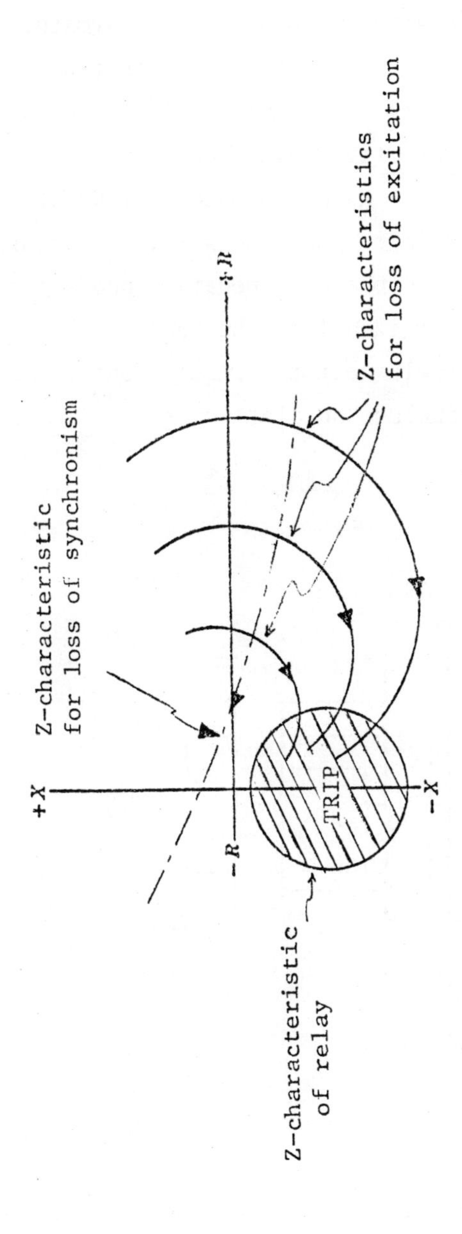

Figure 9.1.9. Impedance characteristics of the system and loss-of-excitation mho relay.

As discussed earlier in Section 6.4 under "Heating Effect of Rotor Currents", standards have been established for the operation of generators with unbalanced armature currents. The time for which the rotor can withstand such an imbalance varies inversely as the square of the negative-sequence current. The protective relay should have a time-current characteristic ($I_2^2t = K$) which matches that of the machine as closely as possible. An inverse-time overcurrent relay operating from the output of a negative phase sequence current filter that is energized from the generator CT's as shown in Figure 9.1.10 should meet the requirements. Some forms of the relay also include an alarm unit.

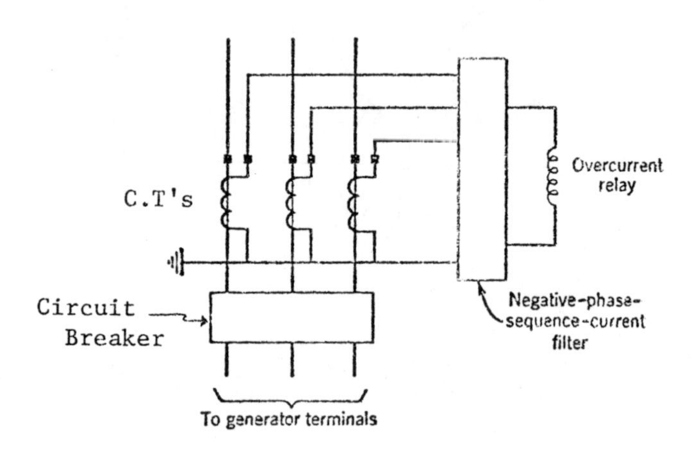

Figure 9.1.10. Schematic connections for negative-phase-sequence-overcurrent-relay for protection against unbalanced stator currents.

(d) Rotor Overheating because of Overexcitation:

Protection against overheating because of overexcitation is not generally provided. Such protection would be indirectly provided by the stator-overheating protective equipment or by the excitation-limiting features of the voltage-regulator equipment.

(e) Vibration

Protective relaying equipment provided against rotor overheating because of unbalanced three-phase stator currents and field ground faults prevents or minimizes vibration under those circumstances. If the vibration-detecting equipment is recommended under field ground-fault protection, it will also provide protection if vibration occurs from a mechanical failure or abnormality. For large steam turbo generators, it is common to provide vibration recorders that can also be used to control an alarm or to trip if desired. However, it is not the general practice to trip.

(f) Motoring:

If the steam supply is reduced sufficiently, the heat caused by windage loss due to the turbulence of the trapped air while the generator is idling or running as a motor can damage turbine blades. Motoring protection is essentially for the benefit of the prime mover or the system, and not for the generator. For a turbine that will not overheat unless its generator runs as a motor, motoring is prevented by sensitive power-directional-relaying equipment which operates on about 0.5% reverse power, its setting depending on the type of steam turbine. Topping turbines require faster settings than condensing turbines. The reverse-power relay usually has a time delay varying from seconds to minutes, also depending on the type of turbine, to prevent undesired operation on transient power reversals such as those occuring during synchronizing or system disturbances.

Reverse-power relays are occasionally used to prevent other types of generators from motoring. Cavitation may occur on low water flow in the case of a hydraulic turbine. Unattended hydraulic turbines are sometimes provided with motoring protection by means of power-directional-relaying equipment capable of operating on motoring current of somewhat less than about 2.5% of the generator's full-load rating. Diesel engine sets usually require motoring protection with a setting of about 15 to 25% reverse power, since the motoring may constitute an undesirably high load on the system and also, there may be danger of fire or explosion from unburnt fuel. In the case of gas turbines, there is usually no turbine requirement for motoring protection; relaying equipment should be chosen based primarily on the undesirability of imposing the motoring load on the system.

In any case, it is desirable that the turbine manufacturer's recommendations be sought.

(g) Overspeed:

Sudden loss of load or opening of a circuit breaker may cause overspeed. The overspeed protection may be furnished as a part of the prime mover, or of its speed governor, or of the generator. It should be responsive to machine speed by mechanical means or equivalent electrical connection; if electrical, the overspeed element should not be adversely affected by the generator voltage. It should operate the speed governor to shut down the prime mover, trip the generator circuit breaker, and also trip the auxiliary breaker, if auxiliary power is taken from the generator leads. The protective equipment should prevent the overfrequency operation of loads connected to the system supplied by the generator, and also prevent possible over-frequency operation of the generator itself from the a.c. system.

Steam and hydro sets are usually provided with mechanical overspeed devices but, because of the slower throttling down of the hydro and gas-turbine sets, overspeed relays become more necessary on the latter. The setting of an overspeed relay may be 115% for steam or 140% for hydro machines. Quick acting relays operating when the wattful power falls relative to the steam pressure have also been used for large steam sets. Out-of-step tripping relays are sometimes used on very large steam sets to anticipate speeding up due to loss of load and cut off the steam when the generator has slipped one pole and is 180° out of synchronism.

(h) Bearing Failure:

The temperature of the white metal or the oil can be monitored by an instrument with alarm contacts. The relay may be actuated by a thermometer-type bulb inserted in a hole in the bearing; a resistance-temperature-detector relay, such as the one used for stator-overheating protection, with the detector embedded in the bearing may be used. Such protection for attended generators is generally only to sound an alarm, while it would shut down the generator in unattended stations where the size and the importance of the generator warrants it. On large machines where the lubricating oil is circulated through the bearing under pressure, failure of the oil cooling equipment may be detected by comparison of the inlet and outlet temperatures of the oil; provision is also usually made for giving an alarm if the oil stops flowing.

(i) Loss of Synchronism:

Abnormal conditions such as the loss of field, motoring, overspeed, bearing failure, and loss of synchronism affect both the rotor and the stator, even though these are discussed here under "Rotor Protection". An out-of-step relay can be provided for detecting loss of synchronism; however, such protection is seldom used on an individual generator because it is very unlikely to run out of synchronism with the system or the other generators unless it loses its field (the case of which has already been discussed) or unless the governor becomes defective. Automatic synchronizing by an electronic relay is a common practice for large machines. If one station loses synchronism with another station, the necessary tripping to remove the generators that are out of step is usually performed in the interconnecting transmission system between them.

Other Miscellaneous Forms of Protection:

Features provided for the protection of the prime mover are generally mechanical and are not usually classified under generator protective-relaying equipment. Except for the protection against motoring and overspeed, the protection of the prime mover and its associated mechanical equipment is not treated in this book.

(a) Auxiliary Failures:

Protective equipment is usually provided for the loss of vacuum and loss of boiler pressure with very large generating units. The loss-of-vacuum relay also provides protection against loss of auxiliaries to some extent, because a fall in vacuum may be due to the station auxiliary failures. It is a common practice to reduce the load until the condition for a fall in vacuum is checked; if the vacuum continues to fall to a dangerous value, a vacuum relay closes its contacts and causes to shut down the set. As a safeguard against a fall in boiler pressure, a steam pressure device may be arranged to remove the load from the turbine. Sometimes the generating unit is shut down automatically on the loss of the induced draught fans.

(b) Voltage Regulator Failure:

Inadvertent tripping of the generator or damage to the rotor may be caused because of the faulty operation of the voltage regulator. Protection needs to be provided to guard against the voltage regulator failure, particularly on large generators using direct cooling of the stator and rotor. A definite time d.c. overcurrent relay energized from a shunt or a d.c. current transformer in the rotor circuit is commonly used; suitable time delay should be set as the rotor is subjected to overcurrent during certain system faults; if, however, the overcurrent condition persists beyond the setting, the relay will operate and switch the excitation to a preselected value.

Protective relaying is sometimes provided to prevent the malfunction of the regulator because of voltage failure. While it should not operate for a normal voltage reduction during system faults, it must respond to the failure of any one fuse on either the h.v. or l.v. side of the voltage transformer. A voltage balance relay which compares the voltage obtained from the instrument transformer with the voltage derived from the voltage regulator transformer, or a current bias voltage relay whose setting increases with an increase in the stator current, may be utilized. It is also usual to supply the regulator reference voltage from a separate voltage transformer so as to minimize the risk of a short-circuit on the secondary wiring causing thereby a fuse failure.

(c) Interlocked Overcurrent Protection"

For reasons of economy, one may in some cases locate protective current transformers only on one side of the circuit breaker. A special overcurrent relay interlocked with the appropriate unit protection needs to be provided to detect the faults occuring between the breaker contacts and the current transformer secondaries.

(d) External Fault Back-up Protection:

If primary relaying should fail, provision should be made not to supply the short-circuit current to a fault in an adjacent system element. The external fault back-up relay is usually energized by current and voltage sources on the low-voltage side of the power transformer for a unit generator-transformer arrangement. Care should be exercised to see that the connections are such that the distance-type units measure distance properly for high-voltage faults. A back-up relay should have characteristics similar to the relays being backed up; a negative phase sequence overcurrent relay is not the best suited for the purpose and also, such a relay would not operate for three-phase faults. Simple inverse-time overcurrent relaying is quite satisfactory for single-phase-to-ground faults. For phase faults, a single-step distance-type relay with definite time delay is preferred; a voltage-restrained or voltage-controlled inverse-time overcurrent relay may also be used; however, inverse-time overcurrent relaying is considered to be inferior and not very reliable for phase-fault back-up protection.

Protective Relaying Schemes:

The IEEE Standard 242-1975 on "Recommended Practice for Protection and Coordination of Industrial and Commercial Power Systems" shows the various protective features to be applied for the generators as in Figure 9.1.11.

Figure 9.1.12 illustrates a coordinated protective package of relays for a large steam-turbine generator, unit step-up transformer, and the excitation system. The relays are identified in Table 9.1.1 and the actions recommended typically by one manufacturer are included therein. Figure 9.1.13 shows a coordinated protective package of circuits for a large steam-turbine generator and its excitation system. Protective circuits generally accomplish protection by limiting or regulating a variable to a pre-established value; from this point of view, circuits added to the exciter can be broadly considered as protective. Such circuits and their key functions are described in Figure 9.1.13. Details given here should be considered as typical, while the particular recommendations by different manufacturers of generators, excitation systems, and relays may vary. Protection by relays and by circuits should be integrated with each other to accomplish a coordinated protective system. Protective relaying furnished with the excitation system usually includes generator field ground, generator field overvoltage, generator volts/hertz over-excitation protection, exciter output unbalanced voltage, and excitation potential circuit ground.

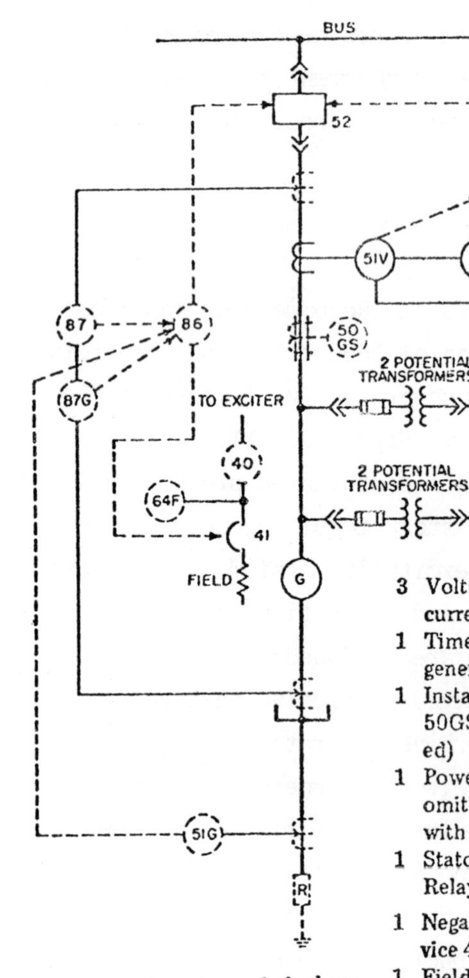

- 3 Voltage-Controlled or Restrained Time Overcurrent Relays (Device 51V)
- 1 Time Overcurrent Relay (Device 51G) (use if generator neutral is grounded)
- 1 Instantaneous Overcurrent Relay (Device 50GS) (use if generator neutral is not grounded)
- 1 Power Directional Relay (Device 32) (may be omitted if protective function is included with steam turbine)
- 1 Stator Impedance or Loss of Field Current Relay (Device 40)
- 1 Negative Phase Sequence Current Relay (Device 46)
- 1 Field Circuit Ground Detector (Device 64F)
- 1 Potential Transformer Failure Relay (Device 60V)
- 1 Lockout Relay (Device 86) (hand reset)
- 3 Fixed or Variable Percent Differential Relays (Device 87)
- 1 Current-Polarized Directional Relay (Device 87G)

NOTE: Devices shown dashed are optional for small or low-voltage machines.

Figure 9.1.11. Protective functions of generators.

[*Reference:* IEEE Standard 242-1975, "Recommended Practice for Protection and Coordination of Industrial and Commercial Power Systems".]

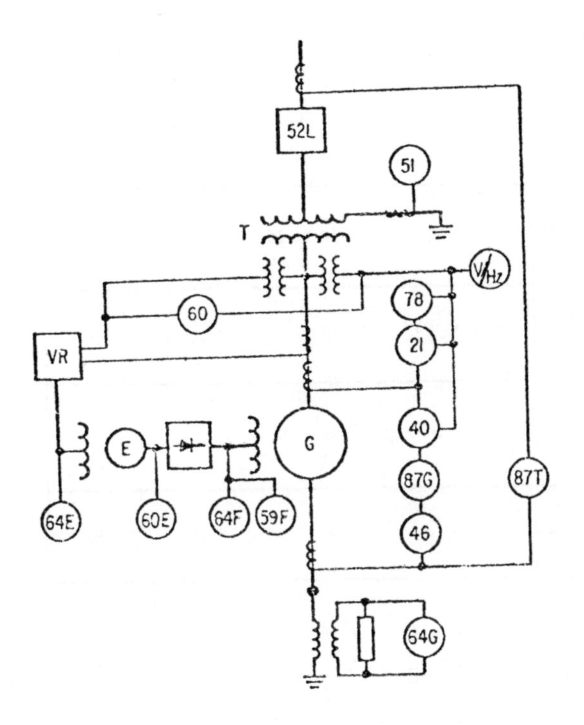

Figure 9.1.12. Protective relaying -- generator and excitation system. (Refer to Table 9.1.1 for device designations).

G - Generator
T - Step-up transformer
E - Exciter alternator
VR - Voltage regulator
52L - Main generator breaker

As shown in Figure 9.1.12, no circuit breaker is provided between the generator and the transformer. The fault detecting relays usually trip the main and field breakers, apply breaking, inject carbon dioxide gas, shut off the steam, and also shut down certain auxiliary equipment. Relays protecting against overload, overheating, overvoltage, negative-sequence current, and overheating due to an external fault may open the main and field breakers only. Conditions operating alarms only may include several abnormal situations,

Table 9.1.1

DEVICE NUMBER	FUNCTION \ ACTION	TURBINE STOP-VALVE TRIP	MAIN BREAKER TRIP	REMOVE EXCITATION	OPERATOR ALARM
	GENERATOR				
21	System Phase Fault Back-up		X		X
40	Generator Loss-of-Excitation*		X	X	X
46	Generator Negative Phase Sequence Current		X	X	X
51	System Ground Fault Back-up		X		X
78	Loss-of-Synchronism		X	X	X
59F	Generator Field Overvoltage	X	X	X	X
60	Generator Voltage Sensing Monitor				X
64F	Generator Field Ground	X	X	X	X
64G	Generator Stator Ground	X	X	X	X
87G	Generator Stator Current Differential	X	X	X	X
87T	Transformer Current Differential	X	X	X	X
V/Hz	Volts/Hertz Overexcitation Protection**			X	X
	EXCITATION SYSTEM				
60E	Exciter Alternator Unbalanced Voltage	X	X	X	X
64E	Exciter Field Ground				X

* recognizes certain cases of loss-of-synchronism
** functions only if main generator breaker is open

[*Reference:* J. Berdy, M. L. Crenshaw and M. Temoshok, "Protection of Large Steam Turbine Generators during Abnormal Operating Conditions", CIGRE Int'l Conf. on Large High Tension Electric Systems, Paper No. 11-05, 1972.]

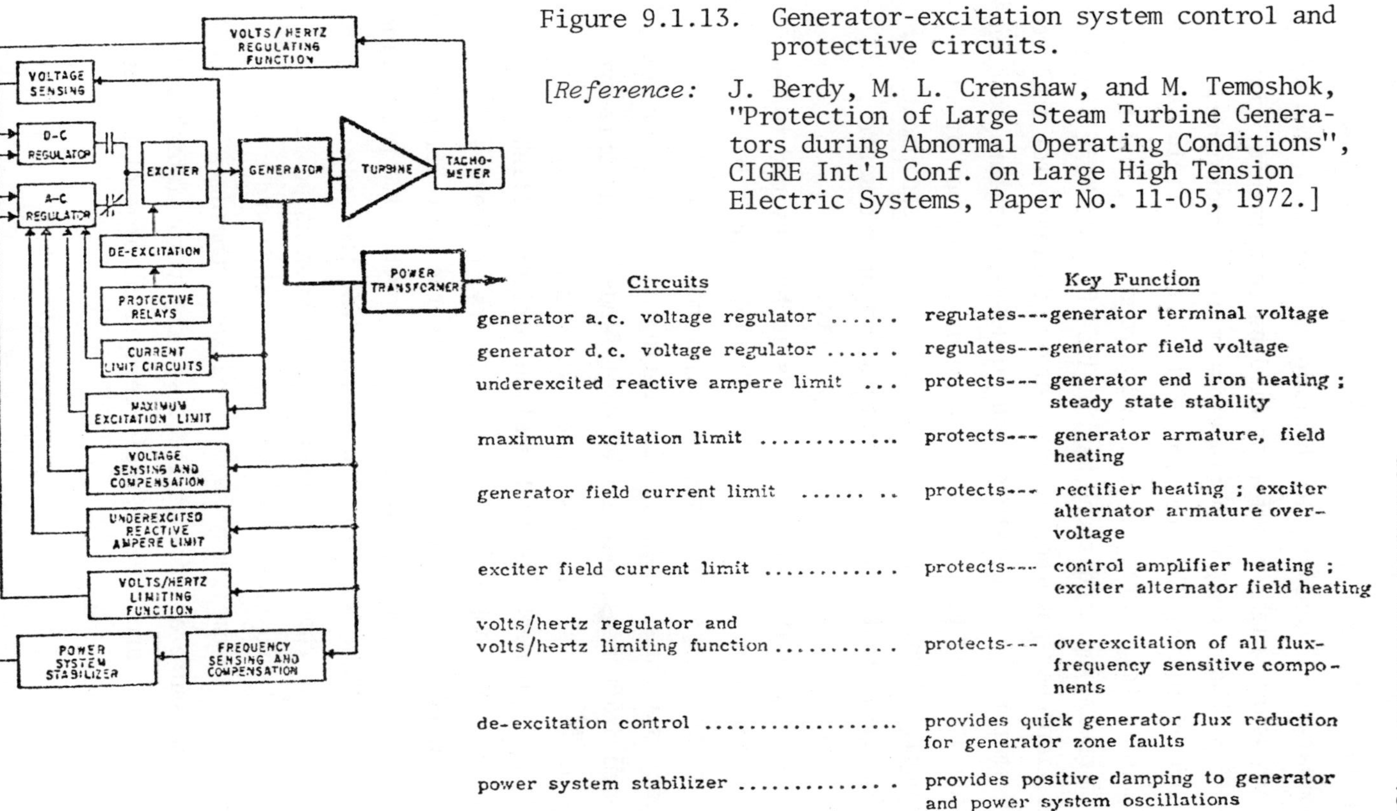

Figure 9.1.13. Generator-excitation system control and protective circuits.

[*Reference:* J. Berdy, M. L. Crenshaw, and M. Temoshok, "Protection of Large Steam Turbine Generators during Abnormal Operating Conditions", CIGRE Int'l Conf. on Large High Tension Electric Systems, Paper No. 11-05, 1972.]

Circuits	Key Function
generator a.c. voltage regulator	regulates---generator terminal voltage
generator d.c. voltage regulator	regulates---generator field voltage
underexcited reactive ampere limit ...	protects--- generator end iron heating ; steady state stability
maximum excitation limit	protects--- generator armature, field heating
generator field current limit	protects--- rectifier heating ; exciter alternator armature over-voltage
exciter field current limit	protects--- control amplifier heating ; exciter alternator field heating
volts/hertz regulator and volts/hertz limiting function	protects--- overexcitation of all flux-frequency sensitive components
de-excitation control	provides quick generator flux reduction for generator zone faults
power system stabilizer	provides positive damping to generator and power system oscillations

such as condenser low vacuum, abnormal hydrogen pressure, temperature or density, low bearing oil pressure, low seal oil pressure, high transformer winding temperature, transformer Buchholz gas, high transformer oil temperature, auto-voltage regulator failure, rotor earth fault, field failure, and low battery voltage. For hydro-generators, additional alarm operating conditions may include high bearing temperature, improper governor oil pressure, cooling water failure, high stator air temperature, and guide vane failure to open. Hydrogen cooled generators may also have a number of additional indicators connected with auxiliary equipment such as defoaming tanks, water detectors, vapor extractors, oil conditioners, d.c. emergency pumps, etc.

Comprehensive protection needs to be provided to the power station auxiliaries. The essential group includes boiler feed pumps, condenser pumps, forced draft fans, primary air fans, stokers, circulating pumps, exciter sets, induced draft fans, pulverized coal feeders, and unit-type coal pulverizers. The non-essential group includes coal handling equipment, central coal pulverizers, clinker grinders, air compressors, coal crushers, conveyors, ventilating fans, and service pumps. While the protection of the two groups is usually similar, more comprehensive protection and full voltage starting are applied for the essential group so that they can be restarted as soon as possible after an interruption of power supply. They are usually switched to an emergency supply and kept running. In case of low pressure or lack of flow, standby pumps (which are commonly d.c. and have thermal overcurrent protection) are sometimes provided.

CHAPTER X

TRENDS IN DEVELOPMENT OF LARGE GENERATORS

Electric power systems continue to grow in their size and complexity. At the present time in the United States, there is a total generating capacity of about 300,000 MW. The total capacity has been doubling about every ten years for the past several decades and shows every sign of continuing at almost the same rate for at least two more decades. Further growth is predicted at a decreasing rate, with a load of about 20 times the present value by the year 2030. The economic pressures of the 1960's continued the drive of the electric utility industry to reach out for ever larger generator unit ratings. The optimum size of a generating unit can be expected to be a certain percentage of the size of the power system on which it is operating. Viewing all of the interconnected systems of the United States as a whole, the largest presently available unit of about 1500 MW is still less than 1/2% of the total network generating capacity. However, in addition to the total network size, the appropriate unit size is controlled by other factors such as the state of the technology of both generation and transmission.

Although the pressures for ever larger fossil-fueled plants will continue, the growth in maximum unit ratings is expected to take place at a more moderate pace. The tendency is especially toward increasing the rating for a given physical size. On the nuclear front the economic incentive to go for larger ratings is more compelling; however, it appears that the electric utility industry would like to get more operational experience at each new level in plant rating before advancing to the next step, while considering the pressures of economy of unit size, of siting, of environmental considerations, and complex system operation.

Reliability and progress are the two key words which express the needs of the electric utility business. It is a tribute to the generator design engineer that, while forced to work always near the frontier of knowledge and experience, he (or she) has been able to meet the stringent requirements of the electric utility industry for high reliability and availability at a reasonable cost. New concepts, research, development, testing, and design studies continue to open up the possibilities permitting further increases in ratings*.

In this chapter we shall be concerned primarily with steam turbine driven generators, although for completeness brief mention would be made of hydraulic turbine driven generators and some discussion regarding superconductive generation will also be presented.

Large Turbo-Alternators:

Steam turbine driven generators are usually of either 2-pole, 3600 rpm or 4-pole, 1800 rpm for 60 Hz systems. Due to the advent of nuclear power the 1960's saw a sharp turn-up in the ratings of 4-pole, 1800 rpm generators. High speeds impose a limit on the permissible rotor diameter on account of the centrifugal forces developed, and then indirectly on the physical size of the unit. In view of this

*See Harrington, D. B. and Jenkins, S. C., "Trends and Advancements in the Design of Large Generators", American Power Conference, Chicago, April 1970.

as well as an economic pressure to conserve materials, and also size constraints imposed by shipping limitations, one is forced to increase the unit rating without a corresponding increase in size thereby increasing the power per unit volume of active material. The larger power density results in higher current densities, greater winding forces, more losses, and more heat generated per unit volume. Also from the viewpoint of the power system, the increased power density tends to increase the generator reactances and decrease the generator rotary inertia, both of which have an adverse effect on stability. Greater use of cooling towers (because of the present concern for the environment and thermal discharges to water) has the effect of reducing the turbine inertia. The thermal transient performance becomes more critical with increased KVA per unit volume of active material. In addition, the necessarily larger physical unit size presents major mechanical and material challenges. The ways in which these challenges have been met will be discussed in this section in some detail.

The approximate range of some of the technical parameters* characterizing the existing large turbine generators is given below:

Rated Power	167 to 1,250 MW
Rated Apparent Power	200 to 1,500 MVA
Rated Voltage (line to line)	18,000 to 26,000V
Rated Current	10,000 to 45,000 A
Excitation Power (Controlled DC)	500 to 4,200 KW

* [Data furnished by D. B. Harrington and A. C. Shartrand of the General Electric Company.]

Excitation Current	1,200 to 6,500 A
Kinetic Rotational Energy	68.5×10^7 to 29.4×10^9 Joules
Stored Magnetic Energy (at rated voltage)	4×10^6 to 1.3×10^6 Joules
Maximum Peripheral Speeds of Rotating Parts (at rated Speed)	185 to 220 meters/second
Total Generator Weight (including exciter)	270 to 785 metric tons

(a) Voltage and Current Levels:

Voltage and current capabilities in generator design have had to grow in step with increased ratings. Generator designers have long advocated the freedom to select the voltage level which results in the optimum generator design; but in view of the overall economy of the station, there remain questions of standardization of voltage for auxiliaries fed from the generator terminals. Innovations in terminal arrangements and winding connections have been utilized for optimum generator designs with suitable voltage levels. Progress in polymer science and development of new insulating systems suggest that voltage levels greater than the 26-KV currently in service could be achieved. Liquid-cooled high-voltage bushings have been developed to accommodate the increased current of the future generators.

(b) Short-Circuit Ratio

For economy in generator physical size and efficiency, the tendency has been toward lowering short-circuit ratio for years. A typical value is around 0.5. It appears that SCR has reached the point of diminishing returns and the design economies once available are not as pronounced today.

While SCR has had a significant influence on rotor design in the past, other considerations and limitations including system stability concerns are now beginning to dominate the design process.

(c) Reactances:

Increased ratings made possible by conductor cooling continue to result in higher values of transient reactance which tend to have an adverse effect on transient stability. However, developments in excitation and relaying technology have more than compensated for the ill effects of the increased generator reactances on stability. In view of the limited range of parameters the designer can work with in developing a reliable design, and restrictions imposed by shipping limitations, it is expected that the trend toward increased values of transient reactance with increased generator unit ratings will continue.

(d) Excitation Systems:

Several excitation systems with their own particular characteristics and very high-speed voltage response have been developed in the recent years for optimum operation with the generator and greater system economy. Innovative excitation system designs will continue to play an increasingly important role in influencing the generator terminal characteristics and power system performance.

(e) Rotor Thermal Transient Capability:

Conductor-cooled generators inherently have higher values of negative-sequence reactance when optimum designs are achieved. Because the higher reactance would limit negative-sequence currents to lower values, the standards for conductor-cooled generators were drawn in the mid-60's to permit $I_2^2t = 10$. In the overall interest of total system economy, it now appears that even lower values such as 7 may be justified by large generators because of the other system limitations such as the effects of fault duration on stability, which are the controlling elements of the system design and may become limiting before the generator negative-sequence current limits become critical. Much research and developmental work has been taking place in the recent years to improve the knowledge of the phenomena involved in the rotor-surface heating due to unbalanced fault currents.

(f) Stator Winding Forces:

The forces exerted on armature windings are proportional to the square of the current. In a typical large generator, the force on a single pair of bars in a stator slot may reach 120 KN (12 tons); for a 60-HZ generator this force is experienced by the bar over 10 million times per day. By increasing the generator length as well as the number of stator slots, it is possible to hold down bar forces; but these in turn increase voltage, requiring more space for insulation. An optimum strategy is being chosen in which bar forces increase more nearly with the first power of rating rather than with the square. In order to reduce and control the stator winding forces, special winding arrangements have been considered. These include the use of "fractional poles per-circuit windings, FP/C" (also known as multiple-circuit-windings which are currently applied in practice), and a '6-phase' connection with two separate 3-phase windings differing in phase by 30°, the voltage being converted to 3-phase by appropriate transformer connections, which has been proposed by Holly and Willyoung in 1970 (included in Bibliography).

Aside from designing new winding circuit arrangements to keep down forces, a parallel effort has been the development of innovative construction features for supporting the winding reliably. To provide the ability to accommodate increasing force levels, enormous improvement has been taking place in the support of the slot and end sections of the winding, as well as its connections and leads. Even though the short-circuit currents have not increased in direct proportion to the unit ratings, (not only because of the inherently higher machine reactances but also because of the unit generator-transformer arrangement in which the generator is connected to its individual step-up transformer by isolated-phase buswork that practically eliminates the possibility of a terminal short circuit) short-circuit forces have increased along with the increased winding forces during normal operation. Very large forces on end windings arising out of high current short-circuits or severe out-of-phase switching have been a challenge to the machine designers for decades. Several analytical studies and extensive model testing have been performed on the end-winding support systems. The insulating system and its mechanical characteristics do play a significant role in these developments.

(g) Cooling:

If any one thing is to be singled out as making possible the dramatic advance in generator ratings of recent years, it would be the improved and more effective methods of cooling, by direct contact of the cooling fluid with the stator and rotor conductors. The insulation has been removed as an element in the heat-flow path. The use of direct water cooling for stator windings of large turbine generators has become universal among manufacturers around the world. The temperature difference between copper and water is practically negligible in case of the water-cooled armature conductors. The consequent reduction in thermal stress adds margin for the required mechanical stresses and thus leads to the possibility of further advances while reliability is sustained.

In the case of the rotor, some manufacturers seem to prefer hydrogen-gas cooling (in which the coolant at an appropriate pressure is made to enter the rotor at several points and diagonal flowpaths are provided through the rotor conductors), while the others feel that liquid-cooled rotors with elimination of hydrogen are more desirable. Advocating the principle of using optimum cooling agents for each machine part, some manufacturers appear to go for total liquid cooling with direct water cooling in the stator and rotor windings, direct oil cooling in the stator core and end-plates, oil-cooled bearings, and water cooling for stator winding air-gap cylinder and end-shields.

(h) Mechanical Utilization:

Improved performance through better mechanical utilization is largely limited by two groups of problems: the available material properties and allowable stresses, particularly in the rotating parts; and the mechanical stability of the machine in which the generator and turbine are to be considered as one unit. Rotors of larger diameter tend to have greater margin for stability and higher critical speeds. Much work is being done to improve the properties of forgings and to study the stability margin of long rotors with their inherently lower critical speeds.

(i) Electromagnetic Utilization:

The electromagnetic utilization influences to a high degree the design and dimensions of the stator core, the armature winding, the field winding and its excitation system. It also affects significantly the operational electrical stability of the machine. Progress in quality grain-oriented core plates with reduced specific magnetization loss and improved permeability has been responsible in decreasing the stator core outside diameter, which has in turn eased the transport problem. However, the resulting reduction of the stator yoke increases the radial core vibration of twice nominal frequency, caused by the rotating magnetic field. Increased unit ratings bring difficult technological problems not only in the active part of the generator but also in other parts, such as the end region in particular. The end-region fields and their effects on the stator core ends, the stator end windings and their shields have acquired considerable importance; during normal operation as well as during abnormal operation such as under-excited or asynchronous running, harmful effects such as increased losses and occurrence of hot spots in the core with the danger of burn-out may result. Innovative end-region configurations are bieng used to overcome the associated problems.

(j) System Performance:

The effect of probable higher reactances and lower inertias of future generators on system performance, particularly stability margins, has been studied in some depth by Concordia and Brown in 1971*. Since a power system with larger generators requires a higher voltage transmission network, and the reactance of the step-up transformer increases with high-side voltage, the situation can be seen to be aggravated further. Aside from the standpoint of stability, other factors such as physical size, shipment, and fault duty may be limiting in the optimum choice of transformer reactance. Concordia and Brown have tried to evaluate the relative degree of stability of a simple system, expressed in terms of the critical clearing time, or the maximum duration of a 3-phase fault at the high-voltage terminals for which the generator is barely stable. An index of stability is mathematically expressed as a very rough rule-of-thumb given by

$$T_c \propto \sqrt{\frac{H}{(x_d' + x_t)(x_d + x_t)}}$$

where T_c is the critical clearing time, H the inertia constant, x_d' the transient reactance, x_d the synchronous reactance given by the reciprocal of the short-circuit ratio, and x_t the transformer reactance. The necessity of a strong transmission network and the possibility of improving stability, if necessary, have been pointed out, and finally, it has been concluded that the achievement of adequate stability margins seems entirely feasible within the range of unit sizes considered for the near future.

The use of series capacitor compensation is one of many ways of strengthening a network. Also, high response

---

*See Concordia, C. and Brown, P. G., "Effects of Trends in Large Steam Turbine Driven Generator Parameters on Power System Stability", IEEE Trans. 71-TP74-PWR, pp. 2211-18, 1971.

excitation systems with special stabilizers utilizing a shaft-speed signal have sometimes been used to improve stability. Under certain conditions, both of these may interact with the torsional natural frequencies of the turbine-generator rotor. Subsynchronous resonance phenomena has been under serious study in the recent years with the popular use of series capacitor compensation. In the case of the excitation system stabilizer problem, the rotor speed may no longer be a simple concept but may be different for each element. More research work will undoubtedly be done for studying torsional oscillations in detail and controlling them adequately.

The capacity of a generator to respond to abnormal system conditions is also affected by uprating. Possibility of overloading of both armature and field, as well as capability of supplying large emergency reactive power demands need to be considered. It has generally been found desirable to specify a generator with a rated power factor of 0.9 or even 0.85 in order to build-in reactive capability, even though the normal operating power factor may be nearly unity. As another case of abnormal operating condition, the rotor thermal transient capability (discussed under item (e) of this Section) needs to be considered to withstand the effects of induced negative phase sequence currents.

The introduction of direct current transmission lines into power systems can expose a turbine-generator to harmonic currents which can have a number of adverse effects (electrical, thermal, and mechanical). These are being studied actively in the recent years and means to control them to acceptable levels are being developed.

(k) Turbine Considerations:

Along with the increase of generator unit size, one has to cope with the corresponding increase in turbine unit size and even boiler size. At the present time it appears that uprating the turbine unit is somewhat more difficult than uprating the unit generator, with several potential problems such as handling the larger steam flows and ensuring rotor stability.

(*l*) Future Trends:

While there are many interrelated factors limiting the growth of generator unit ratings, we have not yet reached the limit of unit rating with the present generator cooling technology, stator winding short-circuit and vibration control, innovative design improvements, improved materials, and available methods of strengthening the transmission network. Conventional generators with a high degree of reliability can be produced in the ratings predicted for the 1980's. It is believed that reasonably predictable extensions of the present technology may suffice up to double the present ratings.

A pattern of broad-based development program involving laboratory and prototype testing will continue to be at the heart of progress. Radically new concepts of generation such as superconductive generation will have to stand on their own feet and compete in capital cost, performance, economy of operation, and especially in reliability and availability with the conventional designs, before they can be brought into commercial operation.

Hydroelectric Generators:

The required ratings of hydroelectric generators have

not been as large as for steam generators, but within recent years a large increase in the maximum ratings available, up to about 500 to 750 MVA, can be noticed. Since such ratings are entirely suitable for the largest hydro-power developments presently contemplated, there appears to be no great incentive for larger ratings in the foreseeable future. Compared to steam generators, hydrogenerators are relatively much larger in physical size because of the much smaller speed such as 100 rpm; much more assembly needs to be done on site in view of the shipping limitations. While the past hydrogenerators have been designed to withstand a great overspeed on sudden loss of load without counting on any corrective action of the control schemes, it seems that more recent large hydrogenerators may have to count on limiting the overspeed from the viewpoint of overall economy. Hydrogenerators in general have higher transient reactance, lower overall inertia because of the extremely small turbine inertia, and lower synchronous reactance. Insofar as power system stability is concerned, the first two changes are adverse while the third one is favorable. Several of the designing as well as constructional aspects of hydrogenerators are quite unique to that class of machinery, and much innovation is undoubtedly needed for their development.

Superconductive Generation:

The present development of technological bases for the production of future cryogenerators indicates that large turboalternators with liquid helium cooled superconducting field windings are feasible, and that the projected advantages should be realizable through further intensive research and development*. However, only the future will show at what rating the economic transition from the present conventional fluid cooling to cryo-cooling lies. Superconductivity does permit the present flux density limit of 2 to 2.5 T in the rotor iron to be exceeded considerably; however, one has to forego the ferromagnetic properties of steel. Current loadings and power density can be increased significantly with practically no change in subtransient and transient reactances, and no deterioration in electrical stability.

Figure 10.1.1 shows a typical basic layout of a superconducting generator. The rotor is a rotating Dewar vessel containing a superconducting field winding of Niobium-Titanium (NbTi) alloy wire which is kept superconductive by liquid helium at 4 to 5 K. The cold rotor interior is thermally insulated in the radial direction by a vacuum space and a radiation shield, and in axial direction by mechanical connections to the shaft with a high thermal resistance. The rotor outer cylinder is an electrical damping screen and a shield protecting the rotor winding from transient influences of the armature. The stator winding is directly water cooled, and its support structure must be of a nonmagnetic material and must resist high short-circuit forces and steady-state fatigue loads. Immediate surroundings must be shielded from the strong magnetic fields; the best rating per unit volume is achieved by using a shield of laminated magnetic iron, while minimum

---

*See Jefferies, M. J. et al, "Prospects for Superconductive Generators in the Electric Utility Industry", IEEE Trans. PAS, pp. 1659-1669, September/October 1973; Smith, J. L., "Superconductors in Large Synchronous Machines", EPRI Research Project Report, June 1975.

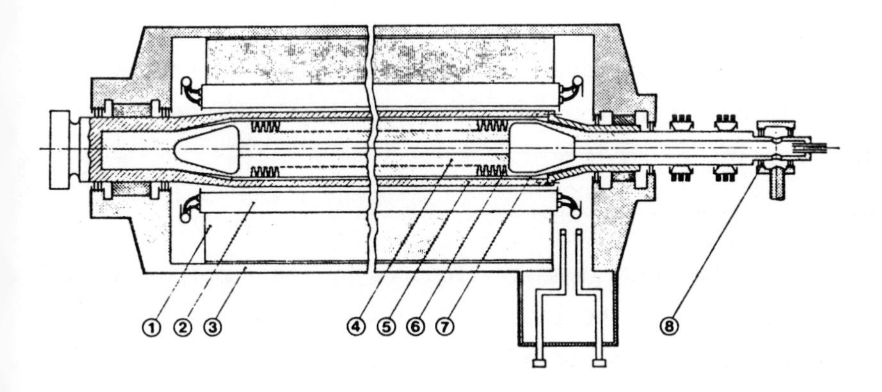

Figure 10.1.1. A typical basic layout of a cryogenerator with superconducting rotor winding

1. Laminated iron shield
2. Armature winding and supporting structure
3. Stator housing
4. Inner rotor
5. Outer rotor shield (damper cylinder)
6. Superconducting winding
7. Rotor thermal insulation (vacuum)
8. Rotor vacuum seals.

[*Reference:* Abegg, K., "The Growth of Turbogenerators", Philosophical Transactions of the Royal Society of London, Vol. 275, No. 1248, August 1973, pp. 51-67].

mass per unit volume results by using a conducting copper screen with a reduced rating per unit volume to about two-thirds.

Potential advantages of superconductive generation appear to be the following as compared to the conventional fluid-cooling technology:

(a) Reduced Size and Weight:

The higher flux density produced by the superconducting field winding, and the increased volume for active armature conductors made possible by the absence of iron and reduced insulation requirements, allow better utilization of space and materials. It is reasonably expected that the weight could be reduced to about one-third, or three times as much power can be obtained from a given frame size.

(b) Higher Efficiency:

Superconducting field windings require almost negligible excitation power. More effective space utilization reduces armature losses, and windage losses are also reduced because of reduction in size. The efficiency is expected to be increased by about 0.4%.

(c) Higher Voltage and Lower Current Levels:

Higher economical terminal voltage with consequent lower terminal current can be employed because of the simplified insulation geometry made possible by the absence of iron. The unit transformer associated with the generator can be made less expensive.

(d) Improved System Performance:

There is practically no steady-state stability limit since the synchronous impedance is so low; the generator can be operated at its full rating to zero power factor underexcited, for typical external system impedances. Improved transient stability results since the lower synchronous reactance yields a smaller torque angle initially and the post-fault machine reactance can be reduced, even though the rotor inertia is smaller. Improved performance in the presence of negative sequence currents (i.e., improved I_2^2t capability) can be expected because more thermal storage can be put into the appropriate rotor members, since space is more readily available and freedom of material selection is greater.

(e) Higher Rotor Critical Speeds:

Substantially higher critical speeds for the rotor will result for a given rating, thereby causing less tendency to develop vibration problems.

(f) Reduced Cost:

Materials for stator construction are the same as those for a conventional turboalternator, but only less is required for a given rating. The absence of magnetic steel core should reduce capital costs. The superconducting rotor will require significantly less material, but will demand more fabrication labor per unit of material. Hoping that the costs of superconductor and refrigeration equipment are modest, it may be possible to make the superconducting machine less expensive and quite reliable through electrical, mechanical, and cryogenic optimization.

Some of the key problem areas presenting serious design challenges are the following:

(a) Electromagnetic Rotor Shield:

A reliable rotor shielding scheme needs to be developed

to satisfy the conflicting requirements for magnetic shielding, electrical damping, mechanical strength to withstand very high short-circuit forces, thermal capacity, cryogenic performance, and desirable terminal characteristics of the machine.

(b) Overspeed Control Requirements:

More severe overspeed control requirements are needed because of the smaller inertia.

(c) Optimized Superconducting Rotor:

Proper balance needs to be achieved amid the various features such as magnetic field intensity, mechanical strength, cryogenic cooling, refrigeration requirement, and transient electrical performance.

(d) Optimized Stator Winding and its Structure:

Because of the much higher flux densities used and the absence of stator teeth, much higher stator conductor forces result. An armature winding and its structure should be optimized to utilize fully the high magnetic field produced by the superconductor, while withstanding the forces. Also, the electric field distribution needs to be managed in order to achieve higher terminal voltage. Since the stator winding is exposed to high tangential and radial fluxes, special measures become necessary to minimize the resulting armature losses.

(e) Reliability, Availability, and Cost:

In the final analysis, relative reliability, availability, and cost become overriding considerations in a practical sense. Any decrease from the present high level of reliability would not be tolerated by the electric utility industry. This naturally leads to an appropriately cautious attitude on the part of the designers, and to the necessity for thorough testing of the new class of cryogenerators, before they can be put into commercial operation on a large scale.

APPENDIX A

UNITS, CONSTANTS, AND CONVERSION FACTORS FOR RATIONALIZED MKS SYSTEM

TABLE A-1

Physical Quantities

Physical Quantity	Symbols	Unit	Abbreviation	In terms of basic units
Length	l, d ...	meter	m	
Mass	m	kilogram	kg	
Time	t, T	second	sec	
Charge	q, Q	coulomb	C	
Current	i, I	ampere	A	C/sec
Frequency	f	hertz	Hz	sec^{-1}
Force	F	newton	N	$kg \cdot m/sec^2$
Energy	U	Joule	J	$N \cdot m = kg \cdot m^2/sec^2$
Power	P	watt	W	$J/sec = kg \cdot m^2/sec^3$
Potential, emf	ϕ, V	volt	V	$W/A = N \cdot m/C = kg \cdot m^2/C \cdot sec^2$
Electric flux	ψ_e	coulomb	C	

TABLE A-1 (Cont'd)

Physical Quantity	Symbols	Unit	Abbreviation	In terms of basic units
Capacitance	C	farad	F	C/V = A · sec/V =
				= C^2 · sec^2/kg · m^2
Resistance	R	ohm	Ω	V/A = kg · m^2/C^2 · sec
Conductance	G	mho	℧	A/V = C^2 · sec/kg · m^2
Magnetic flux	ψ_m	weber	Wb	V · sec = kg · m^2/C · sec
Magnetic flux density	B	tesla	T	Wb/m^2 = V · sec/m^2 =
				= kg/C · sec
Inductance	L	henry	H	Wb/A = V · sec/A =
				= kg · m^2/C^2
Free-space permeability	μ_o	henry/meter	H/m	Ω · sec/m = kg · m/C^2
Conductivity	σ	mho/meter	℧/m	C^2 · sec/kg · m^3
Free-space permittivity	ε_o	farad/meter	F/m	℧ · sec/m = C^2
				= C^2 · sec^2/kg · m^3

TABLE A-2

Constants

Permeability of free space	$\mu_o = 4\pi \times 10^{-7}$ weber/amp-turn meter or (H/m)
Permittivity (capacitivity) of free space	$\varepsilon_o = 8.854 \times 10^{-12}$ coulomb^2/newton-meter^2 or (F/m)
Acceleration of gravity	$g = 9.807$ m/sec^2
Velocity of light	$c = 2.998 \times 10^8$ m/sec

TABLE A-3

Conversion Factors

Length	1m = 3.281 ft = 39.37 in
Mass	1kg = 0.0685 slug = 2.205 lb (mass)
Force	1 newton = 0.225 lb = 7.23 poundals
Torque	1 newton-meter = 0.738 lb-ft
Energy	1 joule (watt-sec) = 0.738 ft-lb
Power	1 watt = 1.341×10^{-3} hp
Moment of inertia	1 $kg \cdot m^2$ = 0.738 slug-ft^2 = 23.7 lb-ft^2
Magnetic flux	1 weber = 10^8 maxwells (lines)
Magnetic flux density	1 weber/m^2 = 10,000 gauss = 64.5 kilolines/in^2
Magnetizing force	1 amp-turn/m = 0.0254 amp-turn/in = 0.0126 oersted

APPENDIX B

IEEE STANDARDS AND TEST CODES FOR SYNCHRONOUS MACHINES

APPENDIX B

IEEE STANDARDS AND TEST CODES FOR SYNCHRONOUS MACHINES

IEEE Standard	Title	Code
1-1969	Temperature Limits in the Rating of Electric Equipment, General Principles for	SH00018
32-1972	Neutral Grounding Devices, Requirements, Terminology, and Test Procedure for	SH40329
43-1974	Testing Insulation Resistance of Rotating Machinery, Recommended Practice	SH20438
56-1958	(Reaff 1971) (ANSI C50.25-1972) Insulation Maintenance for Large AC Rotating Machinery, Guide for	SH02410
67-1972	(ANSI C50.30-1972) Turbine-Generators, Guide for Operation and Maintenance of	SH40675
81-1962	Ground Resistance and Potential Gradients in the Earth, Recommended Guide for Measuring	SH00281
86-1975	Basic Per Unit Quantities for AC Rotating Machines, Definitions of	SH10868
94-1970	Automatic Generation Control on Electric Power Systems, Definitions of Terms for	SH00398
95-1975	Insulation Testing of Large AC Rotating Machinery with High Direct Voltage, Recommended Practice for	SH10959

IEEE Standard	Title	Code
96-1969	Rating Electric Apparatus for Short-Time, Intermittent or Varying Duty, General Principles for	SH00414
97-1969	Specifying Service Conditions in Electrical Standards, Recommended Practice for	SH00422
100-1972	(ANSI C42.100-1972) Dictionary of Electrical & Electronic Terms	SH02881
115-1965	Synchronous Machines, Test Procedure for	SH00554
118-1949	Resistance Measurement, Master Test Code for	SH02477
121-1959	Rotary Speed, Recommended Guide for Measurement of	SH00596
122-1959	Speed-Governing of Steam Turbines Intended to Drive Electric Generators 500 KW and Larger, Recommended Specification for	SH00604
125-1950	Hydraulic Turbines Intended to Drive Electric Generators, Recommended Specification for Speed-Governing of	SH00620
126-1959	Internal Combustion Engine-Generator Units, Recommended Specification for Speed-Governing of	SH00638
136-1959	Aircraft Generator and Regulator Characteristics, Test Procedure and Presentation of	SH00737
138-1960	Aircraft AC Generators, Test Procedure for	SH00752

IEEE Standard	Title	Code
143-1954	Application Guides for Ground Fault Neutralizers, Grounding of Synchronous Generator Systems, Neutral Grounding of Transmission Systems	SH02493
260-1967	(ANSI Y10.19-1969) Letter Symbols for Units Used in Science and Technology	SH01495
268-1973	Units in Published Scientific and Technical Work, Recommended Practice for	SH12682
279-1971	(ANSI N42.7-1972) Protection Systems for Nuclear Power Generating Stations, Criteria for	SH01685
280-1968	(ANSI Y10.5-1968) Letter Symbols for Quantities Used in Electrical Science and Electrical Engineering	SH01693
282-1968	Speed Governing and Temperature Protection of Gas Turbines Intended to Drive Electric Generators, Recommended Specification for	SH01719
286-1975	Measurement of Power-Factor Tip-Up of Rotating Machinery Stator Coil Insulation, Recommended Practice for	SH02865
294-1969	(ANSI C16.44-1972) Noise Temperature of Noise Generators, State-of-the-Art of Measuring	SH01834
315-1971	(ANSI Y32.2-1970) Graphic Symbols for Electrical and Electronics Diagrams (CAS Z99-1972)	SH02337

IEEE Standard	Title	Code
322-1971	(ANSI Z210.2-1972) Rules for Use of Units of the International System of Units, Recommended Practice	SH02386
352-1975	Reliability Analysis of Nuclear Power Generating Station Protection Systems, Guide for General Principles of	SH13524
387-1972	Diesel-Generator Units Applied as Standby Power Supplies for Nuclear Power Generating Stations, Criteria for	SH02964
421-1972	Excitation Systems for Synchronous Machines, Criteria and Definitions for	SH04218
433-1974	Insulation Testing of Large AC Rotating Machinery with High Voltage at Very Low Frequency, Recommended Practice for	SH04330
492-1974	Operation and Maintenance of Hydro-Generators, Guide for	SH04929

American National Standard (ANSI)	Title	Code
C50.5-1955	Rotating Exciters for Synchronous Machines	SH02204

APPENDIX C

TYPICAL CONSTANTS OF THREE-PHASE SYNCHRONOUS MACHINES

APPENDIX C

TYPICAL CONSTANTS OF THREE-PHASE SYNCHRONOUS MACHINES

Explanatory notes for table which appears on following page:

† From <u>Electrical Transmission and Distribution Reference Book</u>, by permission of the Westinghouse Electric Corporation.

* High-speed units tend to have low reactance and low-speed units high reactance.

** x_0 varies so critically with armature winding pitch that an average value can hardly be given. Variation is from 0.1 to 0.7 of x_d''. Low limit is for two-thirds pitch windings.

TYPICAL CONSTANTS OF THREE-PHASE SYNCHRONOUS MACHINES†

Reactances are per unit; values below the line give the normal range of values, those above give an average value.

	x_d (unsat.) (1)	x_q rated current (2)	x'_d rated voltage (3)	x''_d rated voltage (4)	x_2 rated current (5)	x_0** rated current (6)
Two-pole turbine generators	1.20 / 0.95-1.45	1.16 / 0.92-1.42	0.15 / 0.12-0.21	0.09 / 0.07-0.14	$= x''_d$	0.03 / 0.01-0.08
Four-pole turbine generators	1.20 / 1.00-1.45	1.16 / 0.92-1.42	0.23 / 0.20-0.28	0.14 / 0.12-0.17	$= x''_d$	0.08 / 0.015-0.14
Salient-pole generators & motors (with dampers)	1.25 / 0.60-1.50	0.70 / 0.40-0.80	0.30 / 0.20-0.50*	0.20 / 0.13-0.32*	0.20 / 0.13-0.32*	0.18 / 0.03-0.23
Salient-pole generators (without dampers)	1.25 / 0.60-1.50	0.70 / 0.40-0.80	0.30 / 0.20-0.50*	0.30 / 0.20-0.50*	0.48 / 0.35-0.65	0.19 / 0.03-0.24
Capacitors, air-cooled	1.85 / 1.25-2.20	1.15 / 0.95-1.30	0.40 / 0.30-0.50	0.27 / 0.19-0.30	0.26 / 0.18-0.40	0.12 / 0.025-0.15
Capacitors, hydrogen-cooled at $\frac{1}{2}$ lb/in^2 kVA rating	2.20 / 1.50-2.65	1.35 / 1.10-1.55	0.48 / 0.36-0.60	0.32 / 0.23-0.36	0.31 / 0.22-0.48	0.14 / 0.030-0.18

APPENDIX D

LAPLACE TRANSFORMS AND PARTIAL-FRACTION EXPANSION

APPENDIX D

LAPLACE TRANSFORMS AND PARTIAL-FRACTION EXPANSION

D-1. Laplace Transforms

The *Laplace transform* of a function f(t) is defined as

$$F(s) = \mathcal{L}f(t) = \int_0^\infty f(t)\, e^{-st}\, dt \qquad (D.1.1)$$

where t is a real variable; f(t) is a real function of t and is equal to zero for $t < 0$; s is a complex variable and F(s) is a function of s; and e is equal to 2.71828...

The *inverse transform* is defined as

$$f(t) = \mathcal{L}^{-1}[F(s)] = \frac{1}{2\pi j} \int_{\sigma - j\infty}^{\sigma + j\infty} F(s)\, e^{st}\, ds \qquad (D.1.2)$$

where σ is chosen to the right of any singularity of F(s).

The Laplace transform is a mathematical tool which permits solving time-domain problems in the frequency domain. The philosophy behind the Laplace transform is illustrated by the flow diagram shown on the following page.

In *transforming differentiation*, the Laplace transform preserves the transform of the original function except for an algebraic operation; the initial conditions are directly included as a part of the solution:

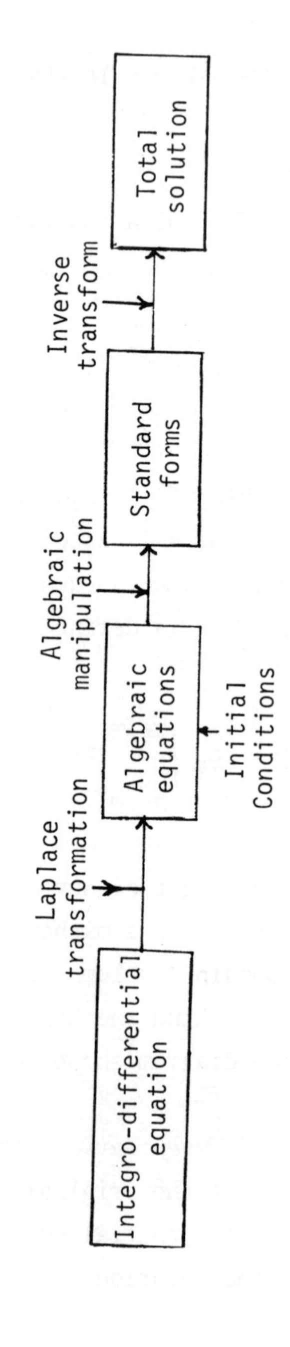

Figure D.1.1. Flow diagram for Laplace transformation methods in solving differential equations.

$$L[i'(t)] = sI(s) - i(o) \tag{D.1.3}$$

$$L[i^n(t)] = s^n I(s) - s^{n-1} i(o) - s^{n-2} i'(o) - \ldots - i^{n-1}(o) \tag{D.1.4}$$

where i(o) is the limit of i(t) as $t \to 0+$ (i.e., time approaching zero from the right); the superscript ' indicates the first derivative d/dt, and the superscript n represents the nth derivative d^n/dt^n; I(s) is the Laplace transform of i(t).

The *Laplace transform of an integral of time* is given by:

$$L[i^{-1}(t)] = L\left[\int_o^t i(t)\ dt\right] = \frac{I(s)}{s} + \frac{i^{-1}(o)}{s} \tag{D.1.5}$$

$$L[i^{-n}(t)] = L\left[\iint \ldots \int i(t)\ dt^n\right] =$$

$$= \frac{I(s)}{s^n} + \frac{i^{-1}(o)}{s^n} + \frac{i^{-2}(o)}{s^{n-1}} + \ldots + \frac{i^{-n}(o)}{s} \tag{D.1.6}$$

where $i^{-1}(o)$ is the initial value of the integral term associated with the energy-storing element just before applying the forcing function.

The Laplace transform of the product of a constant A and a time function i(t) is the constant A multiplied by the Laplace transform of i(t):

$$L[Ai(t)] = A\ I(s) \tag{D.1.7}$$

The Laplace transform of the sum (or difference) of two time functions is the sum (or difference) of the Laplace transforms of the time functions:

$$\mathcal{L}[i_1(t) \pm i_2(t)] = I_1(s) \pm I_2(s) \tag{D.1.8}$$

The *shifting or time-displacement theorem* for real translation is given by

$$\mathcal{L}[f(t - T) \cdot u(t - T)] = e^{-sT} F(s) \tag{D.1.9}$$

in which a time function f(t) is delayed by time T, and u(t) is a unit step function.

The *complex translation theorem* states that if f(t) is Laplace-transformable and has the transform F(s), then multiplication of f(t) by the exponential time function $e^{-\alpha t}$ becomes a translation in the s-domain and vice versa. Expressed mathematically, one has

$$\mathcal{L}[e^{-\alpha t} f(t)] = F(s + \alpha) \tag{D.1.10}$$

The *initial value theorem* states that if the Laplace transform of f(t) is F(s), and f(t) is Laplace-transformable, then

$$\lim_{t \to 0} f(t) = \lim_{s \to \infty} s\, F(s) \tag{D.1.11}$$

if the limit exists. Directly with the transform version of the solution, it is possible to evaluate the initial value of the time-domain solution f(t) without ever having to determine f(t) formally.

The *final value theorem* states that if the Laplace transform of f(t) is F(s), and if the poles of [F(s)] lie inside the left half of the s-plane (i.e., sF(s) is analytic on the imaginary axis and in the right half of the s-plane), then

$$\lim_{t \to \infty} f(t) = \lim_{s \to o} sF(s) \tag{D.1.12}$$

The final value of a time function or the steady-state solution can be obtained directly from the transform solution without ever having to evaluate the complete solution in the time domain. However, the final value theorem is not valid if the denominator of [sF(s)] contains any root whose real part is zero or positive.

The *uniqueness* of the Laplace transformation is such that there cannot be two different functions having the same Laplace transformation, F(s). That being the case, one may use the table of Laplace transforms to find f(t), provided one has the necessary form of F(s) in the table.

D-2. Summary of Properties of Laplace Transformation

(Please refer to the following page.)

Summary of Properties of Laplace Tranformation

Property	Time Function	Laplace Transform
Linearity	$a_1 f_1(t) \pm a_2 f_2(t)$	$a_1 F_1(s) \pm a_2 F_2(s)$
Differentiation	$f'(t)$	$sF(s) - f(o)$
	$f^n(t)$	$s^n F(s) - s^{n-1} f(o) - s^{n-2} f'(o) - \ldots - f^{n-1}(o)$
Integration	$f^{-1}(t)$	$\frac{F(s)}{s} + \frac{f^{-1}(o)}{s}$
	$f^{-n}(t)$	$\frac{F(s)}{s^n} + \frac{f^{-1}(o)}{s^n} + \frac{f^{-2}(o)}{s^{n-1}} + \ldots + \frac{f^{-n}(o)}{s}$
Multiplication by t	$k\, f(t)$	$- \frac{d}{ds} [F(s)]$
Division by t	$\frac{1}{t} f(t)$	$- \int F(s)\, ds$
Time delay or shift	$f(t - T) \cdot u(t - T)$	$e^{-sT} F(s)$

Property	Time Function	Laplace Transform
Exponential translation	$e^{-at} f(t)$	$F(s + a)$
Change of scale	$f(at)$, $a > 0$	$\frac{1}{a} F(\frac{s}{a})$
Initial value	$f(0+)$	$\lim_{s \to \infty} sF(s)$
Final value	$f(\infty)$	$\lim_{s \to 0} sF(s)$ [*sF(s) has poles only inside the left half of the s-plane.*]

D-3. Table of Laplace Transforms

Note that all f(t) should be thought of as being multiplied by u(t), i.e., f(t) = 0 for $t < 0$.

(Please refer to the following page for the table.)

Table of Laplace Transforms

$f(t)$	$F(s)$
$\delta(t)$ (unit impulse or delta function)	1
$\delta(t - T)$	e^{-sT}
$u(t)$ or 1 (unit step function)	$\frac{1}{s}$
$u(t - T)$	$\frac{e^{-sT}}{s}$
$(t - T)\, u(t - T)$	$\frac{e^{-sT}}{s^2}$
$tu(t - T)$	$\frac{(1 + sT)\, e^{-sT}}{s^2}$
t^n (n - integer)	$\frac{n!}{s^{n+1}}$
e^{-at}	$\frac{1}{s + a}$

f(t)	F(s)
$e^{-at} t^n$	$\frac{n!}{(s + a)^{n+1}}$
$\frac{e^{-at} - e^{-bt}}{b - a}$	$\frac{1}{(s + a)(s + b)}$
$\sin \omega t$	$\frac{\omega}{s^2 + \omega^2}$
$\cos \omega t$	$\frac{s}{s^2 + \omega^2}$
$e^{-at} \sin \omega t$	$\frac{\omega}{(s + a)^2 + \omega^2}$
$e^{-at} \cos \omega t$	$\frac{s + a}{(s + a)^2 + \omega^2}$
$\sinh at$	$\frac{a}{s^2 - a^2}$

f(t)	F(s)
$\cosh at$	$\frac{s}{s^2 - a^2}$
$[k_1 e^{-at} \cos \omega t + \frac{k_2 - k_1 a}{\omega} e^{-at} \sin \omega t]$	$\frac{k_1 s + k_2}{(s + a)^2 + \omega^2}$
$\frac{\omega}{\sqrt{1 - a^2}} e^{-a\omega t} \sin \omega \sqrt{1 - a^2}\, t$	$\frac{\omega^2}{s^2 + 2a\omega + \omega^2}$
$\frac{1}{2\omega} t \sin \omega t$	$\frac{s}{(s^2 + \omega^2)^2}$
$\frac{1}{2\omega} (\sin \omega t + \omega t \cos \omega t)$	$\frac{s^2}{(s^2 + \omega^2)^2}$

D-4. Partial-Fraction Expansion

A formalized approach to resolve F(s) into a summation of simple factors is known as the method of partial-fraction expansion, which is based on expanding a rational function of s in terms of the factors of the denominator polynomial. Let us consider a rational function (i.e., one which can be expressed as a ratio of two polynomials)

$$F(s) = \frac{N(s)}{D(s)} \tag{D.4.1}$$

where N(s) denotes the numerator polynomial and D(s) denotes the denominator polynomial. As a first step in the expansion of the quotient N(s)/D(s), we check to see that the degree of the polynomial N is less than that of D. If this condition is not satisfied, divide the numerator by the denominator to obtain an expansion in the form

$$\frac{N(s)}{D(s)} = B_0 + B_1 s + B_2 s^2 + \ldots + B_{m-n} s^{m-n} + \frac{N_1(s)}{D(s)} \tag{D.4.2}$$

where m is the degree of the numerator and n is the degree of the denominator.

The roots of the equation

$$N(s) = 0 \tag{D.4.3}$$

are said to be the *zeros* of F(s); and the roots of the equation

$$D(s) = 0 \tag{D.4.4}$$

are said to be the *poles* of F(s).

The new function $F_1(s)$ given by $N_1(s)/D(s)$ is such that the degree of the denominator polynomial is greater than that of the numerator. The denominator polynomial $D(s)$ is typically of the form

$$D(s) = a_n s^n + a_{n-1} s^{n-1} + \dots + a_1 s + a_0 \tag{D.4.5}$$

By dividing the numerator $N_1(s)$ and the denominator $D(s)$ by a_n, $F_1(s)$ may be rewritten as follows:

$$F_1(s) = \frac{N_2(s)}{D_1(s)} = \frac{N_2(s)}{n + \frac{a_{n-1}}{a_n} s^{n-1} + \dots + \frac{a_0}{a_n}} \tag{D.4.6}$$

The particular manner of evaluating the coefficients of the expansion is dependent upon the nature of the roots of $D_1(s)$ in Eq. (D.4.6). We shall now discuss different cases of interest. The possible forms of the roots are (i) real and simple (or distinct) roots, (ii) conjugate complex roots, and (iii) multiple roots.

Real and Simple (or Distinct) Poles:

If all the poles of $F_1(s)$ are of first-order, Eq. (D.4.6) may be written in terms of a partial-fraction expansion as

$$F_1(s) = \frac{N_2(s)}{(s - p_1)(s - p_2) \dots (s - p_n)} \tag{D.4.7}$$

or

$$F_1(s) = \frac{K_1}{(s - p_1)} + \frac{K_2}{(s - p_2)} + \dots + \frac{K_n}{s - p_n} \tag{D.4.8}$$

where $p_1, p_2, \ldots, p_n$ are distinct, and $K_1, K_2, \ldots, K_n$ are non-zero finite constants. For any k, the evaluation of the residue K_k of $F_1(s)$ corresponding to the pole $s = p_k$ is done by multiplying both sides by $(s - p_k)$ and letting $s \to p_k$.

$$K_k = \lim_{s \to p_k} [(s - p_k) F_1(s)] \tag{D.4.9}$$

Equation (D.4.9) is valid for k = 1, 2, ..., n. Once the K's are determined in Eq. (D.4.8), the inverse Laplace transform of each of the terms can be written down easily in order to obtain the complete time solution.

Conjugate Complex Poles:

It is possible that some of the poles of Eq. (D.4.6) are complex. Since the coefficients a_k in Eq. (D.4.5) are real, complex poles occur in complex conjugate pairs and will always be even in number. Let us consider a pair of complex poles for which case Eq. (D.4.6) may be written as

$$F_1(s) = \frac{N_2(s)}{(s + a + jb)(s + a - jb) D_2(s)} \tag{D.4.10}$$

or

$$F_1(s) = \frac{K_1}{[s + (a + jb)]} + \frac{K_2}{[s + (a - jb)]} + \frac{N_3(s)}{D_2(s)} \tag{D.4.11}$$

The procedure for finding K_1 and K_2 is the same as outlined earlier for unrepeated linear factors or simple poles. Thus we have

$$K_1 = \lim_{s \to (-a-jb)} [(s + a + jb)\, F_1(s)] \tag{D.4.12}$$

and

$$K_2 = \lim_{s \to (-a+jb)} [(s + a - jb)\, F_1(s)] \tag{D.4.13}$$

It can be shown that K_1 and K_2 are conjugates of each other. The terms in time function, $\mathcal{L}^{-1}[F(s)]$, due to the complex poles of $F_1(s)$, are then found easily.

Alternate Representation for Complex Poles:

Complex poles can be combined to yield a quadratic term in the partial fraction expansion. The representation may best be illustrated by considering one real pole and two complex conjugate poles. Let us then consider

$$F_1(s) = \frac{N_2(s)}{(s - p_1)(s + a + jb)(s + a - jb)} =$$

$$= \frac{N_2(s)}{(s - p_1)[(s + a)^2 + b^2]} =$$

$$= \frac{N_2(s)}{(s - p_1)(s^2 + As + B)} \tag{D.4.14}$$

which can be written as

$$F_1(s) = \frac{K_1}{(s - p_1)} + \frac{K_2 s + K_3}{(s^2 + As + B)} \tag{D.4.15}$$

Evaluating K_1 as before, one has

$$K_1 = \lim_{s \to p_1} [(s - p_1) F_1(s)] \tag{D.4.16}$$

It follows from Eqs. (D.4.14) and (D.4.15) that

$$N_2(s) = K_1(s^2 + As + B) + (K_2 s + K_3)(s - p_1) \tag{D.4.17}$$

Since the above equality must hold for all values of s, the coefficients of various powers of s on both sides of the equality must be equal. These equations of equality are then solved to determine K_2 and K_3.

Even if many pairs of complex conjugate poles occur, this procedure may be used remembering that the partial fraction for each complex conjugate pair will be of the form discussed above.

Multiple Poles:

Let us consider that $F_1(s)$ has all simple poles except say at $s = p_1$ which has a multiplicity m. Then one can write

$$F_1(s) = \frac{N_2(s)}{(s - p_1)^m (s - p_2) \ldots (s - p_n)} \tag{D.4.18}$$

The partial fraction expansion of $F_1(s)$ is given by

$$F_1(s) = \frac{K_{11}}{(s - p_1)^m} + \frac{K_{12}}{(s - p_1)^{m-1}} + \ldots + \frac{K_{1m}}{(s - p_1)} + \frac{K_2}{(s - p_2)} + \ldots + \frac{K_n}{(s - p_n)} \tag{D.4.19}$$

When a multiple root is involved, there will be as many coefficients associated with the multiple root as the order of the multiplicity. For each simple pole p_k we have just one coefficient K_k as before.

For simple poles one can proceed as discussed earlier and apply Eq. (D.4.9) to calculate the residues K_k. To evaluate $K_{11}, K_{12}, \ldots, K_{1m}$ we multiply both sides of Eq. (D.4.19) by $(s - p_1)^m$ to obtain

$$(s - p_1)^m F_1(s) = K_{11} + (s - p_1) K_{12} + \ldots + (s - p_1)^{m-1} K_{1m} + (s - p_1)^m \cdot \left[\frac{K_2}{(s - p_2)} + \ldots + \frac{K_n}{(s - p_n)} \right] \qquad \text{(D.4.20)}$$

The coefficient K_{11} may now be evaluated as

$$K_{11} = \lim_{s \to p_1} \left[(s - p_1)^m F_1(s) \right] \qquad \text{(D.4.21)}$$

Next we differentiate Eq. (D.4.20) with respect to s and let $s \to p_1$ in order to evaluate K_{12}:

$$K_{12} = \lim_{s \to p_1} \left\{ \frac{d}{ds} \left[(s - p_1)^m F_1(s) \right] \right\} \qquad \text{(D.4.22)}$$

The differentiation process can be continued to find the kth coefficient:

$$K_{1k} = \lim_{s \to p_1} \left\{ \frac{1}{(k-1)!} \frac{d^{k-1}}{ds^{k-1}} [(s - p_1)^m F_1(s)] \right\},$$

$$\text{for } k = 1, 2, \ldots, m \qquad \text{(D.4.23)}$$

Note that $K_2, \ldots, K_n$ terms play no role in determining the coefficients $K_{11}, K_{12}, \ldots, K_{1m}$ because of the multiplying factor $(s - p_1)^m$ in Eq. (D.4.20).

The alternate representation discussed for the case of complex poles may also be extended for multiple poles by combining the terms in Eq. (D.4.19) corresponding to the multiple root. In an expansion of a quotient of polynomials by partial fractions, it may, in general, be necessary to use a combination of the rules given above.

The denominator of $F_1(s)$ may not be known in the factored form in some cases, and it will then become necessary to find the roots of the denominator polynomial. If the order is higher than a quadratic and simple inspection (or some engineering approximation) does not help, one may have to take recourse to the computer. Computer programs for finding roots of a polynomial equation (using such methods as Newton-Raphson) are available these days in most computer libraries.

D-5. Application of Laplace Transforms to Solving Differential Equations

With the aid of the theorems concerning Laplace transforms and the table of transforms, linear differential equations can be solved by the Laplace transform method. Transformation of the terms of the differential equation yields an algebraic equation in terms of the variable s. Thereafter, the solution of the differential equation is affected by simple algebraic manipulations in the s-domain. By inverting the transform of the solution from the s-domain, one can get back to the time domain. The response due to each term in the partial-fraction expansion is determined directly from the transform table. There is no need to perform any kind of integration. Because initial conditions are automatically incorporated into the Laplace transforms and the constants arising from the initial conditions are automatically evaluated, the resulting response expression yields directly the total solution (i.e., complementary plus particular solution).

APPENDIX E

BLOCK DIAGRAMS, TRANSIENT RESPONSE, FREQUENCY RESPONSE AND STABILITY CRITERIA

APPENDIX E

BLOCK DIAGRAMS, TRANSIENT RESPONSE, FREQUENCY RESPONSE AND STABILITY CRITERIA

E-1. Block Diagrams

The mathematical relationships of control systems are usually represented by block diagrams which show the role of the various components of the system and the interaction of the variables in it. It is common to use a block diagram in which each component in the system (or sometimes a group of components) is represented by a block. An entire system may, then, be represented by the interconnection of the blocks of the individual elements, so that their contributions to the overall performance of the system may be evaluated. The simple configuration shown in Figure E.1.1 is actually the basic building block of a complex block diagram. In the case of linear systems, the input-output relationship is expressed as a transfer function which is the ratio of the Laplace transform of the output to the Laplace transform of the input with initial conditions of the system set to zero. The arrows on the diagram imply that the block diagram has a unilateral property; or in other words, signal can only pass in the direction of the arrows.

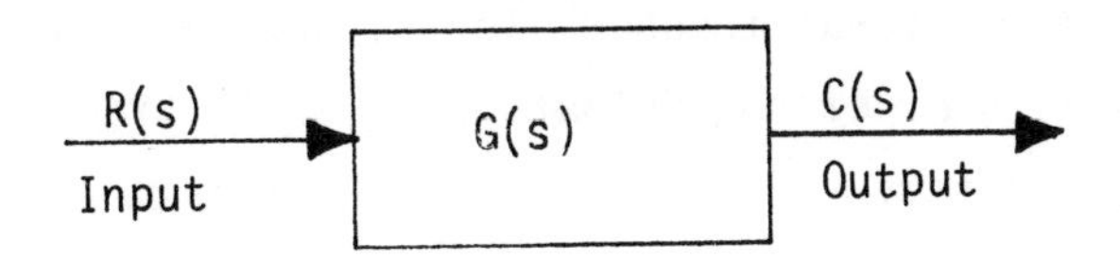

Figure E.1.1. Basic building block of a block diagram.

A box is the symbol for multiplication; the input quantity is multiplied by the function in the box to obtain the output. With circles indicating summing points (in an algebraic sense), and with boxes or blocks denoting multiplication, any linear mathematical expression may be represented by block-diagram notation as in Figure E.1.2 for the case of an elementary feedback control system.

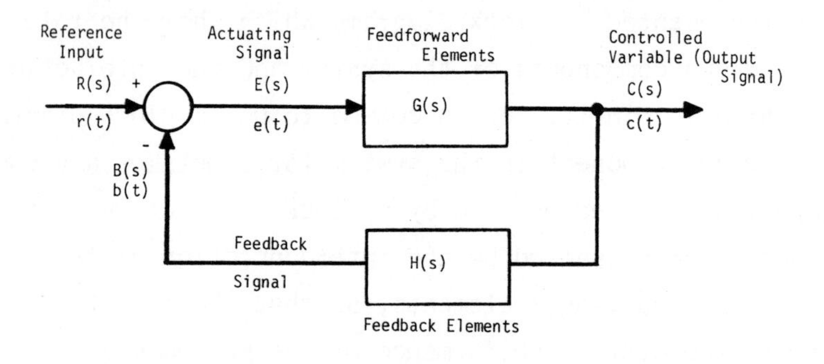

Notation: R(s): Reference Input

C(s): Output Signal (controlled variable)

B(s) = H(s)C(s): Feedback Signal

E(s) = [R(s) - B(s)]: Actuating Signal (error)

G(s) = C(s)/E(s): Forward Path Transfer Function or Open-Loop Transfer Function

M(s) = C(s)/R(s) = [G(s)/1+G(s)H(s)]: Closed-Loop Transfer Function

H(s): Feedback Path Transfer Function

G(s)H(s): Loop Gain

Figure E.1.2. Block diagram of an elementary feedback control system.

The block diagrams of complex feedback control systems usually contain several feedback loops, and they may have to be simplified in order to evaluate an overall transfer function for the system. A few of the block diagram reduction manipulations are given in Table E.1.1; no attempt is made here to cover all the possibilities.

Table E.1.1

Some of the Block Diagram Reduction Manipulations

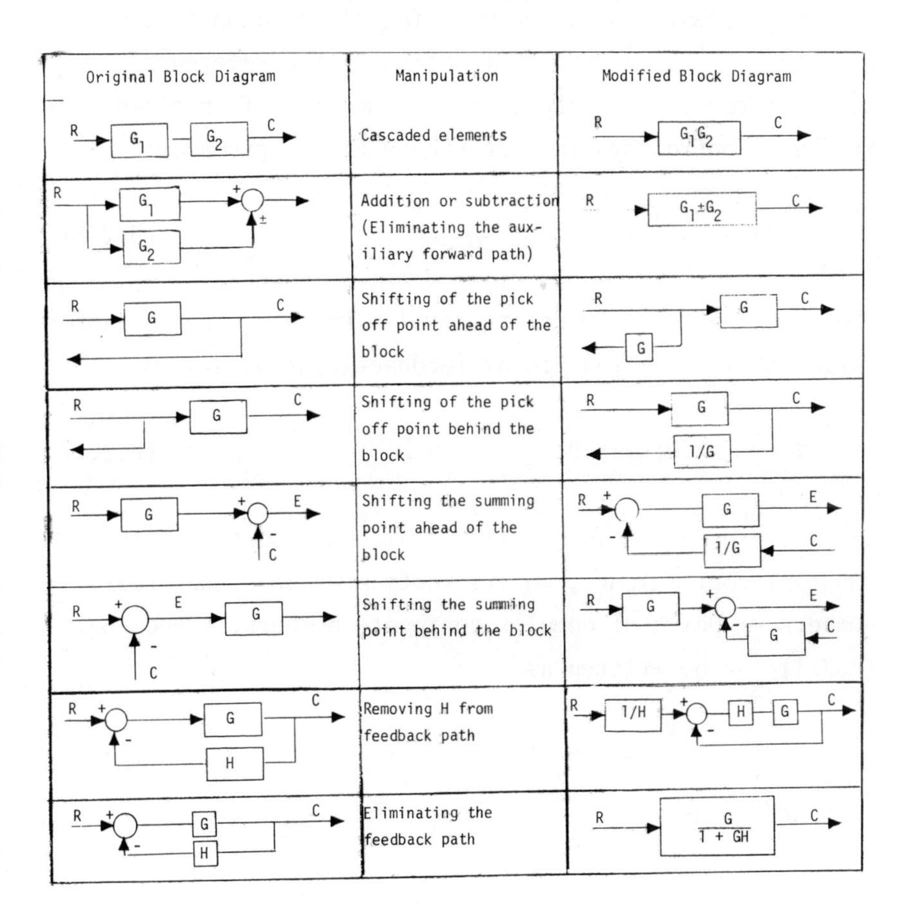

Original Block Diagram	Manipulation	Modified Block Diagram
	Cascaded elements	
	Addition or subtraction (Eliminating the auxiliary forward path)	
	Shifting of the pick off point ahead of the block	
	Shifting of the pick off point behind the block	
	Shifting the summing point ahead of the block	
	Shifting the summing point behind the block	
	Removing H from feedback path	
	Eliminating the feedback path	

E-2. Transient Response

The transient performance of a feedback control system is usually analyzed by using either a unit-step function or a unit-impulse function as the reference input. Consider the closed-loop transfer function of an elementary feedback control system (see Figure E.1.2) given by:

$$M(s) = \frac{C(s)}{R(s)} = \frac{G(s)}{1 + G(s)\,H(s)} \tag{E.2.1}$$

The denominator set equal to zero is known as the characteristic equation of the linear feedback control system.

$$1 + G(s)\,H(s) = 0 \tag{E.2.2}$$

The above equation is usually a rational function with constant coefficients; so its roots must be either real numbers or pairs of complex conjugate numbers. Hence Eq. (E.2.1) may be written as

$$\frac{C(s)}{R(s)} = \frac{k \prod_{j=1}^{\nu} (s + z_j)}{\prod_{i=1}^{n+2m} (s + p_i)} =$$

$$= \frac{K \prod_{j=1}^{\nu} (s + z_j)}{\prod_{i=1}^{n} (s + \sigma_i) \prod_{k=1}^{m} [s + (\alpha_k + j\omega_k)][s + (\alpha_k - j\omega_k)]} \tag{E.2.3}$$

where $-\sigma_i$ denote the real roots, and $(-\alpha_k \pm j\omega_k)$ denote the complex conjugate roots of the characteristic equation. No repeated roots have been assumed here in the characteristic equation. The response of the system to an input signal r(t) is given by

$$c(t) = \mathcal{L}^{-1}[C(s)] =$$

$$= \mathcal{L}^{-1}\left[R(s)\frac{K\prod_{j=1}^{\nu}(s + z_j)}{\prod_{i=1}^{n}(s+\sigma_i)\prod_{k=1}^{m}[s+(\alpha_k+j\omega_k)][s+(\alpha_k-j\omega_k)]}\right] \tag{E.2.4}$$

If the input r(t) is a unit-step function, the output is

$$c(t) = \mathcal{L}^{-1}\left[\frac{K\prod_{j=1}^{\nu}(s + z_j)}{s\prod_{i=1}^{n}(s + \sigma_i)\prod_{k=1}^{m}[s^2 + 2\alpha_k s + (\alpha_k^2 + \omega_k^2)]}\right] \tag{E.2.5}$$

By using the partial-fraction expansion and the Laplace transform table, the output response can be evaluated as a function of time. The constant term in the total solution represents the steady-state response; the transient response is characterized by either the exponential terms, the damped sinusoids, or both. The location of the roots of the characteristic equation in the s-plane uniquely defines the form of the transient response. The real roots σ_i and the real parts of the complex roots α_k appear as exponents; these control the damping of the time response. The imaginary parts of the roots ω_k appear as the frequencies of sinusoidal oscillations of the response. It can easily be verified that for a stable response, the roots of the characteristic equation should not be found in the right half of the s-plane. Roots which are on the imaginary axis correspond to systems with sustained constant amplitude oscillations. The roots which are closest to the $j\omega$-axis are sometimes known as the dominant roots of the characteristic equation.

E-3. Response of a System to an Arbitrary Input

For some input functions r(t), such as an arbitrary input, the transform R(s) is either a very complicated expression or not easily obtained. For such cases, it becomes necessary to find the transient response without knowing the function R(s). The arbitrary input could be approximated by a sum of pulses:

$$r(t) \simeq r_1[u(t) - u(t - t_1)] + r_2(u(t - t_1) - u(t - t_2)] + \\ + \ldots + r_n[u(t - t_{n-1}) - u(t - t_n)] \qquad \text{(E.3.1)}$$

For such a case, the response of the system c(t) at time t is obtained by adding the individual responses due to each pulse.

$$c(t) = c_1(t) + c_2(t) + \ldots \qquad \text{(E.3.2)}$$

where $c_1(t)$ is the response due to the first pulse, etc. Accuracy is increased by using more pulses of a smaller width to approximate the input.

A considerable saving in computational time and effort is realized by the use of the convolution integral method, which yields an exact rather than an approximate solution. The general transformed expression for C(s) is of the form

$$C(s) = \frac{L_m(s)\ R(s)}{L_n(s)} + \frac{I(s)}{L_n(s)} = W(s)\ R(s) + \frac{I(s)}{L_n(s)} \qquad \text{(E.3.3)}$$

The desired time response is given by the inverse transformation:

$$c(t) = \mathcal{L}^{-1}[W(s)\ R(s)] + \mathcal{L}^{-1}\left[\frac{I(s)}{L_n(s)}\right] \qquad \text{(E.3.4)}$$

where the last term is evaluated directly from a knowledge of the initial conditions. For an arbitrary input for which R(s) is not known, the first term may be found by the application of the following convolution integral:

$$\mathcal{L}^{-1}[W(s)\ R(s)] = \int_0^\infty w(\lambda)\ f(t - \lambda)\ d\lambda \qquad \text{(E.3.5)}$$

where

$$w(\lambda) = \mathcal{L}^{-1}\left\{\left[\frac{L_m(s)}{L_n(s)}\right](1)\right\} = \mathcal{L}^{-1}[W(s)] \qquad \text{(E.3.6)}$$

is the unit-impulse response of the system and can be computed directly since W(s) is known. It can be verified that the value of the integral given by Eq. (E.3.5) is zero for values of λ greater than t. Thus the upper limit of integration may be taken as t rather than infinity, i.e.,

$$\mathcal{L}^{-1}[W(s)\ R(s)] = \int_0^t w(\lambda)\ f(t - \lambda)\ d\lambda \qquad \text{(E.3.7)}$$

which can also be shown

$$= \int_0^t f(\lambda)\, w(t - \lambda)\, d\lambda \tag{E.3.8}$$

The complete time response is given by

$$c(t) = \int_0^t w(\lambda)\, f(t - \lambda)\, d\lambda + L^{-1}\left[\frac{I(s)}{L_n(s)}\right] \tag{E.3.9}$$

The impulse response $w(\lambda)$ is also known as the weighting function because it weighs the past values of the input.

For the case of networks, the response to an impulse is determined only by the constants of the network. The impulse response thus characterizes the network just as the transfer function, and may be used for identification purposes. For a two terminal-pair network (i.e., two-port network) with zero initial conditions described by the input and output voltage transforms related by

$$V_2(s) = H(s)\, V_1(s) \tag{E.3.10}$$

the output voltage as a function of time may be obtained as

$$v_2(t) = L^{-1}[H(s)\, V_1(s)] = \int_0^t v_1(\lambda)\, h(t - \lambda)\, d\lambda =$$

$$= \int_0^t v_1(t - \lambda)\, h(\lambda)\, d\lambda \tag{E.3.11}$$

It is apparent that $h(t)$ is the impulse response of the network as well as the inverse transform of the transfer function H(s). Equation (E.3.11) shows that if h(t), the impulse response, is known, only the input voltage $v_1(t)$ need be specified in order to determine the output voltage through convolution. In other words, any input convolved with the unit-impulse response yields the output.

E-4. Frequency Response

The frequency response method is a graphical method through which the description of the system is given in terms of its response to a sinusoidally varying input signal. It provides a convenient means for investigating the dynamic behavior of control systems. For a linear system, the output will also be a sine wave of the same frequency as the input; for a nonlinear system, the output will, in addition, contain higher harmonics, and sometimes subharmonics. The transfer function describing the sinusoidal steady-state behavior of the system can simply be obtained from the transfer function by replacing the Laplace operator s with jω. The sinusoidal transfer function corresponding to the open-loop transfer function G(s) of a certain feedback control system is given by

$$G(j\omega) = [G(s)]_{s=j\omega} = |G| \underline{/\phi_g} \tag{E.4.1}$$

The following methods of plotting G(jω) are usually employed in the analysis and design of feedback control systems:

(a) *Polar Plot:* This is a plot of magnitude vs. phase shift on polar coordinates as ω is varied from zero to infinity. The polar plot is primarily used for the analysis and design of systems by means of the Nyquist

criterion*.

(b) Bode Plot (corner plot): Plots of magnitude in decibel vs. log ω (or ω) and phase angle (in degrees) vs. log ω (or ω) in rectangular coordinates are often referred to as the Bode plot, the corner plot, or the logarithmic plot of G(s). The gain in decibel (db) is given by

$$\text{gain in db} = 20 \log_{10} |G| \tag{E.4.2}$$

The product factors in the expression of G(jω) become additive terms, because logarithms are used. The shape of the Bode plot for the most commonly encountered functions makes it possible to represent approximately the exact function plot by straight line asymptotes.

(c) Magnitude vs. Phase Shift Plot: This is a plot of magnitude in decibel vs. phase shift (in degrees) on rectangular coordinates with frequency as a varying parameter on the curves. This plot can directly be superposed on the Nichols chart** to determine the relative stability of the closed-loop system. The Bode plots are easy to construct; the data necessary for the construction of the polar plot and the magnitude vs. phase-shift plot can be obtained directly from the Bode plot.

---

* See for details any book on Control Systems, such as Automatic Control Systems by Benjamin C. Kuo, Prentice-Hall, Inc., 1962.

** E. A. Guillemin, "Computational Techniques Which Simplify the Correlation Between Steady-State and Transient Responses of Filters and other Networks", Proc. Nat'l. Electronics Conference, No. 9, 1953-1954.

E-5. Frequency Response from the Transient Response

The method described here was developed by Guillemin (see the second footnote on the previous page); it possesses the advantage that the accuracy obtained with only a few terms (thereby minimizing the computational effort) is better than that obtained by other available techniques. The transform for the nth derivative of a function $y(t)$ for which the initial conditions are zero is given by

$$L[y^{(n)}(t)] = \int_0^\infty y^{(n)}(t)\ e^{-st}\ dt = s^n Y(s) \qquad (E.5.1)$$

or

$$Y(s) = \frac{1}{s^n} \int_0^\infty y^{(n)}(t)\ e^{-st}\ dt \qquad (E.5.2)$$

or

$$Y(j\omega) = \frac{1}{(j\omega)^n} \int_0^\infty y^{(n)}(t)\ e^{-j\omega t}\ dt \qquad (E.5.3)$$

Guillemin's procedure is based primarily upon a graphical interpretation of the above equation. A function $y(t)$ may be represented by a straight-line approximation, indicated by $y^*(t)$. The first derivative $y^{*(1)}(t)$ can be seen to be a series of steps; the second derivative yields the train of impulses:

$$y^{*(2)}(t) = \sum_{k=1}^{\nu} a_k\ u(t - t_k) \qquad (E.5.4)$$

where a_k is the area of the kth impulse, $u(t - t_k)$ denotes a unit impulse which occurs at time t_k, and ν is the total number of impulses. Substituting Eq. (E.5.4) into Eq. (E.5.3), one has for n = 2:

$$Y^*(j\omega) = \frac{1}{(j\omega)^2} \int_0^{\infty} \sum_{k=1}^{\nu} a_k\, u(t - t_k)\, e^{-j\omega t}\, dt \qquad (E.5.5)$$

or

$$= \frac{1}{(j\omega)^2} \sum_{k=1}^{\nu} a_k\, e^{-j\omega t_k} \qquad (E.5.6)$$

For a given input f(t) and corresponding output y(t), the frequency response is given by

$$G^*(j\omega) = \frac{Y^*(j\omega)}{F^*(j\omega)} \qquad (E.5.7)$$

Improved accuracy is obtained by approximating the function with a series of parabolas. In such a case, $y^{(1)}(t)$ is determined exactly, and this derivative is then approximated by straight lines. Hence, it follows that the second derivative $y^{*(2)}(t)$ is a series of steps and the third derivative is a train of impulses. The approximation $Y^*(j\omega)$ then becomes

$$Y^*(j\omega) = \frac{1}{(j\omega)^3} \sum_{k=1}^{\nu} a_k\, e^{-j\omega t_k} \qquad (E.5.8)$$

E-6. Transient Response from the Frequency Response

The general operational representation for a differential equation is

$$y(t) = G(D)\, f(t) \qquad (E.6.1)$$

With zero initial conditions and unit-impulse input, the transform is given by

$$Y(s) = G(s) \tag{E.6.2}$$

The inverse transform of the above equation is

$$y(t) = w(t) = \frac{1}{2\pi j} \int_{\sigma-j\infty}^{\sigma+j\infty} G(s)\ e^{st}\ ds \tag{E.6.3}$$

where w(t) represents the impulse response. Substitution of (jω) for s yields

$$w(t) = \frac{1}{2\pi} \int_{-\infty}^{\infty} G(j\omega)\ e^{j\omega t}\ d\omega \tag{E.6.4}$$

Integrating by parts, one can write

$$w(t) = \left. \frac{G(j\omega)\ e^{j\omega t}}{2\pi j t} \right|_{-\infty}^{\infty} +$$

$$+ \frac{1}{2\pi(-jt)} \int_{-\infty}^{\infty} G^{(1)}(j\omega)\ e^{j\omega t}\ d\omega \tag{E.6.5}$$

Noting that G(jω) goes to zero for infinite values of ω for any realizable function, further integration by parts yields the following general expression:

$$w(t) = \frac{1}{2\pi(-jt)^n} \int_{-\infty}^{\infty} G^{(n)}(j\omega)\ e^{j\omega t}\ d\omega \tag{E.6.6}$$

which may be rewritten as follows:

$$w(t) = \frac{1}{2\pi(-jt)^n} \int_{-\infty}^{\infty} [G_R^{(n)}(j\omega) \cos \omega t - G_I^{(n)}(j\omega) \sin \omega t] d\omega +$$

$$+ \frac{j}{2\pi(-jt)^n} \int_{-\infty}^{\infty} [G_R^{(n)}(j\omega) \sin \omega t +$$

$$+ G_I^{(n)}(j\omega) \cos \omega t]\, d\omega \tag{E.6.7}$$

where G_R and G_I denote, respectively, the real and imaginary parts of G. It can be shown that, for even values of n and $t > 0$,

$$w(t) = \frac{-2}{\pi(-jt)^n} \int_{0}^{\infty} G_I^{(n)}(j\omega) \sin \omega t \, d\omega \tag{E.6.8}$$

$$w(t) = \frac{2}{\pi(-jt)^n} \int_{0}^{\infty} G_R^{(n)}(j\omega) \cos \omega t \, d\omega \tag{E.6.9}$$

For odd values of n and $t > 0$, one obtains

$$w(t) = \frac{2j}{\pi(-jt)^n} \int_{0}^{\infty} G_I^{(n)}(j\omega) \cos \omega t \, d\omega \tag{E.6.10}$$

$$w(t) = \frac{2j}{\pi(-jt)^n} \int_{0}^{\infty} G_R^{(n)}(j\omega) \sin \omega t \, d\omega \tag{E.6.11}$$

With the nth derivative given by a train of impulses, Eqs. (E.6.9) and (E.6.11) become

$$w(t) = \frac{2}{\pi} \frac{(-1)^{n/2}}{t^n} \sum_{k=1}^{\nu} a_k \cos \omega_k t, \quad n \text{ even} \tag{E.6.12}$$

$$w(t) = \frac{2}{\pi} \frac{(-1)^{(n+1)/2}}{t^n} \sum_{k=1}^{\nu} a_k \sin \omega_k t, \quad n \text{ odd} \tag{E.6.13}$$

where a_k is the area of the kth pulse and ω_k is the angular velocity at which it occurs. The approach used here is essentially the same as in Section E-5. One can also work with the imaginary part G_I and a set of equations similar to (E.6.12) and (E.6.13) can be obtained from Eqs. (E.6.8) and (E.6.10). Thus, the impulse response can be obtained by working with either the real or the imaginary part of $G(j\omega)$. Once the impulse response is determined, the response to any arbitrary input may be found by the convolution technique presented in Section E-3.

E-7. Stability Criteria

A system is defined as stable if the output response to any bounded input disturbance is finite. Since this implies that all the roots of the characteristic equation must be located in the left half of the s-plane, the problem of determining the stability of a linear system is one of finding the roots of the characteristic equation. It is very tedious and time-consuming to find the roots for polynomials of the third order or higher. Hence, without actually solving for the roots of the characteristic equation, the system stability needs to be evaluated by alternate methods.

The Routh-Hurwitz Criterion:

The general form of the characteristic equation of a linear system is given by

$$F(S) = a_o s^n + a_1 s^{n-1} + a_2 s^{n-2} + \dots + a_{n-1} s + a_n = 0 \tag{E.7.1}$$

In order that there be no roots of the above equation with positive real parts, it is necessary, but not sufficient, that

(a) all the coefficients of the polynomial have the same sign,

and (b) none of the coefficients vanish.

The necessary and sufficient condition that all the roots of an nth-order polynomial lie in the left half of the s-plane is that the polynomial's Hurwitz determinants D_k $(k = 1, 2, \dots, n)$ must all be positive. The Hurwitz determinants of Eq. (E.7.1) are given by

$$D_1 = a_1; \quad D_2 = \begin{vmatrix} a_1 & a_3 \\ a_o & a_2 \end{vmatrix}; \quad D_3 = \begin{vmatrix} a_1 & a_3 & a_5 \\ a_o & a_2 & a_4 \\ 0 & a_1 & a_3 \end{vmatrix}$$

$$\dots D_n = \begin{vmatrix} a_1 & a_3 & a_5 & \dots & a_{2n-1} \\ a_o & a_2 & a_4 & \dots & a_{2n-2} \\ 0 & a_1 & a_3 & \dots & a_{2n-3} \\ 0 & a_o & a_2 & \dots & a_{2n-4} \\ 0 & 0 & a_1 & \dots & a_{2n-5} \\ \vdots & & & & \\ 0 & 0 & 0 & \dots & a_n \end{vmatrix} \tag{E.7.2}$$

in which the coefficients with negative subscripts or with subscripts larger than n are replaced by zeros. The Routh-Hurwitz criterion may be stated as follows:

The necessary and sufficient condition that all the roots of the polynomial $F(s) = 0$ given by Eq. (E.7.1) lie in the left half of the s-plane is that $a_o > 0$, $D_1 > 0$, $D_2 > 0$, ..., $D_n > 0$, where D_1, D_2, ..., D_n are Hurwitz determinants given by Eq. (E.7.2).

The above criterion can be applied without actually working with the higher-order determinants by following the procedure indicated below:

The coefficients of the characteristic equation (E.7.1) are to be arranged into two rows:

$$a_o \quad a_2 \quad a_4 \quad a_6 \quad \cdots$$

$$a_1 \quad a_3 \quad a_5 \quad a_7 \quad \cdots$$

As a next step, form the following array of numbers (known as Routh's tabulation) by following the operations indicated, as shown for a sixth-order system as an example:

Routh's Tabulation for a Sixth-order System

s^6	a_o	a_2	a_4	a_6
s^5	a_1	a_3	a_5	0
s^4	$\frac{a_1a_2 - a_3a_o}{a_1} = A$	$\frac{a_1a_4 - a_oa_5}{a_1} = B$	$\frac{a_1a_6 - a_o \times 0}{a_1} = a_6$	0
s^3	$\frac{Aa_3 - a_1B}{A} = C$	$\frac{Aa_5 - a_1a_6}{A} = D$	$\frac{A \times 0 - a_1 \times 0}{A} = 0$	0
s^2	$\frac{CB - AD}{C} = E$	$\frac{Ca_6 - A \times 0}{C} = a_6$	$\frac{C \times 0 - A \times 0}{C} = 0$	0
s^1	$\frac{ED - Ca_6}{E} = F$	0	0	0
s^o	$\frac{Fa_6 - E \times 0}{F} = a_6$	0	0	0

Now let us investigate the signs of the numbers in the first column of the above tabulation, since the elements in the first column are related to Hurwitz determinants:

$$a_o = a_o; \quad a_1 = D_1; \quad A = D_2/D_1; \quad C = D_3/D_2$$
$$E = D_4/D_3; \quad F = D_5/D_4; \quad a_6 = D_6/D_5 \tag{E.7.3}$$

If all the elements of the first column are positive, the roots of the polynomial are all in the left half of the s-plane. If there are negative signs in the elements of the first column, the number of sign changes indicates the number of roots with positive real parts.

When the first element in any row of the Routh's tabulation is zero, while the other elements are not, simply multiply the polynomial by a factor (s + a) where a is any positive real number, in order to restore the missing power of s, and carry on with Routh's test.

When all the elements in one row of the Routh's tabulation are zero, the equation corresponding to the coefficients just above the row of zeros is called the auxiliary equation, the order of which is always even indicating the number of root pairs that are equal in magnitude but opposite in sign. In order to correct the situation for the application of Routh's test, simply take the first derivative of the auxiliary equation with respect to s, replace the row of zeros with the coefficients of the resultant equation, and carry on. The roots with equal magnitude can be obtained by solving the auxiliary equation.

For a general cubic equation given by

$$a_o s^3 + a_1 s^2 + a_2 s + a_3 = 0 \tag{E.7.4}$$

the application of the Hurwitz criterion yields

$$D_1 = a_1 > 0$$

$$D_2 = \begin{vmatrix} a_1 & a_3 \\ a_o & a_2 \end{vmatrix} = a_1 a_2 - a_o a_3 > 0$$

$$D_3 = a_3 D_2 > 0 \qquad \text{or} \qquad a_3 > 0 \tag{E.7.5}$$

Thus for stability, it follows that

$$a_1 a_2 > a_o a_3 \tag{E.7.6}$$

A frequent use of Routh's criterion is to determine the condition of stability of a linear feedback control system.

The Nyquist Criterion:

The Routh's criterion gives information regarding only the absolute stability of the system, while it does not tell how stable it is or how to improve the system. In addition to the absolute system stability, the Nyquist criterion* indicates the degree of stability of a stable system, and furnishes information as to how the system stability may be improved in cases where necessary. Also, the Nyquist locus gives information concerning the frequency response of the system. The Nyquist criterion makes extensive use of conformal mapping and was originated from the

* See for details any book on Control Systems, such as Automatic Control Engineering by Francis H. Raven, McGraw-Hill Book Company, 1968.

principle of the argument* in complex variable theory. The details and application of the Nyquist criterion are not presented here.

The Root-Locus Technique:

The Bode plot and the Nyquist criterion are useful techniques in designing a feedback control system from the frequency response point of view. In the study of linear systems, however, it is equally important sometimes to consider the time domain specifications. The root-locus method, developed by Evans**, is a graphical method of determining the roots of the characteristic equation by knowing the factored form of the feedforward and feedback elements of a control system. The technique has been greatly developed in the past few years. The advantage of working in the s-plane is that the roots of the characteristic equation not only give direct information concerning the transient response, but also give an indication of the sinusoidal frequency behavior of the system to some extent. For details and application of the root-locus method, the reader is referred to books on control systems.

Nonlinear Systems:

Most of the physical systems can be considered to be linear only within a limited range of operation. The principle of superposition is not valid for a nonlinear system. For the study of nonlinear control systems, the

---

* Advanced Engineering Mathematics, C. R. Wylie, Jr., McGraw-Hill Book Company, 1951.

** Control System Dynamics, W. R. Evans, McGraw-Hill Book Company, 1954.

phase-plane method (which is a graphical technique of solving nonlinear differential equations) is often utilized for first- or second-order systems or higher-order systems which can be approximated by second-order systems. However, it cannot be extended to systems with sinusoidal inputs. The describing-function method developed by Goldfarb and Kochenburger* has been found to be very useful for sinusoidal studies of nonlinear feedback systems of any order. This method is applicable for any nonlinearity which has the characteristic that if the input is a sinusoid, the output is a periodic function. Because of its simplicity and wide range of applicability, the describing-function technique is one of the most versatile methods for investigating the nonlinear effects.

While the phase-plane method offers an adequate graphical procedure for determining the transient response of first- or second-order nonlinear systems, it has become necessary to develop other methods of analysis for third- and higher-order nonlinear systems. One of the most promising methods for investigating the behavior of higher-order systems is the method known as Liapunov's direct method**. Much work has been done in recent years on generating Liapunov functions and applying the same to

---

* R. J. Kochenburger, "A Frequency Response Method for Analyzing and Synthesizing Contactor Servomechanisms", Trans. AIEE, Vol. 69, pp. 270-284, 1950.

** J. LaSalle and S. Lefschetz, <u>Stability by Liapunov's Direct Method</u>, Academic Press, Inc., New York, 1961.

various systems for stability studies*.

---

*F. Williams, S. A. Louie, and G. W. Bills, "Feasibility of Liapunov Functions for the Stability Analysis of Electric Power Systems Having up to 60 Generators", IEEE Trans. PA&S, Vol. PAS-91, pp. 1145-1157, 1972.

T... [illegible] 3. A... [illegible] and [illegible] "Practicality
of Liapunov Functions for the Stability Analysis of
[illegible] Power Systems with up to 6 Generators" [illegible]
[illegible] PAS-91 [illegible] 14... [illegible] 1972.

APPENDIX F

VECTOR RELATIONS

Table F-1

Vector Identities

Algebraic

$\overline{F} \cdot \overline{G} = \overline{G} \cdot \overline{F}$

$\overline{F} \times \overline{G} = -\overline{G} \times \overline{F}$

$\overline{F} \cdot (\overline{G} + \overline{H}) = \overline{F} \cdot \overline{G} + \overline{F} \cdot \overline{H}$

$\overline{F} \times (\overline{G} + \overline{H}) = \overline{F} \times \overline{G} + \overline{F} \times \overline{H}$

$\overline{F} \times (\overline{G} \times \overline{H}) = \overline{G}(\overline{H} \cdot \overline{F}) - \overline{H}(\overline{F} \cdot \overline{G})$

$\overline{F} \cdot (\overline{G} \times \overline{H}) = \overline{G} \cdot (\overline{H} \times \overline{F}) = \overline{H} \cdot (\overline{F} \times \overline{G})$

Integral

$$\oint_s \overline{F} \cdot \overline{ds} = \int_v \nabla \cdot \overline{F} \, dv$$

$$\oint_\ell \overline{F} \cdot \overline{d\ell} = \int_s (\nabla \times \overline{F}) \cdot \overline{ds}$$

Differential

$\nabla(f + g) = \nabla f + \nabla g$

$\nabla \cdot (\overline{F} + \overline{G}) = \nabla \cdot \overline{F} + \nabla \cdot \overline{G}$

$\nabla \times (\overline{F} + \overline{G}) = \nabla \times \overline{F} + \nabla \times \overline{G}$

$\nabla(fg) = f\nabla g + g\nabla f$

$\nabla \cdot (f\overline{F}) = \overline{F} \cdot \nabla f + f(\nabla \cdot \overline{F})$

$\nabla \cdot (\overline{F} \times \overline{G}) = \overline{G} \cdot (\nabla \times \overline{F}) - \overline{F}(\nabla \times \overline{G})$

$\nabla \times (\overline{F}) = (\nabla f) \times \overline{F} + f(\nabla \times \overline{F})$

$\nabla \cdot \nabla f = \nabla^2 f$

$\nabla \cdot (\nabla \times \overline{F}) = 0$

Table F-1 (Cont'd)

$$\oint_s f(\nabla g) \cdot \overline{ds} = \int_v [f\nabla^2 g + (\nabla f) \cdot (\nabla g)]\, dv \qquad \nabla \times (\nabla f) = 0$$

$$\oint_s [f\nabla g - g\nabla f] \cdot \overline{ds} = \int_v (f\nabla^2 g - g\nabla^2 f)\, dv \qquad \nabla \times (\nabla \times \overline{F}) = \nabla(\nabla \cdot \overline{F}) - \nabla^2 \overline{F}$$

$$\nabla \times (f\nabla g) = \nabla f \times \nabla g$$

(Note that f, g are scalars; $\overline{A}$, $\overline{F}$, $\overline{G}$ and $\overline{H}$ are vectors.)

Table F-2

Vector Relations in Orthogonal Curvilinear Coordinate Systems

In rectangular coordinates only, the following relations apply:

$$\nabla = \hat{x}\frac{\partial}{\partial x} + \hat{y}\frac{\partial}{\partial y} + \hat{z}\frac{\partial}{\partial z}$$

$$\nabla \cdot \nabla = \nabla^2 = \frac{\partial^2}{\partial x^2} + \frac{\partial^2}{\partial y^2} + \frac{\partial^2}{\partial z^2}$$

$$\nabla^2 \overline{A} = \hat{x}\,\nabla^2 A_x + \hat{y}\nabla^2 A_y + \hat{z}\nabla^2 A_z$$

Rectangular Coordinates

$$\nabla f = \hat{x}\frac{\partial f}{\partial x} + \hat{y}\frac{\partial f}{\partial y} + \hat{z}\frac{\partial f}{\partial z}$$

$$\nabla \cdot \overline{A} = \frac{\partial A_x}{\partial x} + \frac{\partial A_y}{\partial y} + \frac{\partial A_z}{\partial z}$$

$$\nabla^2 f = \frac{\partial^2 f}{\partial x^2} + \frac{\partial^2 f}{\partial y^2} + \frac{\partial^2 f}{\partial z^2}$$

$$\nabla \times \overline{A} = \hat{x}\left(\frac{\partial A_z}{\partial y} - \frac{\partial A_y}{\partial z}\right) + \hat{y}\left(\frac{\partial A_x}{\partial z} - \frac{\partial A_z}{\partial x}\right) + \hat{z}\left(\frac{\partial A_y}{\partial x} - \frac{\partial A_x}{\partial y}\right)$$

Table F-2 (Cont'd)

Cylindrical Coordinates

$$\nabla f = \hat{r}\,\frac{\partial f}{\partial r} + \frac{\hat{\theta}}{r}\,\frac{\partial f}{\partial \theta} + \hat{z}\,\frac{\partial f}{\partial z}$$

$$\nabla \cdot \overline{A} = \frac{1}{r}\,\frac{\partial}{\partial r}\,(rA_r) + \frac{1}{r}\,\frac{\partial A_\theta}{\partial \theta} + \frac{\partial A_z}{\partial z}$$

$$\nabla^2 f = \frac{1}{r}\,\frac{\partial}{\partial r}\left(r\,\frac{\partial f}{\partial r}\right) + \frac{1}{r^2}\,\frac{\partial^2 f}{\partial \theta^2} + \frac{\partial^2 f}{\partial z^2}$$

$$\nabla \times \overline{A} = \hat{r}\left(\frac{1}{r}\,\frac{\partial A_z}{\partial \theta} - \frac{\partial A_\theta}{\partial z}\right) + \hat{\theta}\left(\frac{\partial A_r}{\partial z} - \frac{\partial A_z}{\partial r}\right) +$$

$$+ \hat{z}\left[\frac{1}{r}\,\frac{\partial}{\partial r}\,(rA_\theta) - \frac{1}{r}\,\frac{\partial A_r}{\partial \theta}\right]$$

Spherical Coordinates

$$\nabla f = \hat{R}\,\frac{\partial f}{\partial R} + \frac{\hat{\theta}}{R}\,\frac{\partial f}{\partial \theta} + \frac{\hat{\phi}}{R \sin \theta}\,\frac{\partial f}{\partial \phi}$$

$$\nabla \cdot \overline{A} = \frac{1}{R^2}\,\frac{\partial}{\partial R}\,(R^2 A_R) + \frac{1}{R \sin \theta}\,\frac{\partial}{\partial \theta}\,(\sin \theta\; A_\theta) + \frac{1}{R \sin \theta}\,\frac{\partial A_\phi}{\partial \phi}$$

$$\nabla^2 f = \frac{1}{R^2}\,\frac{\partial}{\partial R}\left(R^2\,\frac{\partial f}{\partial R}\right) + \frac{1}{R^2 \sin \theta}\,\frac{\partial}{\partial \theta}\left(\sin \theta\,\frac{\partial f}{\partial \theta}\right) +$$

$$+ \frac{1}{R^2 \sin^2 \theta}\,\frac{\partial^2 f}{\partial \phi^2}$$

Table F-2 (Cont'd)

$$\nabla \times \overline{A} = \frac{\hat{R}}{R \sin\theta} \left[\frac{\partial}{\partial\theta} (\sin\theta A_\phi) - \frac{\partial A_\theta}{\partial\phi} \right] +$$

$$+ \frac{\hat{\theta}}{R} \left[\frac{1}{\sin\theta} \frac{\partial A_R}{\partial\phi} - \frac{\partial}{\partial R} (RA_\phi) \right] +$$

$$+ \frac{\hat{\phi}}{R} \left[\frac{\partial}{\partial R} (RA_\theta) - \frac{\partial A_R}{\partial\theta} \right]$$

BIBLIOGRAPHY

The following is a selected list of books and articles that the reader will find useful as a supplementary source of information in the study of this text. The bibliography is by no means complete.

Books

(Arranged alphabetically by author's name)

1. Adkins, B., The General Theory of Electrical Machines, John Wiley & Sons, Inc., New York, 1957.

2. Anderson, P. M., Analysis of Faulted Power Systems, The Iowa State University Press, Ames, Iowa, 1973.

3. Binns, K. J. and Lawrenson, P. J., Analysis and Computation of Electric and Magnetic Field Problems, 2nd Edition, Pergamon Press, Inc., New York, 1973.

4. Byerly, R. T. and Kimbark, E. W., Stability of Large Electric Power Systems, IEEE Press (Selected Reprint Series), New York, 1974.

5. Clarke, E., Circuit Analysis of A.C. Power Systems, Vols. I and II, John Wiley & Sons, Inc., New York, 1943.

6. Concordia, C., Synchronous Machines, John Wiley & Sons, Inc., New York, 1951.

7. Crary, S. B., Power System Stability, Vols. I and II, John Wiley & Sons., Inc., New York, 1945.

8. Del Toro, V., Electromechanical Devices for Energy Conversion and Control Systems, Prentice-Hall, Inc., Englewood Cliffs, N.J., 1968.

9. Elgerd, O. I., Electric Energy Systems Theory: An Introduction, McGraw-Hill Book Company, New York, 1971.

10. Elgerd, O. I., Basic Electric Power Engineering, Addison-Wesley Publishing Company, Reading, MA, 1977.

11. Durney, C. H. and Johnson, C. C., Introduction to Modern Electromagnetics, McGraw-Hill Book Company, New York, 1969.

12. Fitzgerald, A. E., Kingsley, C. Jr. and Kusko, A., Electric Machinery, 3rd Edition, McGraw-Hill Book Company, New York, 1971.

13. Gibbs, W. J., Electric Machine Analysis Using Matrices, Pitman, 1962.

14. Gibbs, W. J., Electric Machine Analysis Using Tensors, Pitman, 1967.

15. Gourishankar, V. and Kelly, D. H., Electromechanical Energy Conversion, 2nd Edition, Intext Educational Publishers, New York, 1973.

16. Hancock, N. N., Matrix Analysis of Electrical Machinery, 2nd Edition, Pergamon Press, Oxford, 1974.

17. Harris, M. R., et al, Per Unit Systems: with Special Reference to Electrical Machines, Cambridge University Press, London, 1970.

18. Hindmarsh, J., Electrical Machines and Their Applications, 2nd Edition, Pergamon Press, Oxford, 1970.

19. IEEE Tutorial Course Text, Modern Concepts of Power System Dynamics, IEEE 70 MG2-PWR, 1970.

20. IEEE Publication, Symposium on Adequacy and Philosophy of Modeling: Dynamic System Performance, IEEE 75-CH 0970-4-PWR, 1975.

21. IEEE Publication, Recommended Practice for Protection and Coordination of Industrial and Commercial Power Systems, IEEE Std. 242-1975, Ch. 14, pp. 287-291, 1975.

22. IEEE Publication, Analysis and Control of Subsynchronous Resonance, IEEE 76-CH 1066-0-PWR, 1976.

23. Johnk, C. T. A., Engineering Electromagnetic Fields and Waves, John Wiley & Sons, Inc., New York, 1975.

24. Jones, C. V., Unified Theory of Electrical Machines, Butterworth, 1967.

25. Kimbark, E. W., Power System Stability, Vols. I, II and III, John Wiley & Sons, Inc., New York, 1948.

26. Knowlton, A. E., Standard Handbook for Electrical Engineers, 8th Edition, Sec. 7, McGraw-Hill Book Company, New York, 1949.

27. Kron, G., Equivalent Circuits of Electric Machinery, John Wiley & Sons, Inc., New York, 1951.

28. Kuo, B. C., Automatic Control Systems, Prentice-Hall, Inc., Englewood Cliffs, NJ, 1962.

29. Lawrence, R. R. and Richards, H. E., Principles of Alternating Current Machinery, 4th Edition, McGraw-Hill Book Company, New York, 1953.

30. Lewis, W. A., The Principles of Synchronous Machines, Illinois Institute of Technology, Chicago, 1959.

31. Liwschitz-Garik, M. and Whipple, C. C., Electric Machinery, Vols. I and II, D. Van Nostrand Company, Inc., Princeton, NJ, 1946.

32. Majmudar, H., Introduction to Electrical Machines, Worcester Polytechnic Institute, Worcester, MA, 1976.

33. Mason, C. R., The Art and Science of Protective Relaying, John Wiley & Sons, Inc., New York, 1956.

34. Matsch, L. W., Electromagnetic and Electromechanical Machines, Intext Educational Publishers, New York, 1972.

35. Meisel, J., Principles of Electromechanical Energy Conversion, McGraw-Hill Book Company, New York, 1966.

36. Nasar, S. A., Electromagnetic Energy Conversion Devices and Systems, Prentice-Hall, Inc., Englewood Cliffs, NJ, 1970.

37. Neuenswander, J. R., Modern Power Systems, International Text Book Company, Scanton, PA, 1971.

38. Peterson, H. A., Transients in Power Systems, Dover Publications, Inc., 1951/1966.

39. Puchstein, A. F., et al, Alternating Current Machines, John Wiley & Sons., Inc., New York, 1954.

40. Routh, E. J., Dynamics of a System of Rigid Bodies, Part II, 6th Edition, Macmillan, London, 1905. (Reprinted as a paperback by Dover, 1961).

41. Say, M. G., Introduction to the Unified Theory of Electromagnetic Machines, Pitman, 1971.

42. Say, M. G., Alternating Current Machines, John Wiley & Sons, Inc., (Halsted Press), New York, 1976.

43. Seely, S., Electromechanical Energy Conversion, McGraw-Hill Book Company, New York, 1962.

44. Skilling, H. H., Electromechanics, John Wiley & Sons, Inc., New York, 1962.

45. Slemon, G. R., Magnetoelectric Devices, John Wiley & Sons, Inc., New York, 1966.

46. Smith, J. L., Superconductors in Large Synchronous Machines, EPRI Research Report (prepared at M.I.T.), June 1975.

47. Stagg, G. W. and El-Abiad, A. H., Computer Methods in Power System Analysis, McGraw-Hill Book Company, New York, 1968.

48. Stevenson, W. D. Jr., Elements of Power System Analysis, 3rd Edition, McGraw-Hill Book Company, New York, 1975.

49. Sullivan, R. L., Power System Planning, McGraw-Hill Book Company, New York, 1977.

50. Wagner, C. F. and Evans, R. D., Symmetrical Components, McGraw-Hill Book Company, New York, 1933.

51. Walsh, E. M., Energy Conversion, Ronald Press, New York, 1967.

52. Warrington, A. R. Van C., Protective Relays: Their Theory and Practice, Vols. I and II, Ch. 9, Chapman and Hall Ltd., Great Britain, 1968.

53. Westinghouse Publication, Electrical Transmission and Distribution Reference Book, 4th Edition, Westinghouse Electric Corporation, East Pittsburgh, PA, 1964.

54. White, D. C. and Woodson, H. H., Electromechanical Energy Conversion, John Wiley & Sons, Inc., New York, 1959.

55. Woodson, H. H. and Melcher, J. R., Electromechanical Dynamics, Vol. I, John Wiley & Sons, Inc., New York, 1968.

Technical Articles

(Arranged Chronologically)

1. Blondel, A., "The two-reaction method for study of oscillatory phenomena in coupled alternators", Revue générale de l'électricité, Vol. 13, pp. 235-251; 515-531, February/March 1923.

2. Doherty, R. E. and Nickle, C. A., "Synchronous machines I, an extension of Blondel's two-reaction theory", AIEE Trans., Vol. 45, pp. 912-926, 1926.

3. Doherty, R. E. and Nickle, C. A., "Synchronous machines II, steady-state power angle characteristics", AIEE Trans., Vol. 45, pp. 927-942, 1926.

4. Doherty, R. E. and Nickle, C. A., "Synchronous machines III, torque-angle characteristics under transient conditions", AIEE Trans., Vol. 46, pp. 1-18, 1927.

5. Stevenson, A. R. and Park, R. H., "Graphical determination of magnetic fields", AIEE Trans., Vol. 46, pp. 112-136, 1927.

6. Wieseman, R. W., "Graphical determination of magnetic fields; practical application to salient-pole synchronous-machine design", AIEE Trans., Vol. 46, pp. 141-154, 1927.

7. Doherty, R. E. and Nickle, C. A., "Synchronous machines IV, single-phase short circuits", AIEE Trans., Vol. 47, pp. 457-487, 1928.

8. Doherty, R. E., "Excitation systems: their influence on short-circuits and maximum power", AIEE Trans., Vol. 47, pp. 944-956, July 1928.

9. Park. R. H., "Two reaction theory of synchronous machines - generalized method of analysis - part I", AIEE Trans., Vol. 48, pp. 716-727, 1929.

10. Doherty, R. E. and Nickle, C. A., "Synchronous machines V, three-phase short circuits", AIEE Trans., Vol. 49, pp. 700-714, 1930.

11. Kilgore, L. A., "Calculation of synchronous machine constants; reactances and time constants affecting transient characteristics", AIEE Trans., Vol. 50, pp. 1201-1214, December 1931.

12. Wright, S. H., "Determination of synchronous machine constants by test", AIEE Trans., Vol. 50, pp. 1331-1350, December 1931.

13. Nickle, C. A., et al, "Single-phase short-circuit torque of a synchronous machine", AIEE Trans., Vol. 51, pp. 966-971, 1932.

14. Shildneck, L. P., "Synchronous-machine reactances: a fundamental and physical viewpoint", General Electric Review, Vol. 35, pp. 560-565, November 1932.

15. Park, R. H., "Two-reaction theory of synchronous machines - II", AIEE Trans., Vol. 52, pp. 352-355, 1933.

16. Lewis, W. A., "Quick-response excitation", Elec. Journal, Vol. 31, pp. 308-312, August 1934.

17. Crary, S. B., "Steady-state stability of composite systems", AIEE Trans., Vol. 53, pp. 1809-1814, 1934.

18. Kingsley, C. Jr., "Saturated synchronous reactance", Elec. Engineering, Vol. 54, pp. 300-305, March 1935.

19. Smith, J. B. and Weygandt, C. N., "Double-line-to-neutral short circuit of an alternator", AIEE Trans., Vol. 56, pp. 1149-1155, 1937.

20. Concordia, C., "Relations among transformations used in electrical engineering problems", General Electric Review, pp. 323-325, July 1938.

21. Clark, E., et al, "Overvoltages caused by unbalanced short circuits - effect of amortisseur windings", AIEE Trans., Vol. 57, pp. 453-466, August 1938.

22. Concordia, C., et al, "Stability characteristics of turbine generators", AIEE Trans., Vol. 57, pp. 732-744, 1938.

23. Boice, W. K., et al, "The direct-acting generator voltage regulator", AIEE Trans., Vol. 59, pp. 149-156, March 1940.

24. Roeschlaub; Von F., "Effect of sequential switching on short-circuit currents in synchronous machines", General Electric Review, pp. 256-261, June 1940.

25. Concordia, C., "Steady state stability of synchronous machines as affected by voltage regulator characteristics", AIEE Trans., Vol. 63, pp. 215-220, 1944.

26. Rankin, A. W., "Per-unit impedances of synchronous machines", AIEE Trans., Vol. 64, (Part I, pp. 569-573; Part II, pp. 839-841), 1945.

27. Harder, E. L. and Valentine, C. E., "Static voltage regulator for rototrol exciter", AIEE Trans., Vol. 64, pp. 601-606, August 1945.

28. Anderson, H. C. Jr., "Voltage variation of suddenly loaded generators", General Electric Review, pp. 25-33, August 1945.

29. Porter, F. M. and Kinghorn, J. H., "The development of modern excitation systems for synchronous condensers and generators", AIEE Trans., Vol. 54, pp. 1020-1028, 1946.

30. Barkle, J. E. and Valentine, C. E., "Rototrol excitation systems", AIEE Trans., Vol. 67, pp. 529-534, 1948.

31. Phillips, A. H., et al, "Excitation improvement: electronic excitation and regulation of electric generators as compared to conventional methods", AIEE Trans., Vol. 69, pp. 338-340, 1950.

32. Bothwell, F. E., "Stability of voltage regulators", AIEE Trans., Vol. 69, pp. 1430-1433, 1950.

33. Concordia, C., "The differential analyzer as an aid in power system analysis", CIGRE Report 311, 1950.

34. Laible, Th., "The effect of solid poles and of different forms of amortisseur on the characteristics of salient-pole alternators", CIGRE Report 111, 1950.

35. Concordia, C., "Effect of buck-boost voltage regulator on steady-state power circuit", AIEE Trans., Vol. 69, pp. 380-384, 1950.

36. Concordia, C., "Synchronous machine damping and synchronizing torques", AIEE Trans., Vol. 70, 1951.

37. Storm, H. F., "Static magnetic exciter for synchronous alternators", AIEE Trans., Vol. 70, pp. 1014-1017, 1951.

38. Dalton, F. K. and Cameron, A. W. W., "Simplified measurement of sub-transient and negative sequence reactances in salient-pole synchronous machines", AIEE Trans., Vol. 71, pp. 752-756, October 1952.

39. Hunter, W. A. and Temoshok, M., "Development of a modern Amplidyne voltage regulator for large turbine generators", AIEE Trans., Vol. 71, pp. 894-900, October 1952.

40. Wust, A. and Dispaux, J., "The operation of turbo-alternators under unbalanced or out-of-step conditions", CIGRE Report 106, 1952.

41. Heffron, W. G. and Phillips, R. A., "Effect of a modern Amplidyne voltage regulator on underexcited operation of large turbine generators", AIEE Trans., Vol. 71, pp. 692-697, August 1952.

42. Laffoon, C. M., "The significance of generator short-circuit ratio", Westinghouse Engineer, Vol. 12, pp. 207-208, November 1952.

43. Lawrence, R. F. and Ferguson, R. W., "Generator negative sequence currents for line-to-line faults", AIEE Trans., Vol. 72, pp. 9-16, February 1953.

44. Ross, M. D. and King, E. I., "Turbine-generator rotor heating during single-phase short circuits", AIEE Trans., Vol. 72, pp. 40-45, February 1953.

45. Alger, P. L., et al, "Short-circuit capabilities of synchronous machines for unbalanced faults", AIEE Trans., Vol. 72, pp. 394-403, June 1953.

46. Pollard, E. I., "Effects of negative sequence currents on turbine-generator rotors", AIEE Trans., Vol. 72, pp. 404-406, June 1953.

47. Anderson, H. C., et al, "System stability limitations and generator loading", AIEE Trans., Vol. 72, Part III, pp. 402-423, June 1953.

48. Farnham, S. B. and Swarthout, R. W., "Field excitation in relation to machine and system operation", AIEE Trans. Paper 53-387, November 1953.

49. Vidmar, M. Jr., "Comparison of four main types of rotary exciters employed in modern practice with regard to the exciting effectiveness", CIGRE Report 138, 1954.

50. Heffron, W. G. Jr., "A simplified approach to steady-state stability limits", AIEE Trans., Vol. 73, pp. 39-44, February 1954.

51. Rothe, F. S., "The effect of generator voltage regulators on stability and line charging capacity", CIGRE Report 321, 1954.

52. Sen, S. K. and Adkins, B., "The application of the frequency response method to electrical machines", Proc. IEE, Vol. 103C, p. 378, 1956.

53. Jones, N. H., et al, "Design of conductor-cooled steam turbine-generators and application to modern power systems", IEEE Trans., PAS-84, No. 2, pp. 131-146, February 1965.

54. King, E. I. and Batchelor, J. W., "Effects of unbalanced current on turbine generator", IEEE Trans., PAS-84, pp. 121-125, 1965.

55. Ahamed, S. V. and Erdelyi, E. A., "Nonlinear theory of salient pole machines", IEEE Trans., PAS-85, pp. 61-70, January 1966.

56. Kimbark, E. W., "Improvement of system stability by switched series capacitors", 31TP65-705, IEEE Trans. PA&S, February 1966.

57. Bobo, P. O., et al, "Performance of excitation systems under abnormal conditions", IEEE Trans., PAS-87, No. 2, pp. 547-553, February 1968.

58. Akers, H. T., et al, "Operation and protection of large steam turbine generators under abnormal conditions", IEEE Trans., PAS-87, No. 4, pp. 1180-1188, April 1968.

59. Erdelyi, E. A., Sarma, M. S. and Coleman, S. S., "Magnetic fields in nonlinear salient pole alternators", IEEE Trans., PAS-87, No. 10, 1968.

60. IEEE Committee Report, "Computer representation of excitation systems", IEEE Trans., PAS-87, No. 6, pp. 1460-1464, 1968.

61. Jackson, W. B. and Winchester, R. C., "Direct and quadrature-axis equivalent circuits for solid-rotor turbine generators", IEEE Trans., PAS-88, pp. 1121-1136, 1969.

62. Oberretl, K., "Magnetic fields, eddy currents and losses taking the variable permeability into account", IEEE Trans., PAS-88, pp. 1646-1657, 1969.

63. IEEE Committee Report, "Proposed excitation system definitions for synchronous machines", IEEE Trans., PAS-88, No. 8, pp. 1248-1258, 1969.

64. Peterson, H. A. and Dravid, N. V., "A method for reducing dead time for single-phase reclosing in EHV transmission", 68TP72-PWR, IEEE Trans. PA&S, April 1969.

65. Harrington, D. B. and Jenkins, S. C., "Trends and advancements in the design of large generators", American Power Conference, Chicago, April 1970.

66. Brown, P. G., et al, "Effects of excitation, turbine energy control, and transmission on transient stability", IEEE Trans., 70TP203-PWR, 1970.

67. Erdelyi, E. A. and Fuchs, E. F., "Nonlinear magnetic field analysis of D.C. machines", IEEE Trans., PAS-89, pp. 1546-1554, 1970.

68. Young, C. C., "The synchronous machine", IEEE 70-M62-PWR, "Modern concepts of power system dynamics", pp. 11-24, 1970.

69. Dougherty, J. J. and Hillesland, T. Jr., "Power system stability considerations with dynamically responsive DC transmission lines", 69TP17-PWR, IEEE Trans. PA&S, January 1970.

70. Sarma, M. S., "Potential functions in electromagnetic field problems", IEEE Trans., MAG-6, No. 3, pp. 513-518, September 1970.

71. Sarma, M. S., "End-leakage of aerospace homopolar alternators", IEEE Trans., AES-6, No. 5, pp. 684-691, September 1970.

72. Concordia, C. and Brown, P. G., "Effects of trends in large steam turbine driven generator parameters on power system stability", IEEE Trans., 71TP74-PWR, pp. 2211-2218, 1971.

73. Chari, M. V. K. and Silvester, P., "Analysis of turbo-alternator magnetic fields by finite elements", IEEE Trans., PAS-90, No. 2, pp. 454-464, 1971.

74. Sarma, M. S., et al, "End-winding leakage of high speed alternators by three dimensional field determination", IEEE Trans., PAS-90, No. 2, pp. 465-477, March/April 1971.

75. Cushing, E. W., et al, "Fast valving as an aid to power system transient stability and prompt resynchronization and rapid reload after full load rejection", 71TP705-PWR, IEEE Trans. PA&S, November 1971.

76. Maliszewski, R. M., et al, "Frequency actuated load shedding and restoration - Part I Philosophy", 70TP672-PWR, IEEE Trans. PA&S, July 1971.

77. Dommel, H. W. and Sato, N., "Fast transient stability solutions", T72-137-3, IEEE Trans. PA&S, January 1972.

78. Meisel, J., et al, "Dynamic control of multimachine power systems based on two-step optimization over admissible trajectories", IEEE Trans., PAS-91, pp. 920-927, May/June 1972.

79. Williams, F., et al, "Feasibility of Liapunov functions for the stability analysis of electric power systems having up to 60 generators", IEEE Trans., PAS-91, pp. 1145-1157, May/June 1972.

80. Yu, Y. N. and Moussa, H. A. M., "Optional stabilization of a multi-machine system", IEEE Trans., PAS-91, pp. 1174-1182, May/June 1972.

81. Linkinhoker, C. L., et al, "Influence of unbalanced currents on the design and operation of large turbine generators", IEEE Trans., 73-012-2, pp. 1597-1604, 1973.

82. Brown, P. G., "Generator I_2^2t requirements for system faults", IEEE PES Winter Power Meeting, 1973.

83. IEEE Working Group Report, "A standard for generator continuous unbalanced current capability", IEEE Trans., 73-128-6, pp. 1547-1549, 1973.

84. Chari, M. V. K., "Finite-element solution of the eddy current problem in magnetic structures", IEEE Trans., T73-320-9, 1973.

85. Concordia, C., "Future developments of large electric generators", Philosophical Trans. of the Royal Society of London, Vol. 275, No. 1248, pp. 39-50, 1973.

86. Abegg, K., "The growth of turbogenerators", Philosophical Trans. of the Roayl Society of London, Vol. 275, No. 1248, pp. 51-68, 1973.

87. IEEE Committee Report, "Excitation system dynamic characteristics", T72-590-8, IEEE Trans. PA&S, January 1973.

88. El-Sherbiny, M. K. and Mehta, D. M., "Dynamic system stability", T73-220-1, IEEE Trans. PA&S, September 1973.

89. Jokl, A. L., et al, "Nonsymmetrical, nonlinear aerospace alternators on nonlinear loads", IEEE Trans., AES-11, No. 3, pp. 298-315, May 1975.

90. Berdy, J., et al, "Protection of large steam turbine generators during abnormal operating conditions", CIGRE 11-05, August 1975.

91. Schulz, R. P., "Synchronous machine modeling", IEEE 75 CH0970-4-PWR, pp. 24-28, 1975.

92. Schleif, F. R., "Excitation system modeling", IEEE 75 CH0970-4-PWR, pp. 29-32, 1975.

93. DeMello, F. P., "Power system dynamics - Overview", IEEE 75 CH0970-4-PWR, pp. 5-15, 1975.

94. Concordia, C. and Schulz, R. P., "Appropriate component representation for the simulation of power system dynamics", IEEE 75 CH0970-4-PWR, pp. 16-23, 1975.

95. Bowler, C. E. J., "Understanding subsynchronous resonance", IEEE 76-CH 1066-0-PWR, pp. 66-73, 1976.

96. Sarma, M. S., "Magnetostatic field computation by finite element formulation", IEEE Trans., MAG-12, No. 6, pp. 1050-1052, November 1976.

97. Demirchian, K. S., Chechurin, V. and Sarma, M. S., "Scalar potential concept for calculating the steady magnetic fields and eddy currents", IEEE Trans., MAG-12, No. 6, pp. 1045-1046, November 1976.

98. Lin, Ing-Shou and Sarma, M. S., "Analysis and control of subsynchronous resonance", 1977 PICA Conf. Proc., May 1977.

99. Kimbark, E. W., "How to improve system stability without risking subsynchronous resonance", IEEE-F-77-125-8, IEEE Winter Power Meeting, February 1977.

100. Umans, S. D., et al, "Modeling of solid rotor turbo-generators", Parts I and II, F 77-160-5, F 77-161-3, IEEE Winter Power Meeting, February 1977.

INDEX